Beginning Microsoft Kinect for Windows SDK 2.0

Motion and Depth Sensing for Natural User Interfaces

Mansib Rahman

Apress®

Beginning Microsoft Kinect for Windows SDK 2.0: Motion and Depth Sensing for Natural User Interfaces

Mansib Rahman
Montreal, Quebec, Canada

ISBN-13 (pbk): 978-1-4842-2315-4   ISBN-13 (electronic): 978-1-4842-2316-1
DOI 10.1007/978-1-4842-2316-1

Library of Congress Control Number: 2017950735

Cover image designed by Freepik

Managing Director: Welmoed Spahr
Editorial Director: Todd Green
Acquisitions Editor: Jonathan Gennick
Development Editor: Laura Berendson
Technical Reviewer: Dwight Goins
Coordinating Editor: Jill Balzano
Copy Editor: April Rondeau
Compositor: SPi Global
Indexer: SPi Global
Artist: SPi Global

Distributed to the book trade worldwide by Springer Science+Business Media New York, 233 Spring Street, 6th Floor, New York, NY 10013. Phone 1-800-SPRINGER, fax (201) 348-4505, e-mail orders-ny@springer-sbm.com, or visit www.springeronline.com. Apress Media, LLC is a California LLC and the sole member (owner) is Springer Science + Business Media Finance Inc (SSBM Finance Inc). SSBM Finance Inc is a **Delaware** corporation.

For information on translations, please e-mail rights@apress.com or visit http://www.apress.com/rights-permissions.

Apress titles may be purchased in bulk for academic, corporate, or promotional use. eBook versions and licenses are also available for most titles. For more information, reference our Print and eBook Bulk Sales web page at http://www.apress.com/bulk-sales.

Any source code or other supplementary material referenced by the author in this book is available to readers on GitHub via the book's product page, located at www.apress.com/9781484223154. For more detailed information, please visit http://www.apress.com/source-code.

Printed on acid-free paper

In loving memory of Uncle Mannan. Entrepreneur for the ages; may your fingerprints taint the pages of this book.

Contents at a Glance

Contents

About the Author

Mansib Rahman is the heinous and uninspiring mastermind behind this slack job of a book (which he is very proud of, by the way). He was formerly a Technical Evangelist at Microsoft, where he encouraged kids to use NoSQL databases because it was "web-scale." And he also tried to get them hooked on Azure. Lots of Azure. His lifelong dream is to one day invent the Kinect Mini, a miniature and easily mountable version of the Kinect meant to be deployed en masse to create interactive projection-mapping experiences and to monitor vital signs in triage scenarios during natural disasters.

Mansib is currently the Research & Development Technical Lead at Desjardins Lab, where he watches BBC documentaries on the HoloLens all day and hacks together holographic telepresence apps on it all night. His other hobbies include canoeing, playing the sax, and opening restaurants. His best friend is his cat, Mittens, although they don't talk much. He lives in Montreal, Canada, and doesn't yet have children to mention in this bio.

You can follow him on Twitter at @gaessaki or reach out to him on LinkedIn. His favorite blog (disclaimer: it's his own blog) is www.ramenscafe.com.

About the Technical Reviewer

Dwight Goins is the Chief Algorithms Officer (CAO) and Lead Emerging Experiences Consultant at THOTH Speed Engineers, located in Coral Springs, Florida. He spends most of his time leading his team in implementing integration solutions using Azure and AWS. He also has built automated frameworks in Chef for DevOps teams with various clients. In his night life, Dwight leads a team of developers in creating medical-based applications using unique implementations with IoT, R, and machine learning. His fascination, however, is with emerging 1-D and 3-D platforms such as bots (conversation as a platform) and augmented reality, virtual reality, and mixed reality using head-mounted displays such as HoloLens. He is an avid Linux user (Gentoo being his favorite flavor) and open source contributor, as well as an InfoSec secure coding instructor. Along with open source development, he also mentors and teaches. Dwight maintains an active MCT (Microsoft Certified Trainer) license and holds the MCPD, MCSD, and Windows 10 UWP Developer certifications, as well as a Certified Data Recovery Professional (CRDP) license. In his spare time, he finds a way to take care of his three beloved daughters, wife, and extended family by constantly keeping them entertained through traveling and family fun.

You can contact the third person Dwight Goins at dngoins@thothspeedengineers.com or read about his first person experiences at http://dgoins.wordpress.com.

Acknowledgments

This book contains over 80,000 words. That's 80,000+ chances to get things wrong. I've been very fortunate that such a responsibility has been vested in me, and I would like to take another thousand words to thank those who made it so.

There is no way I could start to acknowledge anyone without first acknowledging the amazing team at Apress that was responsible for putting this book together. **Jonathan Gennick**, if you hadn't said yes there would be no Kinect 2 book, period. Not just this book—there wouldn't be a single book for the Kinect 2 anywhere on the market. This book was desperately needed, and I was very fortunate to find someone as enthusiastic as yourself to support this project. **Jill Balzano**, you kept me on track and held my hand to the finish line of this epic marathon. I couldn't have imagined a more cooperative editor to work with. Thank you both for having an incredible amount of patience and understanding during what was a very difficult period in my life. There are many others at Apress who worked on this book behind the scenes, and I am no less grateful to them.

To think that a chance encounter in New York years ago has proven to be so instrumental to the completion of this book! Being a student at the time, I could hardly justify the midnight bus fare from Montreal for a Kinect hackathon. When I heard of a Kinect heartbeat-tracking endeavor, I was almost too shy to come and join in. **Dwight Goins**, I've been very fortunate to have one of the world's foremost Kinect experts on my side. This isn't a mere platitude. Given the invariable turnover and separation of concerns within the Microsoft Kinect development teams, Microsoft MVPs such as yourself are the most comprehensive repertoires of Kinect knowledge in the world.

There are very few people who have such an impact on your life that they change the very lens with which you view the world. I can say that **Tommy Lewis**, my former manager at Microsoft, has done this to me twice. Your "Beyond the Tiles" workshop inspired within me an appreciation of design and beauty in everyday things that I hold to this day. By chance, I got to work with you a year later. The little things that you taught me during our weekly syncs accrued over time and ultimately opened my eyes to the true lever of so many human endeavors and reactions: our egos. I haven't been the same person since. Tommy, working with you, in short, was life-altering.

I attribute much of my career acumen to **Susan Ibach**. I want to reiterate my belief that she should write a book about her accumulated wisdom. On another note, I'm not sure why, but I've always pictured her as the poster-child Canadian.

If it weren't for **Sage Franch**, I wouldn't even have a Kinect with which to write this book, nor have gotten the chance to work at Microsoft. I will always fondly remember your surprise giveaway at the MSP summit. Please continue being an inspirational figure to your generation and beyond.

Although I didn't get a chance to work with **Hilary Burt** and **Adarsha Datta** as often, I wanted both of them to know that I've always looked up to their great work ethic and that I aspire to be as productive as them.

Christian Hissibini, vous avez été un ami fiable et généreux et je me sens très chanceux d'avoir rencontré quelqu'un comme vous. Vos talents vous conduiront loin et votre personnalité étonnante vous amènera encore plus loin.

I have to credit **Rudi Chen** for being the catalyst to my career in software. If it weren't for you, I might have been a failed med school applicant right about now. You have a brilliant mind, and I am looking forward to seeing the heights to which it takes you.

My Kinect adventure started with **Marianopolis Brainstorm**, an organization dedicated to cultivating a culture of technological innovation amongst the youth. For many great laughs and unconditional help, I extend my never-ending gratitude to Brainstorm's executives and associates: **Afuad Hossain, Dhanesh Patel, Casey Wang, Jonah Dabora, Patrick Baral, Shohan Rustom, Stephen Thangathurai,** and **Yacine Sibous**.

Of course, I take great pleasure in antagonizing area best friend **Zacharias Ohlin**, which is why I left him out of the previous list. You know as much as anyone that Brainstorm wouldn't have existed without you. You will always be my first editor, and I blame you for much of my writing success.

Most people can write, but very few are fortunate enough to discover it. Even if you never gave me the highest grades, nor always appreciated the humor in my work, I have to thank you, **James Dufault**. It's when you forced me up to do that blasted speech in Grade 7 (for the second time) that I knew I was to write. You were a very engaging teacher to boot, and I hope you take pleasure in correcting my ever-shoddy work.

I am forever grateful to **Mostafa Saadat**. You have always had faith in me, almost to a fault. You kept me going during some of the toughest times in my life, when no one or nothing else did. I couldn't ask for a better co-founder with whom to realize our mutual ambitions. **Qudsia Saadat**, I always said you would be a model one day, and I'm very glad to say that I had a hand in it (see Chapter 2). Hopefully, we'll get mad at each other for many more years to come. And count seminiferous tubule cross-sections in rat testes. And apples.

Shino, **Riwa**, **Yota**, and **Noriko Suzuki**, I am indebted to you all for taking me into your household like family in Japan while I worked on this book. I can't believe there are people as kind as you are in this world. I am certain that you will all be very successful in Canada.

This is always the hardest part. How do you thank the people who gave you everything without being cheesy? I'm pretty sure you can't. Regardless, let it be known that it was **Masum Rahman** whom I would beg for bedtime Gopal Bhai and Liberation War stories and from whom I would learn long English words and that it was **Sanchita Rahman** who bought me all my fairytale and technical books and taught me to read and love them at a very young age. **Marisa**, we grew up under the same roof, but I can't help but feel that we grew up in completely different realities. I hope you can take advantage of our common heritage to face the ever-perilous future, and that you strive to make the world a better place at any cost. You will always have my faith.

Mittens, I've been blessed. The best of times awaits us still. May you be rewarded in turn.

Thank you all,
Mansib

Introduction

If you're reading this, chances are you're looking to integrate the Kinect for Xbox One (colloquially known as the Kinect 2) into a project of yours. Supposing otherwise, let me tell you why you should. Industry titans such as Microsoft, Google, Amazon, Facebook, IBM, and Intel have thrown their hats in the ring and have already starting selling products and services that pertain to the Internet of Things (IoT), Augmented reality (AR), and Virtual reality (VR). With Gartner predicting 20.8 billion connected devices by 2020, it would be incredibly short-sighted for any cutting-edge tech company to not heavily invest in the space. Few doubt that this is the direction in which the tech industry will head towards within the next decade.

We have yet to see wholehearted adoption of IoT and AR/VR in the consumer and enterprise spaces, however. It is still too costly for many, and companies are still somewhat only testing the waters with their solutions. A 2016 study by the Economist Intelligence Unit, co-sponsored by ARM and IBM, cited high infrastructure investment costs as the biggest obstacle to widespread IoT adoption. From an anecdotal standpoint, when I initially started writing this book, the HoloLens was still only available for select developers in Canada and the United States, and at a whopping 3,000 USD a piece (or 5,000 USD for the enterprise edition). It is still very expensive, and while it is somewhat more accessible, even corporate teams are unlikely to possess more than one unit for all of their developers, unless it pertains to their core business (e.g., an AR/VR consulting firm).

Contrast this with the Kinect 2, which everyone pretty much already has in their living room if they own an Xbox One and, if not, can be purchased secondhand on eBay for a fraction of the 150 USD retail price. The HoloLens essentially relies on technology similar to the Kinect; they both make use of a depth camera to paint a picture of the world around them for a computer to interpret, as well as incorporate gesture and voice recognition for interactivity. It should thus come as no surprise that the mastermind behind the HoloLens, Alex Kipman, was also the progenitor of the Kinect.

So, why is this important? Let me reiterate my comments more concisely. The opportunity: a burgeoning market for IoT and AR/VR solutions. The impediment: the cost and availability of aforementioned solutions. The answer: the Kinect.

Wait, the Kinect? Why, yes. The Kinect 2 is essentially a cheap and lo-fi way to create IoT and AR/VR solutions. It collects data from its various sensors and sends it to either a local or a remote computer (the IoT aspect), and it serves as an input device for a digital reality (the AR/VR respect). It is a device that can be adapted to many different scenarios and can be prototyped against before a more integrated and custom solution is devised. In other words, it is the perfect device with which a developer or company can breach the world of IoT and AR/VR.

Let me speak from experience: it was a trivial matter to convince a financial institution to purchase several Kinects, but I have trouble even bringing up the idea of purchasing a second HoloLens, and that is understandable. Two HoloLenses can provide only two connected immersive experiences simultaneously (unless you hook up a Spectator View rig, at the cost of immersion). Twenty-four Kinects can support experiences with over 144 individuals simultaneously (and even more in certain scenarios, such as projection mapping), yet the twenty-four together cost half as much—or about the same, accounting for computer requirements—as a single HoloLens. Wrecking one of those Kinects would only be standard business operating procedure. Lose a HoloLens, and we could anticipate some tough questions being asked by management.

I'll be honest with you. This book has been very late, and Microsoft's interest in developing the Kinect for Windows v2 further is all but non-existent. Some may wonder why we even bothered. This book, in itself, is not going to revitalize the Kinect ecosystem. All eyes are on the HoloLenses and the Oculus Rifts, meanwhile the Kinect has been all but relegated to a footnote in the history of gaming. I wouldn't be surprised if I have drawn snickers from some of my former colleagues at Microsoft. I can't say that I didn't have anxiety throughout the writing of this book.

Then I close the Hacker News and Google News tabs on my browser. I open LinkedIn. I look at my Twitter. I visit hobbyist blogs. I walk into restaurants, banks, and research centers. A grad student in Turkey reaches out to me; he's working on a project to monitor the performance of children walking on a balance beam. A man in Texas is developing a solution to convert American Sign Language to voice. An optical chain wants to reduce their frame inventory and have customers try out frames virtually. A startup wants to capture building-layout models. A marketing team wants to use a 3D printer to print the busts of booth visitors. A restaurant wants to set up a projection-mapping system to detail menu-item availability.

Suddenly, the anxiety fades away, and I know again why I wrote this book. The Kinect is no longer a moonshot technology; it is a mature technology. You won't read about it on TechCrunch or The Verge anymore, but talk to regular people, let their imaginations run wild, and you will have there the market for the Kinect. The Kinect for Xbox 360, let alone the Kinect for Xbox One, is still widely used in enterprise, and its use will only skyrocket in the coming years, even when the current crop of IoT and AR/VR systems are discontinued so that companies can sell you marginally higher spec'd hardware.

Perhaps you're a student looking to create an innovative project for the Microsoft Imagine Cup, or an industry professional developing an AR or IoT solution for a retail chain. I hope you find this book useful. I do my best to discuss all the fundamentals of Kinect development. It's not a replacement for the documentation Microsoft should have provided for all of the Kinect's features, and it wasn't intended to be. Through this book, I only wanted to hold your hand as you navigate the marvelous and limitless world of Kinect development until you feel like you can let go and soar; soar beyond the confines of physical touch-based computing while creating your own realities in the process. May you have as much fun writing your Kinect adventures as I had writing mine.

CHAPTER 1

Getting Started

It would be nice if we could just plug the Kinect in, hash out a quick script on Vim, and execute it on a command line, but, alas, seeing as the Kinect for Windows SDK 2.0 is deeply integrated into the Microsoft developer stack and as there are good development tools available for the Kinect, we'll make a short initial time investment to set up our Kinect properly.

Hardware Requirements

The exact hardware setup required to make use of the Kinect for Windows v2 varies greatly depending on what exactly you want to do. If you want to make use of heavy image processing, it would pay to have a souped-up machine, but for basic usage such as you would find with the Kinect samples' code, the following minimum specs should suffice. All examples in this book have been tested on a Surface Pro 2 (W10, i5, 4GB RAM) and a Surface Book (W10, i5, 8GB RAM, dGPU).

Required Specs

- 64-bit (x64) processor
- Physical dual-core 3.1 GHz processor or faster (recommended i7)
- 4GB of RAM or more
- USB 3.0 controller (recommended built-in Intel or Renesas chipset)
- Graphics adapter with DX 11 support (as always, have the latest drivers installed)
- Windows 8 or 8.1, Windows 8 or 8.1 Embedded Standard, or Windows 10
- Kinect for Windows v2 sensor or Kinect for Xbox One with the Kinect Adapter for Windows and associated USB cabling. It is also recommended that you have the power hub.

If you do not have a dedicated USB 3.0 port available and are using a USB 3.0 host adapter (PCIe), ensure that it supports Gen 2. You might have to disconnect other USB devices, as they may restrict bandwidth on the controller for the Kinect. In fact, this is a very common issue with the Kinect 2. The Kinect transmits a lot of data, and even if your USB appears to be 3.0, in reality it might not be capable of providing enough bandwidth to support your applications.

The Kinect should work with tablet devices as well, as long as they meet the aforementioned criteria. This means you could use it for something like a booth with a keyboard-less Surface Pro type device.

© Mansib Rahman 2017
M. Rahman, *Beginning Microsoft Kinect for Windows SDK 2.0*, DOI 10.1007/978-1-4842-2316-1_1

▪ **Tip** If you want to quickly identify whether a machine supports the Kinect 2 for Windows, use Microsoft's Kinect Configuration Verifier tool, which will check your system for any known issues (Figure 1-1). You can download it at `http://go.microsoft.com/fwlink/p/?LinkID=513889`.

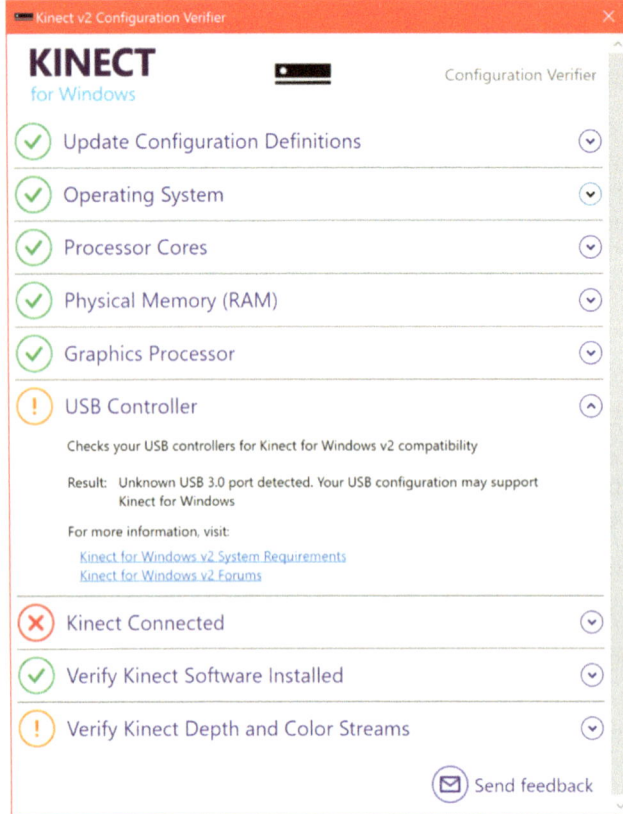

Figure 1-1. *Kinect Configuration Verifier tool*

▪ **Note** The Kinect Configuration Verifier can be misleading at times, specifically in regards to the USB controller. Even if it shows a warning for the USB controller, it is worth testing the Kinect with an actual application before writing the hardware off as unusable.

Software Requirements

- You will need **Visual Studio 2012** or newer. This includes the Express and Community flavors. This does **not** include Visual Studio Code. I personally use Visual Studio Enterprise 2015 (with Update 2), and the examples have been tested with such. You can download a free copy of Visual Studio at `https://www.visualstudio.com/en-us/products/visual-studio-community-vs.aspx`.

- You will need the Kinect for Windows SDK 2.0. You can download it at `https://developer.microsoft.com/en-us/windows/kinect`.

- **Optional**: The speech samples require the **Microsoft Speech Platform SDK Version 11** in order to compile. You can find it at `https://www.microsoft.com/en-us/download/details.aspx?id=27226`.

- **Optional**: The DirectX samples require the **DirectX SDK** in order to compile, which you can grab at `https://www.microsoft.com/en-us/download/details.aspx?id=6812`.

Installing and Setting Up the Kinect

Regardless of whether you purchased a Kinect for Windows v2 or repurposed a Kinect from your daughter's Xbox One, the setup instructions are roughly the same. The key difference is that if you have a Kinect for Xbox One, you will need a Kinect Adapter for Windows (Figure 1-2), which you can purchase on the Microsoft Store at `http://www.microsoftstore.com/store/msusa/en_US/pdp/Kinect-Adapter-for-Windows/productID.308803600`.

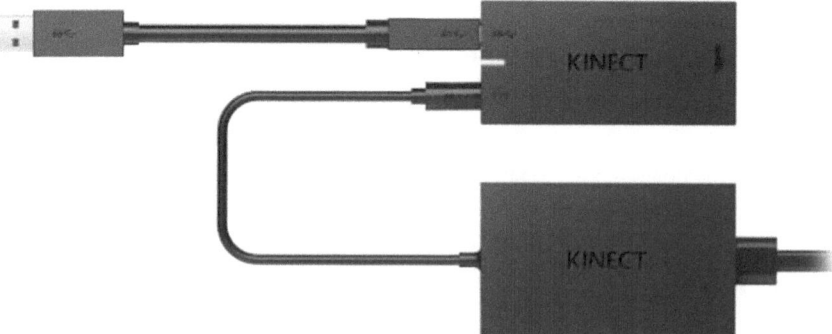

Figure 1-2. *Kinect Adapter for Windows*

Most developers should be able to plug in all the cables of the Kinect on their own without frying anything, but seeing as I very nearly stuck the power code into a USB port, a bit of verbosity in the setup instructions would not be amiss. Before you assemble the hardware, it would be sensible to take care of the software side of things. If you have any other Kinect SDKs and drivers installed on your PC, uninstall them now. This includes open source SDKs such as **libfreenect2** and also any Kinect for Windows v2 Developer Previews that you might have installed at a hackathon or through an insider program. Conflicts are not guaranteed between SDKs, but they have not been tested together, so your mileage might vary.

The next step would be to install the Kinect for Windows v2 SDK, if you have not already. Make sure your Kinect is not connected to your computer and that Visual Studio is not open. Run the installer, accept the terms, and click **Install**. Usually, this will be fairly quick.

■ **Caution**   The technical reviewer felt that it was especially necessary for me to reiterate that you should **not** plug in your Kinect first, and I have to agree. The automated plug and play behavior of Windows can mess with your Kinect and cause you a lot of pain in trying to uninstall and reinstall different drivers and software until you get it right.

To assemble the Kinect, start by attaching the power cord (A in Figure 1-3) to the AC adapter (B). Of the two black bricks that make up the Kinect kit, the adapter is the one that conveniently says "Microsoft AC Adapter" underneath and has a permanently attached power cable. Connect the AC adapter's permanent cable to the designated port on the Kinect for Windows Hub (C). The power port is the one beside the blue Type-B SuperSpeed USB connector port on the Hub. Attach the Kinect's proprietary cable (D) to the other end of the Hub. Finally, attach the USB cable's Type-B connector to the Hub and the other end of the cable to a USB 3.0 port on your PC.

■ **Note**   Your Kinect may have come packaged with two power cords. Normally, this is not a mistake. They are actually two different power cords. One of them is polarized. You can distinguish the polarized cord by the fact that one of its prongs is slightly wider than the other.

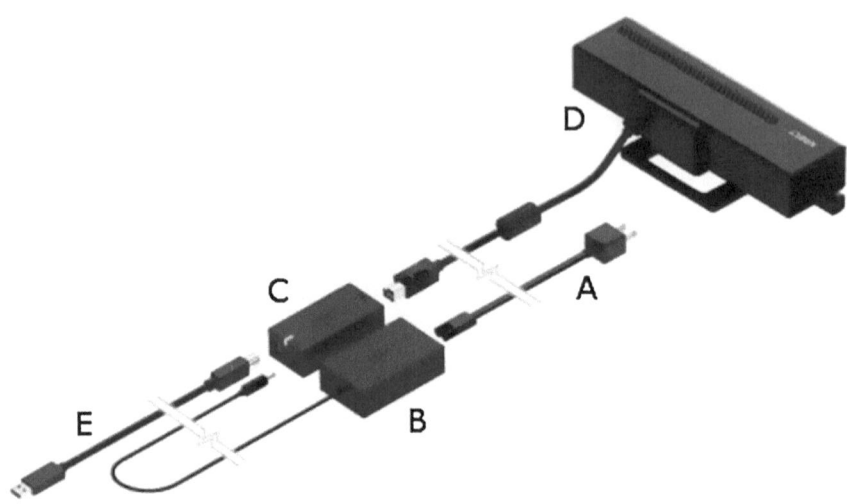

Figure 1-3. *Kinect cabling diagram*

Plug the power cord into a socket, and your PC should start installing the Kinect's drivers. If an *unrecognized driver* error pops up, reconnect the USB to your PC, preferably to a different port, right-click on the Windows icon on the bottom left of your screen, and navigate to the *Device Manager* option on the popup. In the Device Manager window, look for *Kinect sensor devices*, right-click, and then click *Scan for hardware changes.* The drivers should now be installed properly. If this fails, you might not have the proper USB chipset required to support the Kinect. You can take consolation, however, in the fact that I had to attempt this two or three times to get it working with my Surface Book.

Verifying That the Kinect Driver Is Installed Correctly

Start out by checking that the **KinectMonitor.exe** service is running on your machine. Right-click on the Windows icon on the bottom left of your screen (if you are running Windows 8 and have no icon, right-click the bottom left corner of the taskbar) and select *Task Manager* from the drop-down. If there is a *More details* button on the bottom left of the Task Manager window, click it. Scroll down on the Background Processes section and check to see that **KinectMonitor.exe** (Figure 1-4) is one of the entries. You might also notice services such as **KinectService.exe** and **KStudioHostService.exe**. These are actually legacy services that were included in the preview version of the SDK. They have been folded into KinectMonitor.exe and are individually no longer required for the Kinect to operate. You can keep them for backward compatibility, but otherwise you can go ahead and remove them.

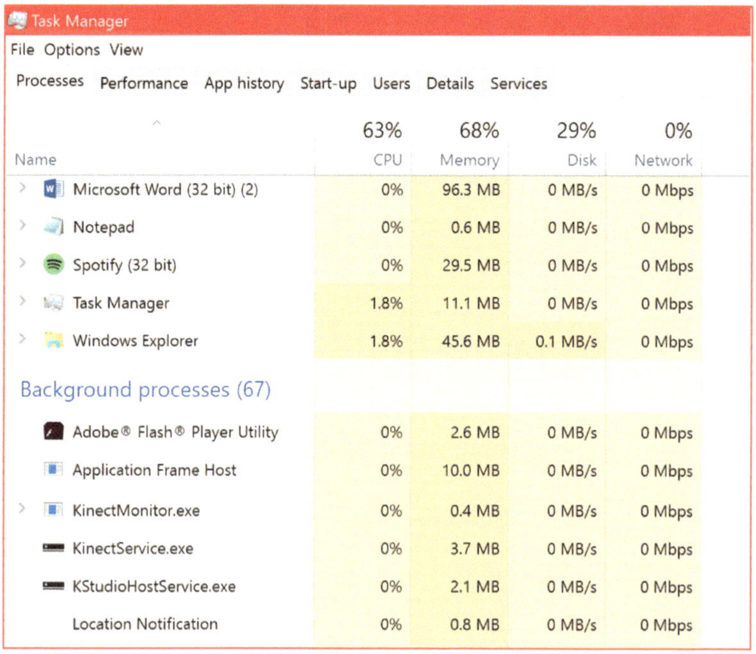

Name	63% CPU	68% Memory	29% Disk	0% Network
> ■ Microsoft Word (32 bit) (2)	0%	96.3 MB	0 MB/s	0 Mbps
> ■ Notepad	0%	0.6 MB	0 MB/s	0 Mbps
> ■ Spotify (32 bit)	0%	29.5 MB	0 MB/s	0 Mbps
> ■ Task Manager	1.8%	11.1 MB	0 MB/s	0 Mbps
> ■ Windows Explorer	1.8%	45.6 MB	0.1 MB/s	0 Mbps
Background processes (67)				
■ Adobe® Flash® Player Utility	0%	2.6 MB	0 MB/s	0 Mbps
■ Application Frame Host	0%	10.0 MB	0 MB/s	0 Mbps
> ■ KinectMonitor.exe	0%	0.4 MB	0 MB/s	0 Mbps
■ KinectService.exe	0%	3.7 MB	0 MB/s	0 Mbps
■ KStudioHostService.exe	0%	2.1 MB	0 MB/s	0 Mbps
Location Notification	0%	0.8 MB	0 MB/s	0 Mbps

Task Manager — File Options View — Processes Performance App history Start-up Users Details Services

Figure 1-4. *Kinect services in Task Manager, notably KinectService.exe*

Next, take a look at your Kinect sensor. Whenever you turn on a Kinect application, there should be a white light turned on at the right end of the device. This white light is the privacy light that indicates that the Kinect is potentially recording. Additionally, if your application requires the services of the Kinect's depth camera or skeletal detection, there should be three red LEDs illuminated at the center of the device, as shown in Figure 1-5.

Figure 1-5. *The Kinect's three red LEDs and white power light. In Xbox One versions, the white light will be shaped liked the Xbox One logo*

Finally, confirm that the Kinect hardware shows up in Device Manager (Figure 1-6). Right-click the Windows icon on the bottom left of your screen on the taskbar and select *Device Manager*. Confirm that the following entries are all present:

- Under *Audio inputs and outputs*, you should see *Microphone Array (Xbox NUI Sensor)*.

- Under *Kinect sensor devices*, you should see *WDF KinectSensor Interface 0*.

- Under *Sound, video and game controllers*, you should see *Xbox NUI Sensor*.

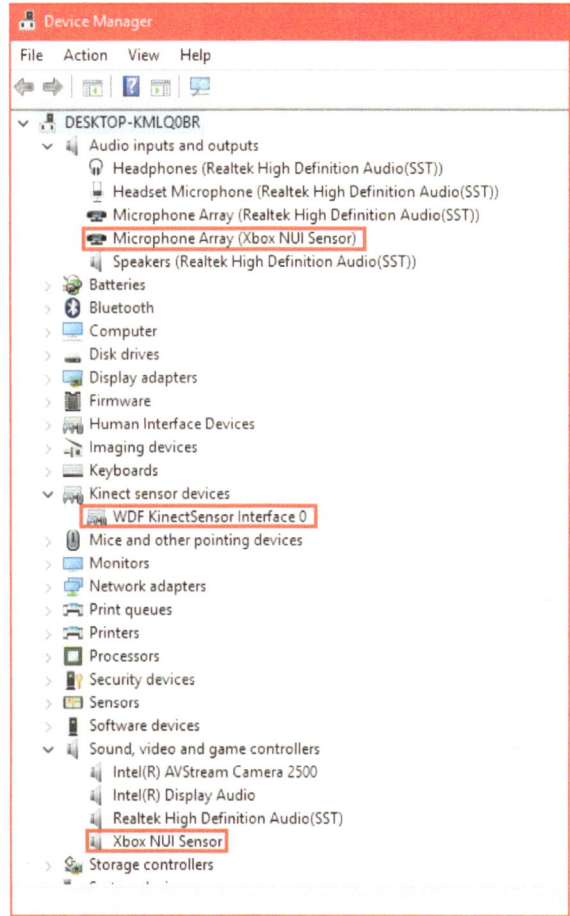

Figure 1-6. *Kinect hardware entries in Device Manager*

Do not fret if the entries are not spelled exactly as shown here; they only need to be similar. If any of the aforementioned criterions failed to manifest, uninstall the Kinect for Windows SDK 2.0 and reinstall it, then restart your machine. Sometimes things just don't fall into place, for undocumented reasons. This might have happened because the Kinect's firmware was being updated or for some other reason along this line.

Positioning the Kinect, and Other Considerations

While the Kinect is relatively versatile, you want to set it up so that it acquires the most accurate representation of the environment surrounding it. The better it can see you, the less data it has to computationally infer about your position and gestures, and the more seamless the resulting user experience is. If I could summarize the content found in this section in one instruction, it would be to position the Kinect in such a way that it can consistently see your whole body and listen to you without interference.

You should strive to have the Kinect positioned in a way similar to what you would see in Microsoft's marketing materials. This means 0.6m to 1.8m (2 to 6 feet) off the ground and ideally (but not necessarily) 15cm (6 inches) from the screen. Have the Kinect centered and on the edge of whatever flat, solid platform it is sitting on and tilted downward so that it can see the floor (see Figure 1-7). Any users need to be within 0.5m to 4.5m (1.6 to 14.7 feet) of the Kinect, but should ideally be within 0.8m to 3.5m (2.6 to 11.5 feet). The Kinect detects audio input from ± 50 degrees from the center of the device and visual input from ± 35 degrees horizontally, thus you should keep any users of the device within this range. The Gen 1 Kinect had a tilt motor, but the Kinect for Windows v2 needs to be adjusted manually. Simply hold the base of the Kinect with one hand and the back of its body with the other, then lightly tilt the base to make any adjustments. This is not necessary from the get-go, but you can use this to calibrate the vertical field of view once you start streaming the Kinect's camera.

Figure 1-7. *One of my development Kinect setups, positioned above my monitor, tilted downward*

While using the Kinect, avoid wearing black and baggy clothes and avoid putting the Kinect directly in front of sunlight. Never touch the face of the Kinect. Always handle the Kinect by holding the sides. The Kinect tries to keep its internal temperature within a narrow range for optimal operation, and you can help it by not covering any of its vents and by keeping the area around it relatively empty. The Kinect should be able to detect bodies in a dark room, but you should keep the room well-lit to make use of its RGB data (read: color camera).

■ **Note** Since the Kinect can have issues with black clothes, it would not be out of the question for us to think that it could potentially have difficulty with certain dark skin pigments. There have been claims that the Kinect does not detect darker-skinned individuals properly. This myth has little merit to it and has been conclusively debunked by the likes of *Consumer Reports* and the *New York Times*, among others. The Kinect uses infrared to detect individuals rather than visible light, so skin color has no effect on whether it can detect a person. So, why does black clothing affect the Kinect? Certain black clothing dyes tend to absorb infrared light, hampering the Kinect's ability to properly track a user.

The Kinect sits well on any hard, flat, stable surface such as a TV stand, but seeing as you are building desktop apps with the Kinect, it is likely that you are using it in a situation where you need it to be hovering over a booth screen or a projector. Fortunately, the Kinect v2, unlike its predecessor, comes with a 1/4-20 UNC thumbscrew–compatible receptacle, the kind you find on a camera in order to mate with tripods. This means you can use the Kinect with a tripod or any other instrument meant to stabilize and elevate cameras and other photographic equipment. There is an official Kinect mount made by a third party that you can check out on the Microsoft Store at `http://www.microsoftstore.com/store/msusa/en_US/pdp/Kinect-TV-Mount-for-Xbox-One/productID.295240200`. Note that the Xbox 360 Kinect mount won't work on the Kinect for Xbox One.

A final word on the topic: as developers, it is more than likely that we will simply have the Kinect strewn on the desk, shelf, or bed next to where we are hacking. Mine is currently sitting on a sofa. To be completely honest, this will work. Just make sure to test your solutions with a proper setup before you commercialize them or give them to others to use. Also, avoid carelessly dropping the Kinect. The Kinect has sensitive parts that will give erroneous measurements if they have been compromised in any way. Fortunately, Microsoft spent quite of time drop-testing the Kinect and putting it through shock tests to make sure these parts refrain from budging.

Starting a Kinect Project in Visual Studio

If you have developed on the Microsoft platform before (WPF, WinForms, UWP, . . . err, MFC?), you could probably skip through to the "Completing & Launching Your Kinect Project in Visual Studio" section. Just know to add the `Microsoft.Kinect` reference to your Visual Studio project through the DLL installed on your computer by the SDK or through NuGet. Seeing as many Kinect developers tend to be students, hobbyists, and researchers who do not typically get to play with the Microsoft stack (for better or for worse), I will take the time to spell out the steps properly.

The Kinect has three APIs: a native C++ API, a managed .NET API, and a WinRT API. There is strong parity between the three APIs, so we will mainly focus on the .NET API, though you can also use C++, HTML/JavaScript, Python, and other languages as well, should you use third-party Kinect SDKs. I will be using C# throughout this book, and I suggest you do too if you are new to development on the Microsoft platform. It is syntactically similar to Java and conceptually to an extent as well.

If you have not downloaded and installed Visual Studio (and not Visual Studio Code) on your computer yet, go do so now, as Kinect development revolves around Visual Studio (though you could probably pull it off with MonoDevelop and its ilk). When that is done, open Visual Studio and click on *New Project* on the Visual Studio Start page, as shown in Figure 1-8. Alternatively, you can click on *File* ➤ *New* ➤ *Project. . .* through the top-left menus. In the New Project dialog, select *Visual C#* templates in the tree menu found to the left. The center column of the New Project dialog will list several C# project types that Visual Studio is configured to support. Choose *WPF Application* from this list. At the bottom of the dialog, you can set a name for your project, as well as a folder location. Once you are satisfied with these options, click OK.

■ **Tip**   WPF, or Windows Presentation Foundation, applications are the typical desktop apps we have seen on Windows systems for many years now. It consists of a graphical subsystem for rendering UIs on Windows and utilizes DirectX (the implementation of which is abstracted from developers). The UI is designed using an XML-based language called XAML. The other major desktop app technology for .NET apps, WinForms (think of AWT in Java), is currently in maintenance mode, and it is recommended that beginners start with WPF or the similar, but more recent Universal Windows Platform (UWP).

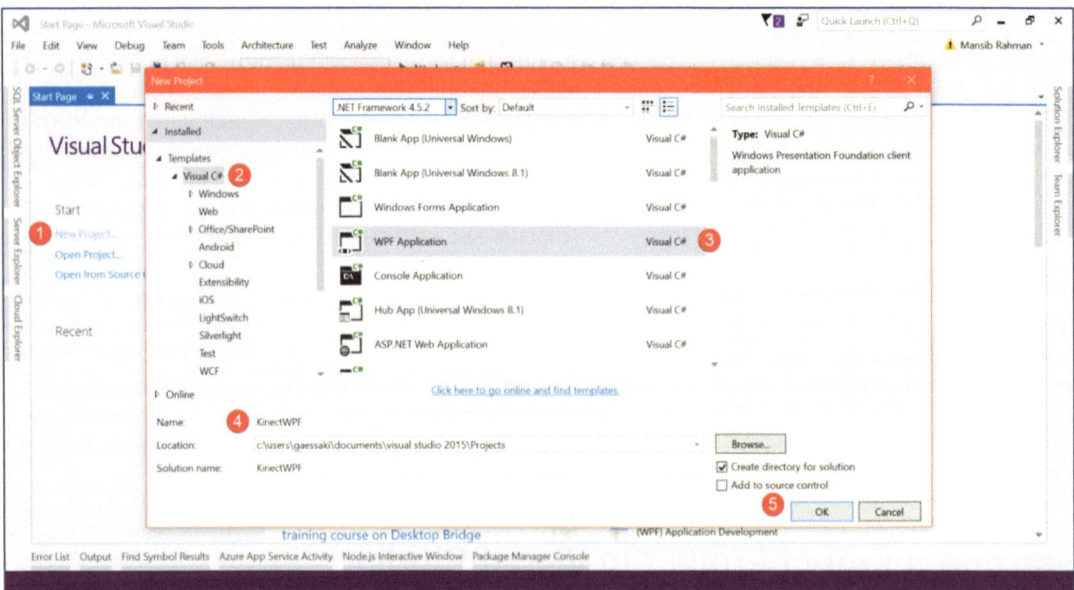

Figure 1-8. *Steps to create a new Kinect project*

After you have created a new project, the next step is to add Kinect references to your project. This ensures that your code has access to the Kinect API that Microsoft built, and it must be done every time you create a new Kinect project. There are two different methods for adding the references to your project. You can browse for DLLs installed on your computer and reference them, or you can add the references through NuGet, a package manager for the Microsoft development platform (think npm for Node.js). The method recommended by the Kinect team is NuGet. This is advantageous if you are working with a team, as you can keep your binary data out of source control, but for most amateurs, either method should suffice. I will go over them both.

Adding the Kinect Reference Through a Local Reference

To add the Kinect reference, repeat the following steps:

1. Find the *References* entry in Solution Explorer (Figure 1-9). Right-click and choose *Add Reference. . . .* If you accidentally closed Solution Explorer, you can bring it up again by pressing **Ctrl + W**, followed by the **S** key. Alternatively, you can click on the View menu at the top-left corner of the window to find it.

10

2. In the Reference Manager dialog, click *Extensions* under the *Assemblies* tree menu on the left.

3. Scroll across the center column of the Reference Manager dialog until you find the `Microsoft.Kinect` entry. Make sure it is checked. Note the other Kinect entries underneath. You do not need them at the moment, but you may need them later, and the method to add them is the same.

4. Click OK.

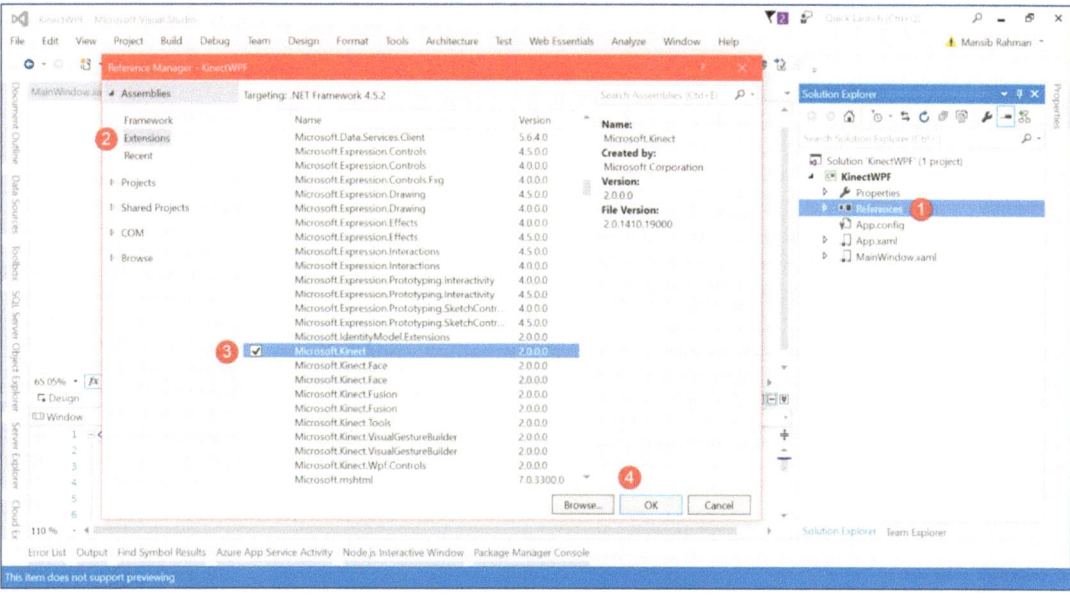

Figure 1-9. *Steps to add the Kinect reference to a Visual Studio project*

Adding the Kinect Reference Through NuGet

To add the Kinect reference, do the following:

1. Find the *Tools* menu at the top left of your Visual Studio window, select *NuGet Package Manager* from the drop-down, and click *Package Manager Console*.

2. In the NuGet console window that appears, type `Install-Package Microsoft.Kinect` and press Enter (Figure 1-10).

The Kinect API should now be referenced. If you encounter an error, check to make sure you are connected to the Internet.

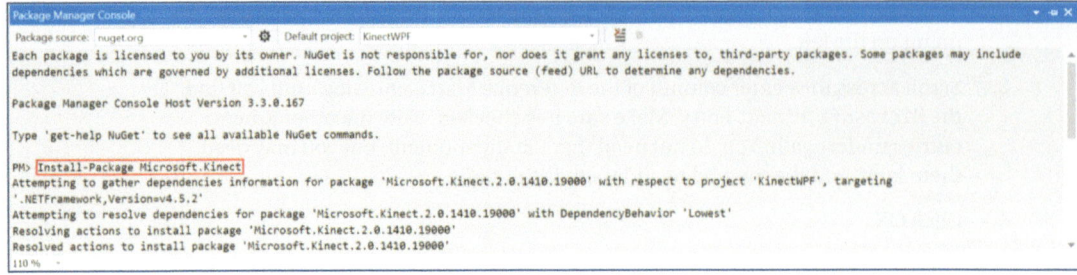

Figure 1-10. *Installing the Kinect reference through NuGet*

Completing and Launching Your Kinect Project in Visual Studio

Now that we have referenced the Kinect API in our project, we can utilize it to interface with the Kinect. Let us start with a trivially simple project. We will create a program that detects whether the Kinect is connected to our computer and operational and, if so, displays "Hello World! Kinect is Ready!"

Start off by navigating to the Solution Explorer and opening the `MainWindow.xaml.cs` file, which should be found under `MainWindow.xaml`. **These are not the same files.** The former is the code-behind, whereas the latter contains the user interface.

I will give you the source code for our app right off the bat in Listing 1-1, but no need to worry—I will go over the code line by line. The following code will end up replacing the stock code given to us by Visual Studio in `MainWindow.xaml.cs`. Once again, I will indulge in some verbosity for the sake of amateur developers or readers unfamiliar with C#/.NET. More experienced .NET developers can probably skim through this section.

Listing 1-1. MainWindow.xaml.cs

```
using System.Windows;

using Microsoft.Kinect;
using System.ComponentModel;

namespace KinectWPF
{
    /// <summary>
    /// Interaction logic for MainWindow.xaml
    /// </summary>
    public partial class MainWindow : Window, INotifyPropertyChanged
    {

        private KinectSensor kinect = null;
        private string statusText = null;

        public MainWindow()
        {
            this.kinect = KinectSensor.GetDefault();
            this.kinect.IsAvailableChanged += this.Sensor_IsAvailableChanged;
```

```csharp
            this.StatusText = this.kinect.IsAvailable ? "Hello World! Kinect is Ready!" :
            "Goodbye World! Kinect is Unavailable!";
            this.DataContext = this;

            InitializeComponent();

            this.kinect.Open();
        }

        public event PropertyChangedEventHandler PropertyChanged;

        private void Sensor_IsAvailableChanged(object sender, IsAvailableChangedEventArgs e)
        {
            this.StatusText = this.kinect.IsAvailable ? "Hello World! Kinect is Ready!" :
            "Goodbye World! Kinect is Unavailable!";
        }

        public string StatusText
        {
            get
            {
                return this.statusText;
            }

            set
            {
                if (this.statusText != value)
                {
                    this.statusText = value;

                    if (this.PropertyChanged != null)
                    {
                        this.PropertyChanged(this, new PropertyChangedEventArgs("StatusText"));
                    }
                }
            }
        }

        private void MainWindow_Closing(object sender, CancelEventArgs e)
        {
            if (this.kinect != null)
            {
                this.kinect.Close();
                this.kinect = null;
            }
        }
    }
}
```

First, we remove all the using System Windows.*; statements, except for using System.Windows; itself, as we will not need those libraries. After our only using statement, we add using Microsoft.Kinect; followed by using System.ComponentModel; on the next line (Listing 1-2). This tells the compiler that we will be using the Kinect API namespace in this class along with the System.ComponentModel namespace, which I will discuss momentarily.

Listing 1-2. MainWindow.xaml.cs Using Statements

```
using System.Windows;

using Microsoft.Kinect;
using System.ComponentModel;
```

Next, we declare our own namespace, KinectWPF, and our class, MainWindow, as shown in Listing 1-3. This code should have already been generated by Visual Studio with whatever name you gave to your project, and **you should not change it to KinectWPF.** The namespace organizes our code under one scope. It is not too important for our little project, but it is used extensively in large projects to reduce naming conflicts.

The triple slashes are just comments that do not affect the code. Typically, you would write comments with double slashes, but the triple slashes generated by Visual Studio here are XML comments for use with IntelliSense.

We will make a small addition to our class declaration. We will add the INotifyPropertyChanged interface at the end, which we access from the System.ComponentModel namespace mentioned earlier. This lets our UI elements know that they can subscribe to the PropertyChanged event and update themselves accordingly. In our app's case, we will trigger an event whenever the computer detects that the Kinect has been plugged in or unplugged that will pass a string for our UI to print.

Listing 1-3. MainWindow.xaml.cs Namespace and Class Declaration

```
namespace KinectWPF
{
    /// <summary>
    /// Interaction logic for MainWindow.xaml
    /// </summary>
    public partial class MainWindow : Window, INotifyPropertyChanged
    {
        //... rest of the code will go here
    }
}
```

Inside our class declaration (Listing 1-4), we will initialize two variables. KinectSensor kinect represents any Kinect sensor; we will soon set it as the Kinect attached to our computer. statusText is a string we will use internally to store the status of our Kinect (either "Hello World! Kinect is Ready!" or "Goodbye World! Kinect is Unavailable!").

Listing 1-4. MainWindow.xaml.cs Class Constructor and Variable Initialization

```
private KinectSensor kinect = null;
private string statusText = null;

public MainWindow()
{
    this.kinect = KinectSensor.GetDefault();
    this.kinect.IsAvailableChanged += this.Sensor_IsAvailableChanged;
```

```
    this.kinect.Open();
    this.StatusText = this.kinect.IsAvailable ? "Hello World! Kinect is Ready!" : "Goodbye
    World! Kinect is Unavailable!";
    this.DataContext = this;

    InitializeComponent();
}

public event PropertyChangedEventHandler PropertyChanged;
```

In our class constructor, `public MainWindow()`, we start by calling the `GetDefault()` method of `KinectSensor`. This assigns any available Kinect sensor as the Kinect from which we will be streaming data. We have an event on `kinect` called `isAvailableChanged` that is raised whenever our computer detects that our Kinect's availability has changed. We assign it (with the += operator) an event handler method called `Sensor_IsAvailableChanged`. This method will let our program know what to do when an `isAvailableChanged` event is raised by the machine. We will define it later.

`IsAvailable` is a property that tells us whether the Kinect is available. We are setting the initial status of the Kinect with this line. If it returns true, we assign "Hello World! Kinect is Ready!" to `StatusText`, and otherwise we set it to "Goodbye World! Kinect is Unavailable!" `StatusText` is a public getter and setter for our `statusText` string.

The `Open()` method of our `kinect` object physically starts our real-world Kinect. As shown in Figure 1-5, the Kinect's privacy light should turn on when the device is in operation. This method is responsible for turning on that light.

`DataContext` is a WPF property that allows our data in the code-behind to interact with our user interface. I will not go too deeply into how exactly it is used, as it would be more appropriate in a book about WPF. In our case, we are declaring that our UI may use the properties of our `MainWindow` class.

`InitializeComponent()` is a Visual Studio–generated piece of code used to bootstrap the UI.

`PropertyChanged` is an event we will raise whenever the computer raises its own event saying that our Kinect's availability has changed. We will raise this event to execute a method to alter our UI.

The rest of our back-end code consists of a couple of methods and a property (Listing 1-5). Our first method, `Sensor_IsAvailableChanged`, is the one we configured earlier to be called whenever the Kinect's availability changes, and it raises an event. In the method body, we have a line similar to one seen earlier that sets the `statusText` based on the Kinect's availability.

Listing 1-5. MainWindow.xaml.cs Method Definitions

```
private void Sensor_IsAvailableChanged(object sender, IsAvailableChangedEventArgs e)
{
    this.StatusText = this.kinect.IsAvailable ? "Hello World! Kinect is Ready!" : "Goodbye
    World! Kinect is Unavailable!";
}

public string StatusText
{
    get
    {
        return this.statusText;
    }
```

```
    set
    {
        if (this.statusText != value)
        {
            this.statusText = value;

            if (this.PropertyChanged != null)
            {
                this.PropertyChanged(this, new PropertyChangedEventArgs("StatusText"));
            }
        }
    }
}

private void MainWindow_Closing(object sender, CancelEventArgs e)
{
    if (this.kinect != null)
    {
        this.kinect.Close();
        this.kinect = null;
    }
  }
}
```

The getter for our statusText string is simple, but the setter is slightly more involved. First, we assign its value if it is different from the incumbent (value is a reserved C# keyword for the data assigned to a setter). Then, we proceed to raise our own event, letting the UI know that it should change the text of any UI element (otherwise referred to as a *control*) bound to the StatusText property.

Our final method, MainWindow_Closing, is a bit of good practice. All we are doing here is turning off our Kinect by calling Close() and disassociating from Close() by declaring the kinect object as null. This method will be called by our UI whenever we decide to exit our program.

Creating the User Interface

Are we done? Not quite—just one last bit. Switch over to MainWindow.xaml in Solution Explorer. You should see our UI in a designer window, as well as a bit of XAML code on the bottom. We are going to make some adjustments so that the UI can display the Kinect's status with some text. The final product for the XAML code should look like Listing 1-6.

Listing 1-6. MainWindow.xaml User Interface

```
<Window x:Class="KinectWPF.MainWindow"
        xmlns="http://schemas.microsoft.com/winfx/2006/xaml/presentation"
        xmlns:x="http://schemas.microsoft.com/winfx/2006/xaml"
        xmlns:d="http://schemas.microsoft.com/expression/blend/2008"
        xmlns:mc="http://schemas.openxmlformats.org/markup-compatibility/2006"
        xmlns:local="clr-namespace:KinectWPF"
        mc:Ignorable="d"
        Title="MainWindow" Height="350" Width="525"
        Closing="MainWindow_Closing">
```

```
<Grid>
    <TextBlock x:Name="textBlock" Text="{Binding StatusText}" HorizontalAlignment="Left"
        Margin="100,100,0,0" TextWrapping="Wrap" VerticalAlignment="Top" FontSize="36"/>

</Grid>
</Window>
```

We are only making two changes to the stock Visual Studio XAML. First, we are adding the `Closing="MainWindow_Closing"` attribute to the `<Window .../>` tag. If you recall from our code-behind, `MainWindow_Closing` is a method that we call when closing the app to properly turn off our Kinect. Here, we are letting the UI know to call it when the user exits the application.

In the `<Grid>` element, we have added a `TextBlock` element. This element will simply display text as per our customizations. Its only essential attribute is `Text="{Binding StatusText}`. Earlier, we set up a `DataContext`, and here all we are doing is letting the `Text` property of `TextBlock` be bound to the `StatusText` property in `MainWindow.xaml.cs` so that it is updated whenever the property changes in the back end.

Keep in mind that as in our back-end code, if you named your project something other than `KinectWPF`, you should keep those names as they appear in the XAML (e.g., in `<Window x:Class="KinectWPF. MainWindow">`).

Whew. If you have not programmed much in your life, this might have been a boatload for you to swallow and if you are an experienced .NET dev, you might have been rolling your eyes throughout (or perhaps it is the other way around?). Regardless, I hope the point you will take away from this is that programming for the Kinect is not conceptually difficult. The rest of the Kinect's sensors actually work in a way similar to its availability detector. If the Kinect has data of any kind, it raises an event, which triggers the associated event handlers, and then you can do what you want with the data. We will see this pattern throughout this book.

Firing Up the Kinect and Taking It for a Spin

As long as your Kinect is plugged in and connected to your computer, any application that wants to use it will be able to access it directly, so you do not have to worry about turning it on or having the user perform any laborious tasks to make use of it. In our case, we will simply run the program in Visual Studio to get started. Press the Start button in Visual Studio, as shown in Figure 1-11. Alternatively, simply press the **F5** key.

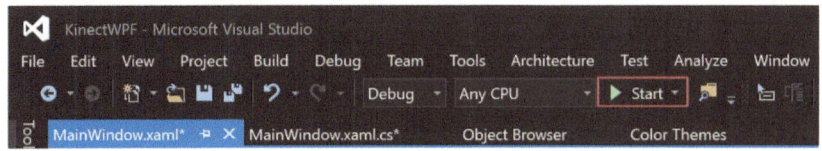

Figure 1-11. *Typical location for the Start button in Visual Studio*

Ready? Get set . . . GO! Ermm, not too many fireworks yet, eh? Do not fret; we will get to all the fun stuff shortly. For now, you should have a little white window on your desktop that displays "Hello World! Kinect is Ready!" as in Figure 1-12. Of course, this is only if your Kinect is connected. Otherwise, it should say "Goodbye World! Kinect is Unavailable!" Go ahead and pull the Kinect's USB cable out of your PC. Nothing happened? Patience is key here. Your computer takes a few seconds to realize that the Kinect has been disconnected. At this point, the "Hello" should change to "Goodbye."

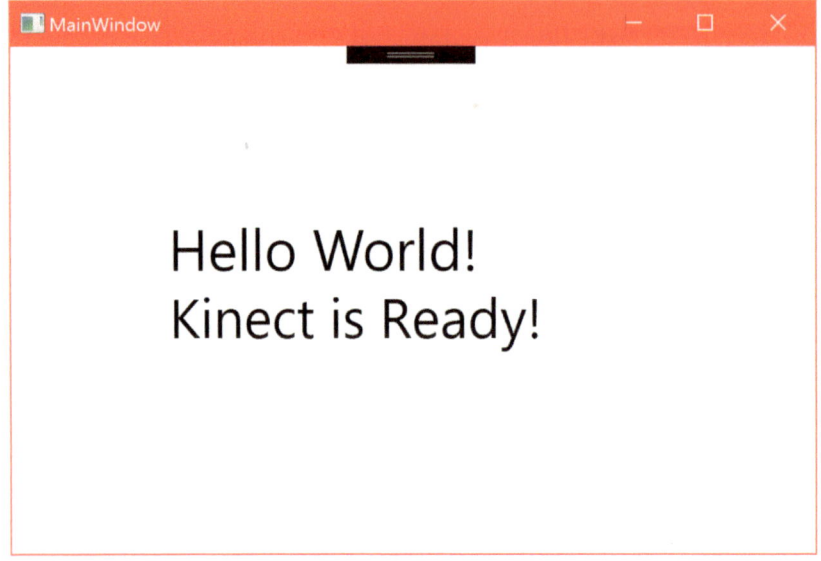

Figure 1-12. *Our barebones Kinect Hello World! app*

Now, of course, you should never yank your Kinect's USB like that. At least, not while it is in use. There are no documented ill effects, as the Kinect does support hot swapping and disconnecting, but, depending on your computer hardware, you may get the rare Blue Screen of Death. The Kinect's cable will get yanked out in the normal course of things, however. Maybe a client at your booth will accidently trip on it from the excitement of getting to try your game out. Maybe you will pull it out in frustration. Either way, you need to know how to deal with such a situation programmatically, the basis for which we demonstrated here.

Kinect Evolution

Before we start programming an entire Kinect app from the ground up with lasers and all (did someone say lasers? Oh, just the infrared lasers in the Kinect), let us take some cues from one of Microsoft's Kinect samples. The new samples are not as creative as the ones they had for the Kinect for Windows v1, but they get the job done. We will take a look at the Kinect Evolution sample. You can find it by searching for *SDK Browser (Kinect for Windows) v2.0* on your PC. Once you open it, scroll all the way to *Kinect Evolution-XAML* on the list, as shown in Figure 1-13, and click the *Install From Web* button. Download the project and extract it to a location of your choice. Compile the project and run it.

Figure 1-13. *Kinect Evolution in the SDK browser*

■ **Note** If you get an error about missing Windows 8.1 Tools, you can choose to install them, or, alternatively, you can download the Kinect Evolution app from the Windows Store for free and run that instead.

Once the app is running, you should see a summary of all the different sensor streams from the Kinect (Figure 1-14). There's a fair chance that your Kinect is positioned awkwardly because you have it attached to your desktop or laptop and thus you do not have enough space, but if you give the camera some distance, you should be able to see the Body Joints and Block Man streams as well. If you click the large left window, you can cycle through the raw visual streams of the Kinect (e.g., infrared and color). The audio stream should show you from which direction the loudest noise is coming, as well as its intensity.

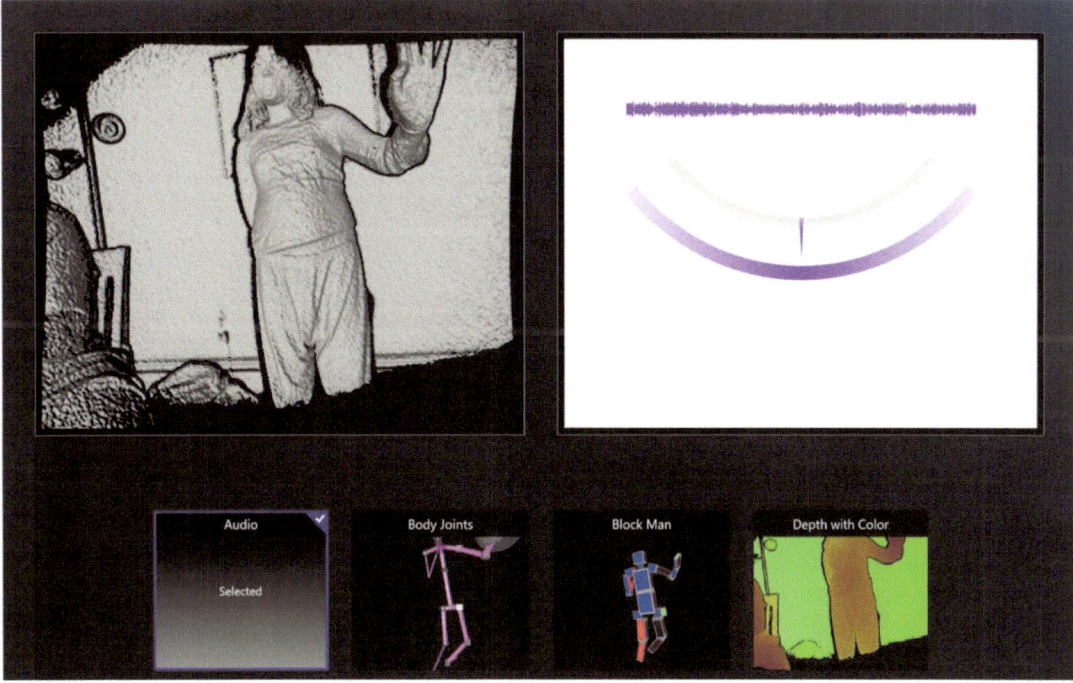

Figure 1-14. *Kinect Evolution: A summary of the Kinect's various sensor streams*

All the streams are interpretations of data coming from the Kinect rather than a direct stream. For example, in the Depth with Color stream, we are programmatically coloring the depth pixels from the regular depth stream to our specifications. The Kinect API does not naturally provide this rendered stream in a one-liner. In the Block Man stream, we are rendering meshes with DirectX based on the joint locations and orientations. Again, the Kinect does not provide much more than raw data from which we must build our worlds. It is interesting to note that the red boxes represent interpolations the Kinect has made about the user's joint locations, as they are obstructed or otherwise not very distinguishable through the Kinect's camera.

Summary

In this chapter, you learned how to install your Kinect and deploy a very basic app. We have not explored Kinect data sources in depth, but an overarching pattern for working with them was introduced.
All of Kinect's sensory data comes in data sources, and we manipulate them by raising events and giving instructions via event handlers. In the next chapter, we will uncover exactly what data sources the Kinect has to offer, along with what other features the Kinect has available.

CHAPTER 2

■ ■ ■

Understanding How the Kinect Works

Although it is conceivable that we can learn to build a house without an education in physics, learn to be a chef without taking a course in food chemistry, and learn about programming without learning how a computer fundamentally relies on transistors and machine code, there is a reason that engineering schools, cooking academies, and computer science programs typically teach the theory that begot the profession before they teach how to actually take part in the profession. We may know to sear steak above a certain temperature for a certain time to hit medium rare, but if we know the Maillard reaction, we are on our way toward knowing how to cook hundreds of foods with perfect browning and flavor, even ones we have never before encountered. Similarly, in this chapter we will cover how the Kinect fundamentally works from an engineering perspective, followed by an overview of its software interface.

Exploring the Hardware

The Kinect is a quite revolutionary piece of hardware, if I may say so myself (though I admit that I may potentially harbor the slightest of biases). It is not so much that the Kinect's features are unprecedented, but rather the fact that Microsoft was able to package all of them in a neat little black plastic box for a comparatively cheap price. Commercial systems with only a subset of the Kinect's capabilities ran in the thousands of dollars only a few years ago. Microsoft was only able to retail the Kinect for consumer use so cheaply because of its having development budgets in the hundreds of millions (the advertising budget for the Kinect for Xbox 360 alone was an estimated 500 million USD), elite scientists at Microsoft Research offices across the globe, and a strong market position in the multi-billion-dollar gaming industry.

While the new Kinect for Xbox One largely has the same or better capabilities than the original Kinect, the hardware behind it has actually been rebuilt from the ground up. It was not simply a matter of adding higher-resolution cameras; the actual technology behind the Kinect v2 is fundamentally different. The original relied on a *structured lighting* approach, which involved plastering a distinct pattern of infrared dots from its laser, somewhat like a QR code, on everything within its field of view. Knowing that objects farther away would have more distorted infrared dot patterns, the Kinect v1's infrared camera captured this distortion through triangulation and determined the depth of any scenery within its field of view.

© Mansib Rahman 2017
M. Rahman, *Beginning Microsoft Kinect for Windows SDK 2.0*, DOI 10.1007/978-1-4842-2316-1_2

Microsoft initially licensed this technology from an Israeli company called PrimeSense (which was later acquired by Apple). During the incubation period of the first Kinect, Microsoft also had a competing team working on a Kinect that relied on technology from two other companies, Canesta and 3DV. By 2010, both of these companies had been acquired by Microsoft, the price being 35 million USD for 3DV and an undisclosed amount for Canesta. While it was initially assumed that these acquisitions were for some form of patent protection, it turned out that the technology from these two companies found its way into the Kinect for Xbox One. Unlike the Kinect v1, the new Kinect employs a *time-of-flight*, or ToF, camera. ToF cameras function by measuring the time it takes for light emitted by a Kinect to reach an object and come back. Knowing the speed of light in a typical living room atmosphere, the Kinect can calculate the distance between itself and the object. Since the Kinect must be able to work with the speed of light, its IR sensor had to have the precision required to measure in picoseconds, or trillionths of a second.

All this results in a depth image with three times the fidelity, a quarter of the motion blur, less image degradation from ambient lighting, better tracking, and a much larger field of view than the original Kinect. Essentially, Microsoft looked at every issue that developers had qualms about in the first Kinect and sought to quench them.

Inside the Kinect

You should never try to take apart your Kinect (whether that is an implicit license to take it apart, I cannot say). To save you the trouble of having to figure out what is in that little black box, I have taken the time to show you the innards right here. Figure 2-1 shows a Kinect without its protective cover. All the coolest parts of the Kinect are on the front face. All the way to the left, we have our regular color camera. The smaller circle in between that and the middle is the depth camera. The depth camera does not show up on the front face of the casing like the color camera does, because the casing is transparent but tinted. An interesting effect of the separation between the color camera and the depth camera is that they see the world from slightly different perspectives. That is to say, their respective images do not correspond to each other, and if they were overlaid on top of each other, they would show discrepancies. This can be corrected through the SDK. Right beside the depth camera we have three white rectangles, which are the infrared (IR) emitters. The last thing of interest is the black bar at the bottom. That contains the directional microphone array, which consists of no less than four microphones. The microphone array relies on the differences in phase of the sound waves arriving at each individual microphone to determine the position of the sound, a method that is similar to that used by our own ears. The microphone array is also capable of filtering out sounds that are not voices. You can see their exact positioning by looking at the foam slits underneath the microphone bar.

In addition to the Kinect unit, you will notice the much thicker and more advanced USB 3.0 wiring. The Kinect has really been pushed to the limit this time and has a much larger bandwidth throughput. The upgrade to 3.0 was because USB 2.0 could no longer transport the copious amounts of data from the Kinect, which now typically transfers at around 2.1 Gbit/s. There is 5GB of data on a frame being processed through the Kinect's motherboard before being whittled down for the USB ports.

Figure 2-1. *A Kinect without its plastic casing*

Comparing the Kinect for Windows v1 to the Kinect for Windows v2

The Kinect for Windows v2 is vastly superior to the Kinect for Windows v1, but you would not know it if you just read the first result off Google. My hard drive is teeming with Microsoft Kinect marketing materials that I obtained at various events and from the net, and they all highlight different things! Let us go through the gamut of improvements in the Kinect for Windows v2 and see where they really made changes. The Kinect is probably capable of even more than what I have listed in Table 2-1. Unfortunately, a lot of these functionalities are hard to find because Microsoft has not advertised them cohesively.

Quick Comparison

Table 2-1. *A Side-by-side Comparison of the Kinect for Windows v1 and the Kinect for Windows v2*

Feature	Kinect for Windows v1	Kinect for Windows v2
Color Camera	640 x 480 x 24 bits per pixel 4:3 @ 30 Hz RGB (VGA), 640 x 480 x 16 bits per pixel 4:3 @ 15 Hz YUV	1920 x 1080 x 16 bit per pixel 16:9 YUY2 @ 30 Hz (15 Hz in low light, HD)
Depth Camera	320 x 240 x 16 bits per pixel @ 30 Hz 11-bit IR depth sensor	512 x 424 x 16 bits per pixel 16-bit ToF depth sensor IR can now be used at the same time as color
Range	Practical: 0.8m to 4.5m (2.6 ft.–14.7 ft.) **Default** 0.4m to 3m (1.4 ft.–9.8 ft.) **Near Mode** Absolute: 0.4m to 8m (1.4 ft.–26.2 ft.)	Only one configuration: 0.5m to 8m (1.6 ft.–26.2 ft.) Quality degrades after 4.5m (14.7 ft.)
Angular Field of View	57.5° Horizontal – 43.4° Vertical	70° Horizontal – 60° Vertical
Audio	16-bit per channel with 16 kHz sampling rate	16-bit per channel with 48 kHz sampling rate
Skeletal Joints	20 joints tracked	25 joints tracked; the additional joints are Neck, left and right Thumbs and Hand Tips
Skeletons Tracked	2 with joints, 6 in total	6 with joints (renamed to Bodies)
Vertical Adjustment	Tilt motor with ±27 degrees of freedom	Manual, also ±27 degrees of freedom
Latency	~100ms	~50ms
USB	2.0	3.0

Windows Store Support

Windows Store support has long been requested by Kinect developers, and the Kinect for Windows v2 now enables this. The APIs are pretty much the same as the ones for WPF. While the market for general Window Store Kinect consumer apps is admittedly bleak, such apps are perfect for special use cases like booths or commercial use because of their accessibility. A great example of a Windows Store app that uses Kinect is 3D Builder, which allows you to scan objects and prepare them for 3D printing.

■ **Note** UWP is not fully supported at the time of writing, but there are plans to do so. This book contains an appendix with details on how to get started with UWP and Kinect.

Unity Support

Much-anticipated Unity support is now available for the Kinect for Windows v2. Most developers should be familiar with Unity by now, but for those still programing their games in BASIC, you should take a serious look at Unity. Unity makes it exceedingly easy for any developer to make nearly AAA-quality games for dirt cheap. With the new Kinect plugin, we can now make complex games with the Kinect without having to maintain an entire development team. In a few easy steps, we can add skeletal tracking to Unity games and do cool stuff like avateering or green-screening. While a paid license was initially required, after the release of Unity 5, anyone can use the Kinect plugin with their apps for free.

Unity apps can be deployed to both the desktop and the Windows Store. A thorough discussion on Unity is available in Chapter 7.

■ **Note**   There is also Unreal support available for the Kinect. It is arguably easier to use than the Unity plugin, but is not free. This plugin will not be covered in this book.

Hand-Gesture Support

There is now support for hand-gesture tracking. The Kinect can let us know whether a user's hands are in any one of three recognized gestures or two non-gestures. The three recognized gestures, as shown in Figure 2-2, are *Open*, when a user opens their palm flat; *Closed*, when a user clasps their hand in a clenched fist; and *Lasso*, a gesture that is oft-mistaken for flipping the bird but is really a pointing gesture with two fingers (the middle and index). The non-gestures are *Unknown* and *NotTracked*. Recognizing these hand gestures in our applications enables better user experiences by facilitating user interactions with prompts (e.g., holding Closed grip to signify that the user confirms the selection).

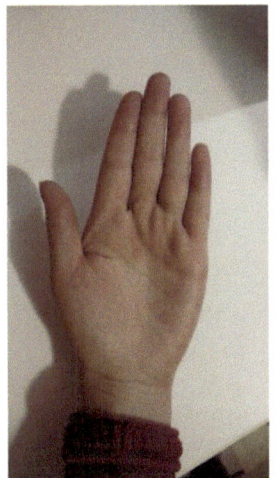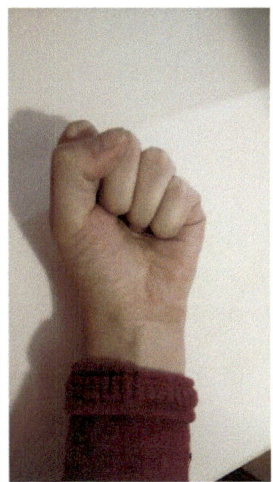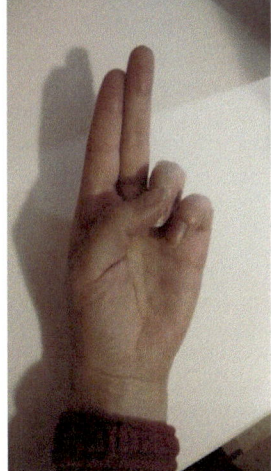

Figure 2-2. *Open, Closed, and Lasso hand gestures*

Visual Gesture Builder

There was no official tool with which to integrate gestures into apps with the Kinect for Windows v1, which meant that the average developer had to essentially track gestures with an assortment of conditional statements. This was very prone to error, not to mention irritating to program. The Kinect for Windows v2 comes with Visual Gesture Builder, a tool that uses machine-learning techniques to help you train gestures and integrate them into your app. The tool is very easy to use and requires no prior knowledge of machine learning.

Face & HD Face API

The original Kinect SDK tracked up to 40 points on a user's face. As with skeletal gestures, we had to make assumptions based on the positional heuristic of face joints to see if a user was making a certain facial gesture. The kind folks at Microsoft went ahead and included a handful of facial gesture–detection algorithms with the Kinect for Windows v2 SDK. These gestures include closed eyes, eyes looking away/not interested, open mouth, moving mouth, happy expression, and neutral expression. Additionally, the Kinect can detect if the user is wearing glasses. The use for this Face API is immediately obvious: we can now determine if a spectator is engaged with our product or not.

There is also an additional HD Face API. HD Face supports a whopping 613 HD face points. It is 3D, whereas the regular Face API is 2D. HD Face can be used for advanced character rigging and character facial animation, among other things.

Kinect Fusion

Kinect Fusion enables 3D object scanning and model creation using the Kinect. The original Kinect Fusion was pretty nifty, but because of throughput issues, low camera resolution, and weakly optimized code, the results were somewhat lackluster. The new Kinect Fusion, depicted in Figure 2-3, has been drastically improved and is much more performant.

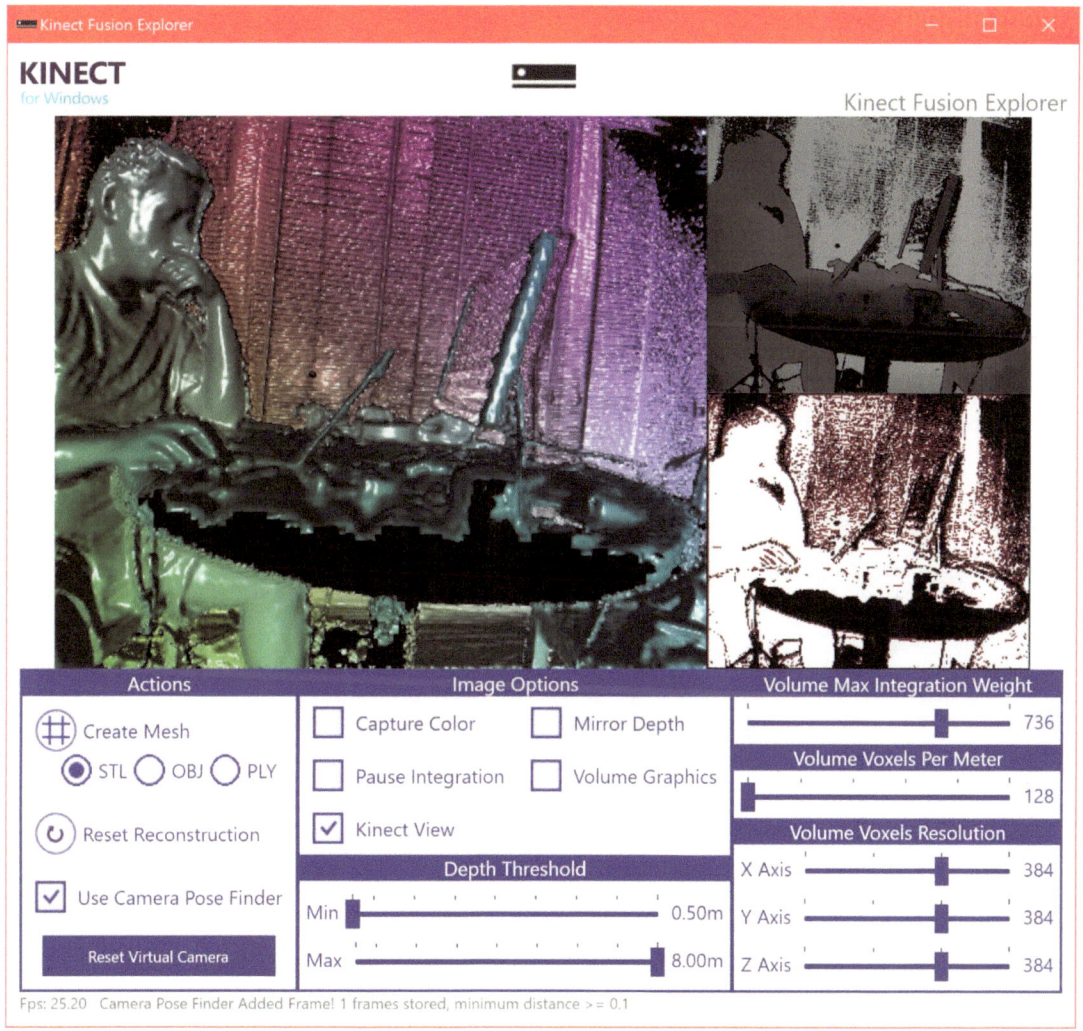

Figure 2-3. *Kinect Fusion reconstruction of the author working on his book*

Kinect Studio

If you have worked with Kinect Studio which was introduced in Kinect for Windows SDK v1.7 onward, you will be glad to know that Kinect Studio makes a comeback in SDK v2. Kinect Studio is an application that allows you to record data with your Kinect and replay it for your apps so as to avoid having to physically debug each time around. It is much more comprehensive than its earlier incarnation, and you now have the capability to toggle the data streams you want to capture. A rundown of the tool will be provided later in this chapter.

Kinect Ripple

The Kinect Ripple is one of the coolest Kinectabilities (can this be a word?) that I have had a chance to play with. Unfortunately, it was never advertised anywhere but within the core Kinect developer circles and in India, and thus never received all the attention that it deserved. Basically, the concept is that you have a projector displaying an image on a wall or screen and another projector displaying an image on the floor. The Kinect is positioned underneath the wall projection, facing away, like in Figure 2-4. The floor projection will become interactive, and you can step on it to activate buttons or tiles. The one on the wall can show a presentation or content based on the interaction with the floor projection. The whole thing can be programmed in JavaScript.

Figure 2-4. *A Kinect Ripple system set up at Microsoft's Hyderabad office*

Kinect RoomAlive Toolkit

Another one of the Kinect's coolest capabilities is RoomAlive. True to its name, it brings your living room to life, and it achieves this by enabling you to do immersive, dynamic projection mapping. The toolkit takes care of calibrating the combined Kinect and projector units (dubbed *procams*), and any surface covered by these units receives a projection-mapping treatment.

In Figure 2-5, we have a real example of a person playing with a RoomAlive app. Creatures are able to erupt from any surface and attack the user. The user can in turn retaliate by hitting them with their limbs or by firing their rifle. You can see where the HoloLens, in particular with its robot shooting game, got its inspiration.

Figure 2-5. *A whack-a-mole type game with a toy rifle (© Microsoft)*

Setting up the system is beyond the scope of this book, but if you want to learn more, you can visit `www.github.com/Kinect/RoomAliveToolkit`. (A future blog post by the author maybe in the works however.

Multi-App Support

The Kinect for Windows v2 now allows for the ability to have multiple apps using the Kinect at the same time. It is a bit of a niche capability, but a very much welcome one in several circumstances. If we created a *Dance Dance Revolution*–type game, for example, we could set up two monitors with two instances of the game running at the same time for two participants competing against each other, while only running one Kinect. Another use case could be having a therapist run a Kinect exercise suite app for one of her patients while at the same time running a general body-analysis app. The analysis app might need to be utilized for many different exercises, so it saves having to integrate it with every new exercise app that is developed.

Exploring the Software Interface

The Kinect for Windows v2 SDK has a robust API that makes extracting any data of interest a cinch. All the sensors in the Kinect have similar design patterns and developer experiences. This pattern generally involves starting the Kinect, delineating which data types we want, and then performing our data manipulation and analysis on frames from a stream of a specific data type. At the forefront of all of this are Data Sources, an API construct that allows us to access data of a specific type.

Understanding Data Sources

Most of the Kinect's data can be conveniently accessed through *Data Sources* defined by the API. Those of you who have tinkered with the Kinect for Windows v1 will know them to be similar to *Streams* (the name change was the result of a stronger adherence to the Windows Design Guidelines at the time of the API's development, which mandated a different definition for the concept of *stream*). Each Data Source focuses on one type of data provided by the Kinect, such as depth data for DepthFrameSource; the full list is described in Table 2-2. Each Data Source provides us with metadata about itself (e.g., whether this Data Source is being actively used by the Kinect) and enables us to access its data through *readers*, which are covered next.

Table 2-2. *The Different Data Sources Available for the Kinect*

Data Source	Description
AudioSource	Supplies audio from the Kinect's microphone array bar. Audio is captured from all directions, but is focused in the form of beams that emphasize sounds from a specific direction. Sounds are split into frames that sync with the camera feeds.
BodyFrameSource	Exposes all data about humans in view of the Kinect sensor. Provides skeletal joint coordinates and orientations for up to six individuals. Start point for any activity requiring the tracking of users.
BodyIndexFrameSource	Yields information on whether a pixel corresponding to a depth image contains a player. Can determine which player occupies a pixel. Useful for green-screening and similar activities.
ColorFrameSource	Provides image data from the Kinect's 1080p HD wide-angle camera. Can be accessed in multiple color formats, such as RGB and YUV. Useful for standard camera-filming and picture-taking operations.
DepthFrameSource	Provides depth data derived from the Kinect's depth camera. Depth distance is given in millimeters from the camera plane to the nearest object at a particular pixel coordinate. Can be used to make custom tracking algorithms and size measurements (do not rely on its accuracy for critical applications).
InfraredFrameSource	Exposes an infrared image from the Kinect's 512 x 424 pixel time-of-flight (ToF) camera. Suitable for computer vision algorithms, green-screening, tracking reflective markers, and imaging in low-light situations.
LongExposureFrameSource	Enables long-exposure infrared photography using the same ToF infrared sensor as InfraredFrameSource. Generates sharp images of stationary features at the expense of blurring bright, fast-moving objects (an effect which can be desired).
FaceFrameSource	Provides recognition of five points on a face in two dimensions (X andY coordinates). This data source also provides data on facial expressions, such as happy and sad. Suitable for facial tracking, facial-expression recognition, and real-time face masking.
HighDefinitionFaceFrameSource	Provides recognition of 36 standard facial points and over 600 more vertices of non-standard facial points in three dimensions (X, Y, and Z coordinates). Suitable for facial recognition, facial micro-expression recognition, and character facial animation.

Readers and Frames

Readers

Most of the Kinect's data is read through readers. The reader's job is to read from the Data Sources and fire an event each time there is a data *frame* available for use, which I will explain briefly.

Readers can access frames in either one of two methods: *Events* and *Polling*. If you have worked with .NET before, Events should be nothing new to you, but for those of you have not, the gist is that a class can use events to notify other classes that something of interest has occurred. In the case of the Kinect SDK, readers raise events that let us know whenever there is a new frame available for us to use. For your typical WPF or Windows Store application, events are the way to go.

The other method, Polling, is a reverse Event of sorts. If you have dabbled with Microsoft's XNA, this should be familiar to you. Instead of waiting for the specific class to drive an interrupt, our application code continuously calls or **polls** the class asking for the current status of whatever you are interested in. So, for example, if you want to know if the Jump button was pressed, your application loop checks to see the status of Jump on each loop, and if it has been pressed you execute the code to handle this. The end result generally is that while you will find out faster whether something of interest has occurred, the system will be taxed unnecessarily as many to most of the calls will return a negative. Of greater concern is the added complexity for any application other than those with the smallest of scopes. Typically, you will go with polling if you are working with game loops or something of the sort.

Accessing Data Sources with readers is as simple as simple can get. Listing 2-1 demonstrates how we would access `ColorFrameSource` and open a `ColorFrameReader`.

Listing 2-1. Opening a Data Source with a Reader

```
private ColorFrameReader colorFrameReader = null;
...
this.colorFrameReader = this.kinect.ColorFrameSource.OpenReader();
```

When a reader is not being utilized, it is best to pause it to minimize the creation of unnecessary frames. If we reach a point in the application where we no longer need to use the Kinect, we can shut off the Kinect sensor altogether, as we saw in Listing 1-5 in Chapter 1. However, if we only need to temporarily pause the collection of data from one Data Source, we are better off pausing the readers, as shown in Listing 2-2.

Listing 2-2. Pausing a Reader

```
this.colorFrameReader.IsPaused = true;
//to re-enable
this.colorFrameReader.IsPaused = false;
```

If we are done with the reader for good, it is best practice to dispose of it, as shown in Listing 2-3.

Listing 2-3. Disposing of a Reader

```
if (this.colorFrameReader != null)
{
    // ColorFrameReader is IDisposable
    this.colorFrameReader.Dispose();
    this.colorFrameReader = null;
}
```

Unlike in the first Kinect SDK, we can now have multiple readers per source. This permits us to modularize our code. An example of this could be for a first-person sports game in which we could have the majority of the screen dedicated to an activity—say, a user trying to hit a target with a bow and arrow—and on the corner of the screen we could have an optional 3D avatar visualization of the person that could be turned off and on. Both the main game and the 3D avatar visualization use the `BodyFrameReader`, but since we want to be able to toggle the 3D avatar visualization and use it for other sports activities like dancing, we put it in its own code module and give it its own reader that can be turned off individually.

> ■ **Note** In addition to the Data Sources listed in the previous section, there is an additional reader that allows us to get all or some of the data (excluding audio and Face API data) at the same time and in sync—the `MultiSourceFrameReader`.
>
> Those who have worked with the Kinect for Windows v1 SDK will find this similar to the `AllFramesReady` event, which also allowed us to access all the streamed data *relatively* synchronously.

Frames

The frame contains all our precious, precious data, in addition to metadata such as dimensions or formatting. Frames are obtained from `FrameReferences`, which are in turn obtained from readers. `FrameReferences` exist mainly to help you keep your frames in sync. Other than the frame itself, `FrameReferences` contain a `TimeSpan` (.NET structure that describes an interval of time) called `RelativeTime`, which informs us when the frame was created. The start time of `RelativeTime` is unknown to us, so we compare it with the `RelativeTimes` of other types of frames to see if they were produced at around the same time.

To access the frames, we add an event handler to our reader, as in Listing 2-4, and whenever our reader raises an event, we will be provided with the most recently produced frame.

Listing 2-4. Adding an Event Handler to a Reader

```
this.colorFrameReader.FrameArrived += this.Reader_ColorFrameArrived;
```

To access the frame, we should make use of the *Dispose* pattern. Kinect data is very memory intensive, and it can easily eat up all our system resources if we are not careful. We do this by relying on the `using` keyword, which will automatically dispose of unneeded resources when we are finished. Accessing frames with the use of this pattern is demonstrated in Listing 2-5.

Listing 2-5. Accessing a Frame from an Event Handler

```
private void Reader_ColorFrameArrived(object sender, ColorFrameArrivedEventArgs e)
{
    using (ColorFrame colorFrame = e.FrameReference.AcquireFrame())
    {
        if (colorFrame != null)
        {
            ...
            //do work with Frame
            ...
        }
    }
}
```

Once the event handler is called, we can obtain the `FrameReference` from the event arguments e and then call its `AcquireFrame()` method to extract the frame. Whenever we receive a frame, we should endeavor to make a local copy or to directly access its data buffer to avoid holding up the system.

■ **Note** Are you not getting new frames for some reason? Then stop being so greedy with them! The Kinect SDK will refrain from giving you new frames if it realizes that you are not getting rid of the old ones. This is because the SDK aims to not hog up all your memory and is helping you do the same. Copy the data you need and chuck the rest.

Working with Kinect Studio

In the course of testing and debugging your Kinect application, you will invariably get tired of performing the same gestures repeatedly. I recall working on a project that involved analyzing respiratory rates with the Kinect for Windows v1 and having to take quick and deep breaths to gauge the accuracy of my algorithms. As you would expect, this lead to hyperventilation, and ultimately I passed out! Thus, you can understand the importance of being able to replay a particular recording to your Kinect application instead of having to redo the gestures or the scene footage.

Enter Kinect Studio. Kinect Studio is an application bundled with the SDK that allows you to capture and replay data collected by the Kinect. You can pretty much film a scene with the Kinect, including all the various Data Sources, and watch it in something sort of like a video player. This data can be fed to your Kinect applications instead of your having to generate footage each time around. Not only does this reduce repetitive, tiring gestures, but it also allows you to set up controls for your algorithmic experiments so that you can determine exact differences in their performance.

Using Kinect Studio

Before starting Kinect Studio, make sure `KinectMonitor.exe` is running on your machine and that your Kinect is turned on. To run Kinect Studio, either find its executable in the Tools folder of your Kinect for Windows SDK 2.0 or search *kstudio* or *Kinect Studio* on your computer (**Windows** key + **S** key).

At first you will be brought to the *Monitor* page. The Monitor page allows you to view all the data being streamed from your Kinect's sensors. You can turn off certain sensors and view the scene from a 3D perspective.

The first thing to do is to click the Connect to Service button (Figure 2-6) at the top left (under *File*). As soon as you click it, you should see a colored depth stream from your Kinect being played back. If at any time you want to stop, you can simply click the Stop button right beside it.

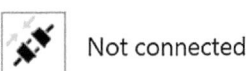

Figure 2-6. The Connect to Service button

The panel on the left (Figure 2-7) contains the switches for all the streams. You must stop the streams before you can toggle any of them. To turn off a stream, simply click the checkmark icon. For convenience, you can hide the streams from appearing on the panel with the eye icon. The topmost eye icon toggles whether hidden streams are shown.

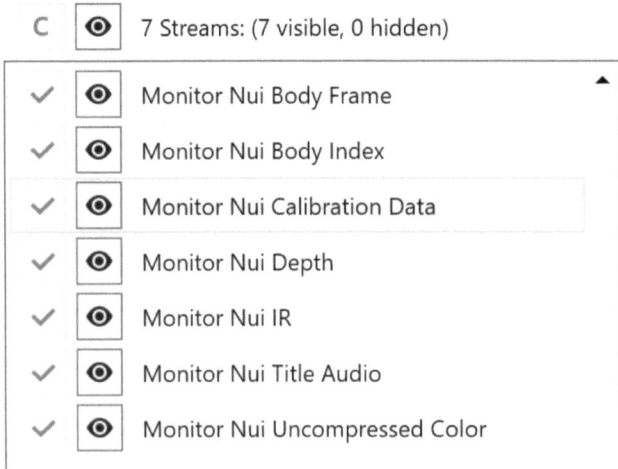

Figure 2-7. *Toggling Data Sources in Kinect Studio*

If you click the *Monitor 3D View* tab, you should be able to see a 3D depth stream from your Kinect (Figure 2-8). The red frustum represents everything that falls in the Kinect's reliable field of view. You might be able to see objects beyond it depending on circumstances, but the quality will be degraded. Its smaller rectangular face is where a cutting plane separates itself from a pyramid's apex. The original pyramid covers the entire field of view of the Kinect starting from its camera through a point-origin perspective, but since the Kinect does not accurately sense near itself, the field of view is truncated. The apex of this original pyramid is colored gray.

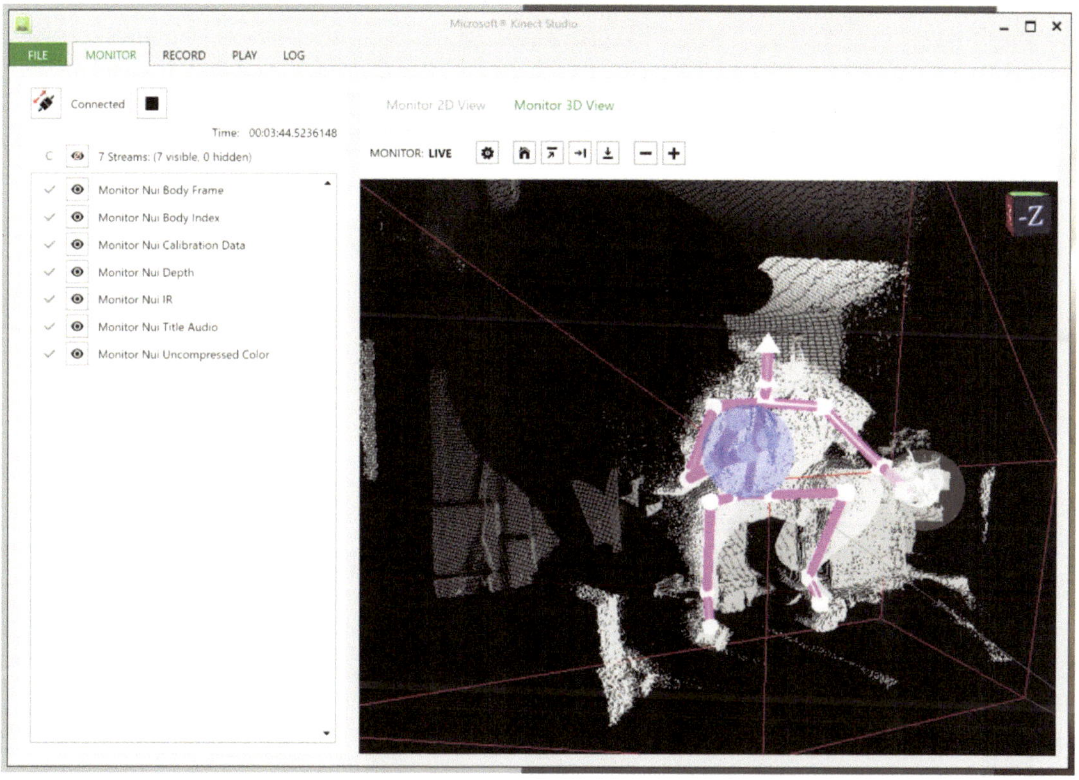

Figure 2-8. *A 3D depth stream from the Kinect, including a body and its skeleton. The blue circle represents a pointing hand (or lasso).*

You are not limited to a colored depth stream in 2D or a gray point cloud in 3D. To access the other streams, click the *Settings* icon, as shown in Figure 2-9. From there you can configure various plugins (Figure 2-10).

Figure 2-9. *The Settings icon, situated above the monitor view*

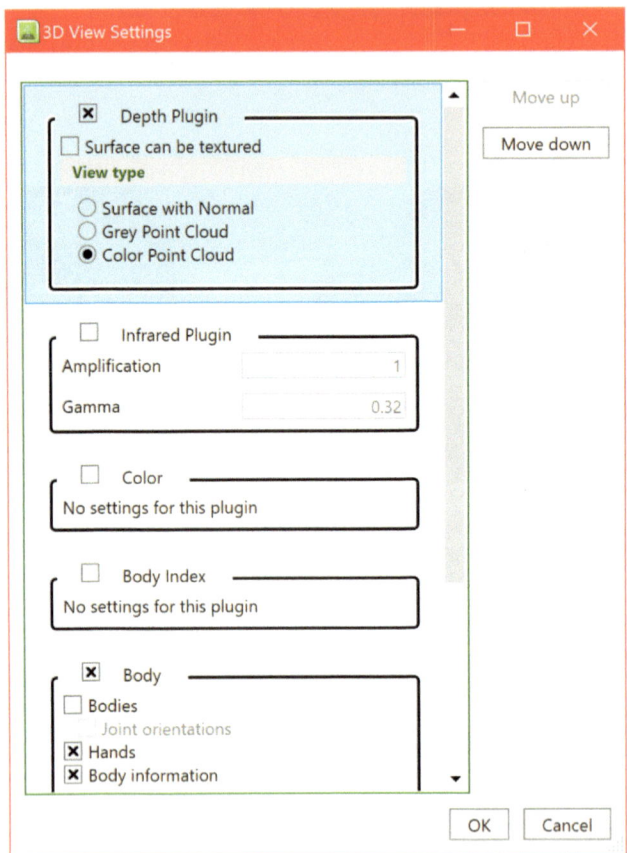

Figure 2-10. *The plugins page where Data Sources can be configured*

It is best that you play around with the options to see what suits you, as there are many configurations. You can alter the Data Sources to visualize what your Kinect sees differently. For example, you can exchange the gray point cloud for a color one, like the one in Figure 2-11. To get color or infrared in 3D, you will have to check the *Surface can be textured* checkbox and then turn on the color or infrared plugin respectively. To the right of the Settings icon in 3D view, you can also click the default view (house), front view, left view, and top view icons to view the 3D image from different planes. You also have the option to zoom in and out, though you will find it much quicker to achieve the same effect by holding **CTRL** while scrolling with your mouse wheel. Left-clicking and dragging your mouse will enable you to pan the camera, while scrolling will enable you to translate up and down on whichever axis is vertical on the screen. Look at the cube at the top-right of the view window to find out which axis is vertical; when left-view is selected, it is the Y-axis.

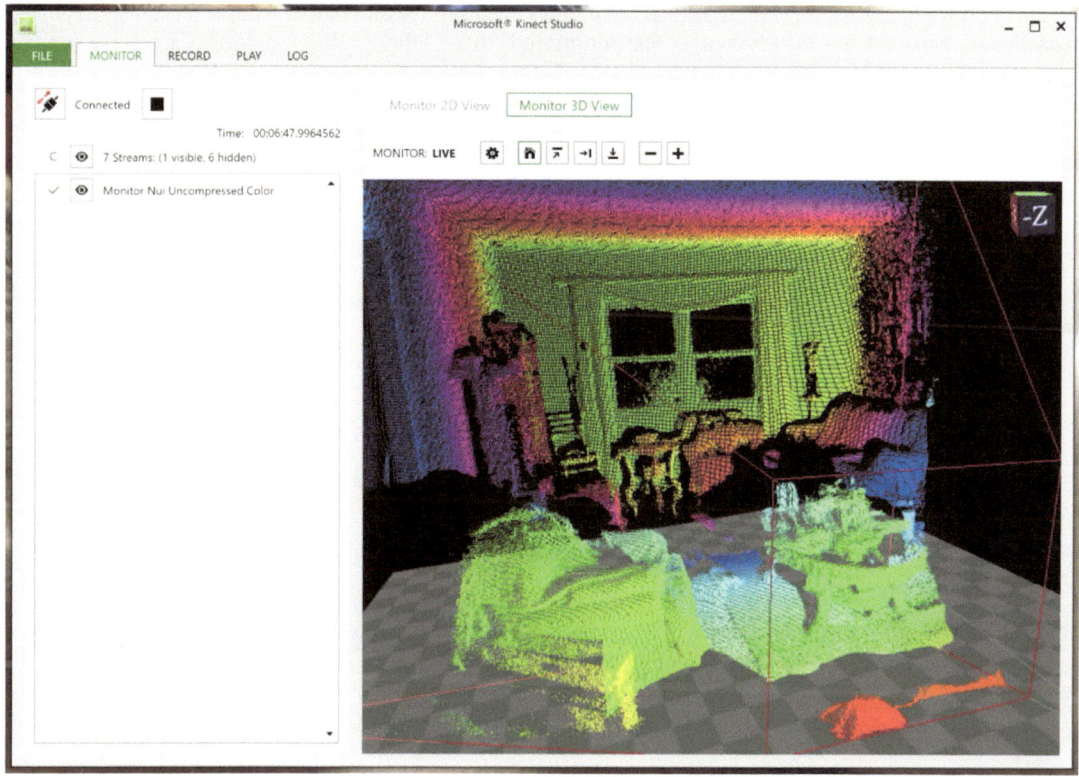

Figure 2-11. *Colored point cloud*

■ **Note** You have to keep the depth plugin and Data Source turned on in 3D view or you will not be able to see color or infrared. This is because there is no point cloud to project on without depth. You also cannot see the infrared and color streams at the same time.

Recording and Playing Back

While the Monitor feature of Kinect Studio is pretty dandy, being able to record and replay scenes from our Kinect takes the cake. Not only is it useful for debugging, but also the *Visual Gesture Builder* tool makes use of the recordings as well. To get started, visit the *Record* tab of Kinect Studio. Conveniently, it looks very much like the Monitor page. The one difference is that there happens to be a few more streams that you can toggle in the left panel. These are not available for view in Kinect Studio itself; rather, the data is recorded for any applications you plan to replay for.

■ **Tip** Only toggle what you need. The recordings grow larger for each Data Source being tracked. A one-minute recording with all Data Sources active will take up a whopping 10 GBs. Remember to delete your recordings after you have finished with them.

To get started, make sure you are connected to the Kinect service and have the desired Data Sources toggled, and then click the *Start Recording File* button (red circle). Film whatever interactions you need, such as gestures or plain video footage, then click the Stop button when you are done. There is no Pause button, so you will have to film whatever you need in one go.

Once you have clicked Stop, Kinect Studio will transfer you to the *Play* page (Figure 2-12). Your recording should have been saved as a .xef file in *Documents ➤ Kinect Studio ➤ Repository*. If you have to reopen your file, go to *File ➤ Open for Edit*, then pick the file you want.

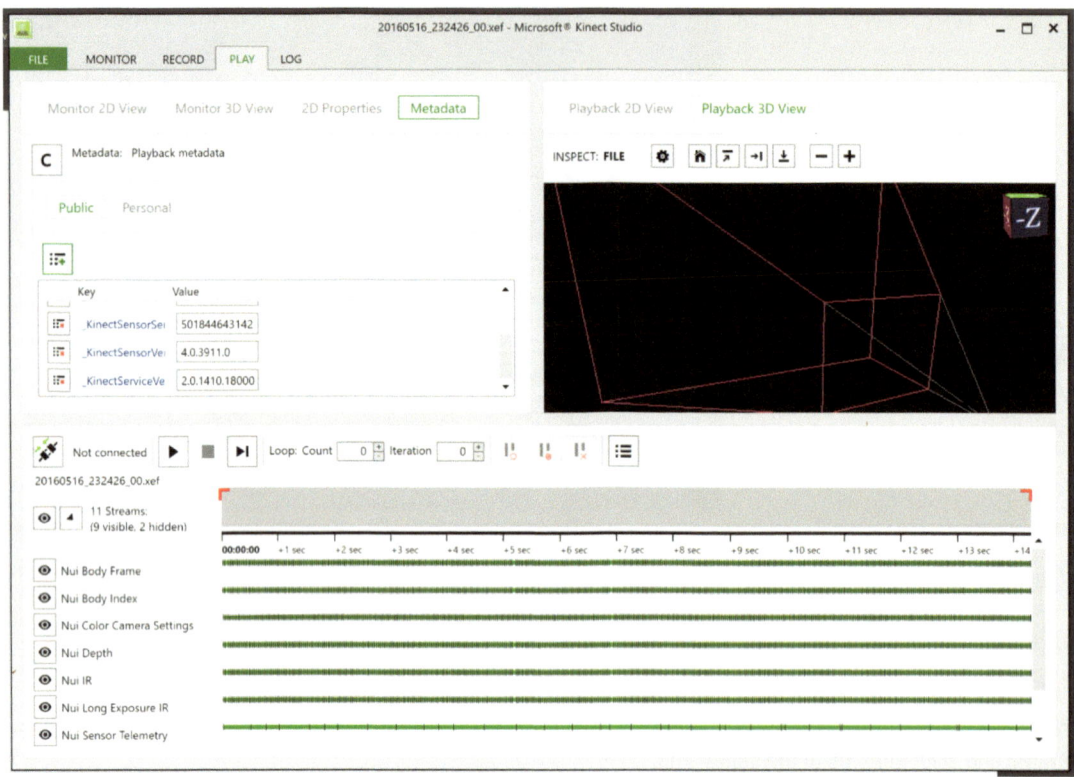

Figure 2-12. *The Play page*

The Play page has a variety of features to help you analyze your footage. The most straightforward and essential ability is naturally the *Play File* button. Clicking it replays the saved footage in the Playback 2D View and 3D View windows at the top right. These work the same as on the Monitor page. Unlike the Monitor page, however, you can actually view frame data on the Play page. In the top-left window, click the *2D Properties* tab; you will be presented with data such as depth in millimeters, RGB values, and infrared intensity about each pixel in Playback 2D View. To view data about a specific pixel, simply hover your mouse over it in the 2D view in the top-right window.

On the bottom half of the application you have a timeline manipulation panel similar to those found in standard video manipulation suites. To view a specific point in time, click on any of the green bars at the desired timestamp (and not the larger gray bar).

If you work with an IDE like Visual Studio or other developer tools such as the Chrome browser dev tools, you will be familiar with the concept of breakpoints. Basically, they allow you to pause execution at a certain line of code. In Kinect Studio, there are breakpoints, or *pause points*, that pause your playback whenever they are reached. You can add them by clicking on the gray bar that spans the bottom half of the window; a red circle should appear, as shown in Figure 2-13. Likewise, you can also add *markers*, which hold metadata about a specific time. This can be used to write notes about a specific time, such as indicating the start of a gesture. Markers are added in the bottom half of the gray bar and appear as blue squares. To create and/or edit metadata, click the *Metadata* tab at the top left portion of the window.

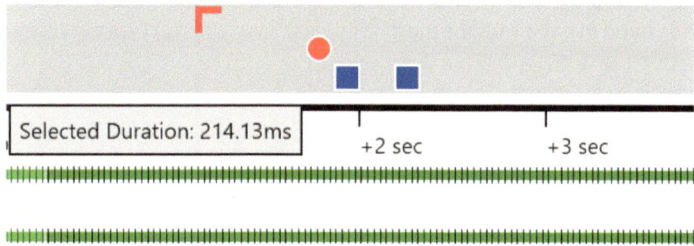

Figure 2-13. *Pause points, markers, and an In point*

You should see two little red brackets at either end of the gray bar. These are *In* and *Out points*, which allow you to set when the playback starts and finishes in the clip. Simply drag them to where you would like, or right-click the desired location on the bar and press *Set In/Out Point*. A final nifty feature in the playback window is the Loop tool. Next to the Play button area, you should see two textboxes that say *Loop: Count* and *Iteration*. You can set a value in Count, and the playback will be repeated this many times. Iteration lets you know how many times the loop has run. You could use this while debugging your application to avoid having to go back and click Play again and again.

Feeding a Recording into a Kinect Application

Start by unplugging your Kinect from your computer. We do not want to stream the Kinect to our application instead of the recording. On the Play page, ensure that you are still connected to the Kinect service. It should say *Connected* next to the Play button. Then, make sure all the Data Sources you are interested in are enabled on the bottom half of the Play page. Do not play the video yet.

Since we have not created a full-fledged Kinect application yet, we are going to use one of the sample applications. Search for *SDK Browser v2.0* on your computer and run it. Find the *Color Basics-WPF* sample and install it to a location of your choice. Open the solution file (.sln) in Visual Studio and then compile the project (green arrow or **F5**). The video window should be black and should say "Kinect not available!" on the bottom. Now, return to Kinect Studio and click the Play button. It should still say "Kinect not available!" on the bottom, but the video feed should work as if you had a real Kinect connected to it. If it is not working, make sure you recorded your Kinect Studio footage with the color Data Source enabled and that it is also enabled on the playback page.

Summary

The Kinect is a device that collects a number of sensory inputs to build a visualization of the world around it. The new Kinect for Windows v2 built on many of the lessons learned from the development of the Kinect for Windows v1, which grants it an unprecedented ability to be used in interactive multimedia applications. Despite all the data it can provide, the programming model for the Kinect is quite simple. Readers acquire data from Data Sources in the form of frames, which we must be sure to handle carefully to avoid bogging down the system. Debugging is also not very difficult, thanks to the time-saving capabilities of Kinect Studio.

Knowing all of this, we are now properly armed to tackle Kinect development. There are many quirks to figure out, and with this knowledge we can now do so.

In the next chapter, we will start working hand in hand with the most basic Data Source: `ColorFrameSource`.

CHAPTER 3

Working with Image Data Sources

Above all else, the Kinect is an overpriced camera (half kidding), and it is imperative that we learn how to work with its image data before we can understand how other features like gestures work. For most amateur applications, you will probably want to give the user some visual feedback anyway. In this chapter, we will explore the peculiarities of working with the various image data sources.

ColorFrameSource: Color Image Data

In Chapter 2, we very briefly mentioned `ColorFrameSource` as one of the many Kinect Data Sources. We went over the general pattern for recovering data from `ColorFrameSource`, but we will now proceed to fill in the blanks. We will build an application to stream the Kinect's video feed from scratch. The code is similar to the samples in **SDK Browser v2.0**, so do not hesitate to follow along with them if you prefer.

ColorFrameSource in WPF

Start off by creating a C# WPF project in Visual Studio. Next, add your `Microsoft.Kinect` reference. This time, we will be including two additional references, `System.Windows.Media` and `System.Windows.Media.Imaging`, which will aid us in turning the Kinect color data into proper images for display and saving.

■ **Tip**   If you do not recall how to start a new project in Visual Studio, or how to add references, revisit Chapter 1. You can use the **Ctrl-Shift-N** key combination to quickly start a new project. You do not need to add the `System.Windows.Media` and `System.Windows.Media.Imaging` namespaces through NuGet or the Visual Studio References manager, as they are a part of WPF's core assemblies, which are by default included by Visual Studio. Simply mention them in the code with the other `using` statements. Many project templates in Visual Studio will include them by default.

Right underneath our class declaration, we will initialize a few variables, as seen in Listing 3-1.

Listing 3-1. MainWindow.xaml.cs Variable Initialization for Color Data

```
public partial class MainWindow : Window
{
    private KinectSensor kinect = null;
    private ColorFrameReader colorFrameReader = null;
    private WriteableBitmap colorBitmap = null;
    ...
```

© Mansib Rahman 2017

M. Rahman, *Beginning Microsoft Kinect for Windows SDK 2.0*, DOI 10.1007/978-1-4842-2316-1_3

In Listing 3-1, in addition to our `KinectSensor` object, we are declaring a `ColorFrameReader` and a `WriteableBitmap`. As we discussed in Chapter 2, the `ColorFrameReader` will be reading frames from the Kinect. `WriteableBitmap` is a WPF object that allows us to modify a set of pixels dynamically. Instead of creating and destroying a new bitmap for every frame of data we receive from the Kinect through use of a `BitmapImage`, our `WriteableBitmap` will allow us to flexibly update the relevant pixels in our image without having to reallocate memory each time around.

In Listing 3-2, we bootstrap our Kinect and its color Data Source within our class constructor.

Listing 3-2. MainWindow.xaml.cs Data Source Bootstrapping and Variable Assignment for Color Data

```
public MainWindow()
{
    this.kinect = KinectSensor.GetDefault();
    this.colorFrameReader = this.kinect.ColorFrameSource.OpenReader();

    this.colorFrameReader.FrameArrived += this.Reader_ColorFrameArrived;

    FrameDescription colorFrameDescription = this.kinect.ColorFrameSource.CreateFrame
    Description(ColorImageFormat.Bgra);

    this.colorBitmap = new WriteableBitmap(colorFrameDescription.Width,
    colorFrameDescription.Height, 96.0, 96.0, PixelFormats.Bgr32, null);

    this.DataContext = this;
    this.InitializeComponent();

    this.kinect.Open();

}
...
```

We start off by creating a new reader from our Kinect with `this.kinect.ColorFrameSource.OpenReader();` and assigning it an event handler with `+= this.Reader_ColorFrameArrived;`. We then declare a `FrameDescription` object using the `ColorFrameSource`'s `CreateFrameDescription` method. This contains various dimensional metadata about our image frames. In this case, it generates the metadata based off the image format given as an input. The format is given as an enumeration of `ColorImageFormat` from the Kinect's API. We are going to go with BGRA for demonstration purposes, but we also have Bayer, RGBA, YUV, and YUY2 for use, in addition to "None."

We then create a `WriteableBitmap` using some standard inputs. The first two inputs should be pretty straightforward; they are just dimensions based off the Kinect's image format. The next two give the pixel density: 96 dpi by 96 dpi (dots per inch). The last two are the pixel format, which is BGRA like our Kinect image format, and the bitmap palette, which is unimportant for our use case.

■ **Tip** If you do not have experience with color formats, just know that they represent different ways of digitally describing colors. Pixel density here refers to how many pixels fit within a fixed space. A higher pixel density will mean that there is more detail in a given area than if there were a lower pixel density.

The remaining code in our constructor is just scaffolding, as explored in Chapter 1. In Listing 3-3, we go over the properties and methods required to complete our code.

Listing 3-3. MainWindow.xaml.cs Methods and Properties for Color Data

```
public ImageSource ImageSource
{
    get
    {
        return this.colorBitmap;
    }
}

private void Reader_ColorFrameArrived(object sender, ColorFrameArrivedEventArgs e)
{
    using (ColorFrame colorFrame = e.FrameReference.AcquireFrame())
    {
        if (colorFrame != null)
        {
            FrameDescription colorFrameDescription = colorFrame.FrameDescription;

            using (KinectBuffer colorBuffer = colorFrame.LockRawImageBuffer())
            {
                This.colorBitmap.Lock();

                if ((colorFrameDescription.Width == this.colorBitmap.PixelWidth) &&
                (colorFrameDescription.Height == this.colorBitmap.PixelHeight))
                {
                    colorFrame.CopyConvertedFrameDataToIntPtr(
                        this.colorBitmap.BackBuffer,
                        (uint)(colorFrameDescription.Width * colorFrameDescription.Height * 4),
                        ColorImageFormat.Bgra);

                    this.colorBitmap.AddDirtyRect(new Int32Rect(0, 0, this.colorBitmap.
                    PixelWidth, this.colorBitmap.PixelHeight));
                }

                this.colorBitmap.Unlock();
            }
        }
    }
}

private void MainWindow_Closing(object sender, System.ComponentModel.CancelEventArgs e)
{
    if (this.colorFrameReader != null)
    {
        this.colorFrameReader.Dispose();
        this.colorFrameReader = null;
    }

    if (this.kinect != null)
    {
        this.kinect.Close();
        this.kinect = null;
    }
}
```

43

We have a total of two methods and one property in this section of the code. First, we have a getter, `ImageSource`, for our `WriteableBitmap`, `colorBitmap`. This can be used by our XAML front end to access the `ImageSource` bitmap and display it to the image on the user interface.

The meat of our program resides in `Reader_ColorFrameArrived`. The concept is simple enough: every frame (which occurs 30 times in a second, or 15 in low-light settings) replaces our `WriteableBitmap`'s data with that of the new frame. The first step is to get the frame from the `FrameReference` in our event args: `using (ColorFrame colorFrame = e.FrameReference.AcquireFrame())`. Remember that to ensure that the frames are disposed of automatically when we are done with them we wrap our code around a `using` block. We then check whether the frame contains any data. Sometimes the frame is invalid, and the Kinect will return a null result, so always account for this.

We have another `using` code block, `using (KinectBuffer colorBuffer = colorFrame.LockRawImageBuffer())`, which locks down `colorFrame`'s underlying image data buffer so that we can appropriate it for our use. If we do not lock it, the Kinect might send us new data before we are finished reading it. The data buffer will be disposed of by the end of the method so that the Kinect can use it to give us new data.

We lock our `WriteableBitmap` with `colorBitmap.Lock()`, which reserves its data buffer for us to manipulate in the back end and blocks the UI from updating the image. The integrity of our data is then checked by comparing the width and height of our color data frame to the format slated for use by our `WriteableBitmap` in `if ((colorFrameDescription.Width == this.colorBitmap.PixelWidth) && (colorFrameDescription.Height == this.colorBitmap.PixelHeight))`.

Once this final check is performed, we can proceed to copy the `ColorFrame` data to our `WriteableBitmap`. We use `ColorFrame`'s `CopyConvertedFrameDataToIntPtr` method, which takes as inputs an `IntPtr` to the buffer we want to copy to, that buffer's size in bytes, and its image format. In our case, the data buffer we want to copy to is that of the `WriteableBitmap`, and thus we reference its pointer: `colorBitmap.BackBuffer`.

■ **Note**   Not familiar with `IntPtr`? `IntPtr` is simply an integer representation of a pointer, or an address to a memory location. Its size corresponds to that of a pointer on a specific platform; thus 32 bits on 32-bit hardware and 64 bits on 64-bit hardware.

Since we are using BGRA32 as our color format, we use 8 bits for each of the Blue, Green, Red, and Alpha (opacity) channels in an individual pixel. This translates to 4 bytes in total per pixel. Thus, to obtain the size of our buffer, we multiply 4 bytes per pixel by the number of pixels (width × height): `(uint) (colorFrameDescription.Width * colorFrameDescription.Height * 4)`.

Afterward, we specify which part of our `WriteableBitmap` we will be changing with the `AddDirtyRect` method. This takes a rectangle designating the changed area as an input. Since each frame update from the Kinect might contain a color image completely unlike the last, we use a rectangle the size of the whole image for this. To finish off our `Reader_ColorFrameArrived` event handler, we unlock the `WriteableBitmap` so that our UI can make use of it.

Our `MainWindow_Closing` method is similar to that of Chapter 1. We simply added the `ColorFrameReader` disposal code from Chapter 2 to ensure that resources are released back to the system.

The XAML code for our UI, as shown in Listing 3-4, is relatively simple. It contains scaffolding for the window and an `Image` element to display our feed.

Listing 3-4. MainWindow.xaml

```
<Window x:Class="Microsoft.Samples.Kinect.ColorFrameSourceSample.MainWindow"
        xmlns="http://schemas.microsoft.com/winfx/2006/xaml/presentation"
        xmlns:x="http://schemas.microsoft.com/winfx/2006/xaml"
        Title="ColorFrameSource Sample"
        Height="440" Width="700"
        Closing="MainWindow_Closing">
```

```
<Grid Margin="10">
  <Grid.RowDefinitions>
    <RowDefinition Height="Auto" />
    <RowDefinition Height="*" />
    <RowDefinition Height="Auto" />
  </Grid.RowDefinitions>

    <Viewbox Grid.Row="1" HorizontalAlignment="Center">
      <Image Source="{Binding ImageSource}" Stretch="UniformToFill" />
    </Viewbox>

  </Grid>
</Window>
```

The image is bound to the ImageSource property in our code-behind through the DataContext, which accesses our WriteableBitmap. We set the Stretch to UniformToFill so that the video feed will be as big as possible while maintaining its original aspect ratio.

Compile the code to check your results. They should look something along the lines of Figure 3-1, minus Redmond bear.

Figure 3-1. *A Kinect application displaying a video feed, featuring Redmond bear*

Well, there you go: your first application streaming live data directly from the Kinect. It is a bit bland for now, not even as featured as the default camera-viewing app provided with Windows was. Let us try to extend it by bundling some additional functionality. At the very least, every camera app should be able to take a screenshot, right?

Taking a Screenshot

Some of you have probably already thought of ways to grab a screenshot from the application. If not, take a couple of moments to think through how you could potentially do this. It seems simple enough, and there is more than one right way to achieve the desired result. The most common first guesses on how to do this are the following: grabbing and saving data from a frame, taking a snapshot of our UI, or saving our WriteableBitmap. Out of these options, the third one is the practical one and is how Microsoft does it in their samples.

Grabbing and saving from a frame would not be ideal, at least not with how our application is currently structured. When the screenshot button is clicked on the UI, an event is raised, and we manage to encode a picture file in its handler. We would have no convenient access to the ColorFrameSource in there, as all that logic and data is located in the ColorFrame event handler. Even if we did, there is no guarantee that there is a frame available for us to pick from at the instant the screenshot button is clicked. This is where polling could have been useful had we not gone with an event-driven model. We could poll for a frame to use as the basis for our screenshot. We could choose to cache a frame in the event handler and call on that, but seeing as our image already has a copy of the data, that would be redundant.

While it is possible with WPF to take a snapshot of our UI element, it is much more straightforward to simply save an image from its original source, the WriteableBitmap. It is already a bitmap, so we can encode it into a picture file with relative ease.

The code for adding screenshot functionality requires very few modifications to the application. Specifically, we will be adding a namespace reference, an event handler method, as shown in Listing 3-5, and a button on the UI, as shown in Listing 3-6. The namespace that we need to add is System.IO, which will provide us with methods with which to save an image file to the drive.

Listing 3-5. MainWindow.xaml.cs Screenshot Event Handler

```
//add to the Using section
using System.IO;

//Directly after the MainWindow_Closing method
private void ScreenshotButton_Click(object sender, RoutedEventArgs e)
{
    if (this.colorBitmap != null)
    {

        BitmapEncoder encoder = new PngBitmapEncoder();

        encoder.Frames.Add(BitmapFrame.Create(this.colorBitmap));

        string photoLoc = Environment.GetFolderPath(Environment.SpecialFolder.MyPictures);

        string time = System.DateTime.Now.ToString("d MMM yyyy hh-mm-ss");

        string path = Path.Combine(photoLoc, "Screenshot " + time + ".png");
```

```
    try
    {
        using (FileStream fs = new FileStream(path, FileMode.Create))
        {
            encoder.Save(fs);
        }
        //provide some confirmation to user that picture was taken
    }
    catch (IOException)
    {
        //inform user that picture failed to save for whatever reason
    }
  }
}
```

None of the code involved in capturing and saving an image is from the Kinect API. We are using code from the System.Windows.Media.Imaging and System.IO namespaces. After checking whether our WriteableBitmap has any data in it, we create a new PngBitmapEncoder, to which we add our WriteableBitmap as a BitmapFrame in encoder.Frames.Add(BitmapFrame.Create(this.colorBitmap));.

▪ **Note**   Why does BitmapEncoder encoder have a Frames property that seems like it could accept more than one frame, when a PNG could only possibly have one? BitmapEncoder can also encode file types such as GIF that naturally have multiple frames. Adding more than one frame to the encoder for a file type such as PNG or JPEG will result in the subsequent frames being discarded. You can use this functionality to create your own GIF camera with the Kinect if you so desire.

After grabbing the location of our *My Pictures* folder and a unique timestamp to discourage our screenshot from overwriting prior ones, our encoder saves the screenshot to our desired path using .NET's FileStream class. As with any input/output operation, this is wrapped in a try/catch block to handle the casual error (e.g., lack of disk space, no write access, and so on).

In MainWindow.xaml we will add the XAML for the Button UI element directly after the Viewbox UI element but inside the Grid UI element with which we will take our screenshots.

Listing 3-6. MainWindow.xaml Screenshot Button Markup

```
<Grid Margin="10">
  ...
  </Viewbox>

  <Button Grid.Row="2" Content="Screenshot" Height="Auto" HorizontalAlignment="Left"
  VerticalAlignment="Center" Margin="10" Click="ScreenshotButton_Click" />

</Grid>
```

Most of the Button attributes are merely decorative. The important thing is to assign the event handler from our code-behind for when the button is clicked, which we accomplish with Click="ScreenshotButton_Click".

If you open our app now, you should be able to click the Screenshot button. There's no picture-taken confirmation (though you can easily include this), so to find out if it worked, go visit your *My Pictures* folder. Now you can go share your beautiful Kinect picture on social media or the like. (Maybe on Snapchat.... oh wait, never mind; I forgot they took down 6snap.)

ColorFrameSource in Windows Store Apps

While the Kinect APIs are very similar for both platforms, the platform APIs are not, so there are some minor differences in how the apps are built. In the case of displaying `ColorFrameSource` data, we have to work around the Windows Store App version of `WriteableBitmap`, which is **not** the same as that of WPF.

The overall code structure for the Windows Store Application will look similar. Instead of the `System.Windows.Media` namespaces, we will have to use the `Windows.UI.Xaml.Media` and `Windows.UI.Xaml.Media.Imaging` namespaces, which have their own version of the `WriteableBitmap` class. Additionally, at the time of writing, instead of the `Microsoft.Kinect` reference we have been using all along, we will have to rely on the `WindowsPreview.Kinect` namespace. Do not let the name fool you, as the API is still pretty complete.

▪ **Note**   If you are following along with the ColorBasics-XAML sample in the SDK Browser, you might be unable to compile. Chances are you got a `System.IO.FileNotFoundException` on the `GetDefault` method of `KinectSensor`. Do not fret, as this is just a result of the app retargeting Windows 8.1 and can be easily remedied. Visit the Reference Manager (can be accessed in Solution Explorer) and then the *Extensions* section in the menu labeled *Windows 8.1*. You will see a yellow caution sign next to *Microsoft Visual C++ Runtime Package*. You have to unselect that and choose *Microsoft Visual C++ 2013 Runtime Package for Windows*, which at version 12 is one version number ahead of the package it is replacing. The program should now compile.

Our initial bootstrapping code will only have a one-line difference (Listing 3-7).

Listing 3-7. MainPage.xaml.cs Data Source Bootstrapping and Variable Assignment in Windows Store Apps

```
public MainPage()
{
    this.kinect = KinectSensor.GetDefault();
    this.colorFrameReader = this.kinect.ColorFrameSource.OpenReader();

    this.colorFrameReader.FrameArrived += this.Reader_ColorFrameArrived;

    FrameDescription colorFrameDescription = this.kinect.ColorFrameSource.CreateFrame
    Description(ColorImageFormat.Rgba);

    this.colorBitmap = new WriteableBitmap(colorFrameDescription.Width,
    colorFrameDescription.Height);

    this.kinect.Open();

    this.DataContext = this;
    this.InitializeComponents();
}
...
```

In contrast to what we saw in Listing 3-2, Listing 3-7 shows how the Windows Store App version of WriteableBitmap's constructor only takes two inputs—a width and a height. This means it is dumb in regards to the image's format and quality, unlike WPF's WriteableBitmap. I have included the rest of the constructor's code to demonstrate the parity between both APIs.

The remaining brunt of the code differences are found in the Reader_ColorFrameArrived event handler.

Listing 3-8. MainPage.xaml.cs ColorFrameReader Event Handler in Windows Store Apps

```
private void Reader_ColorFrameArrived(object sender, ColorFrameArrivedEventArgs e)
{
    using (ColorFrame colorFrame = e.FrameReference.AcquireFrame())
    {
        if (colorFrame != null)
        {
            FrameDescription colorFrameDescription = colorFrame.FrameDescription;

            if ((colorFrameDescription.Width == this.colorBitmap.PixelWidth) &&
            (colorFrameDescription.Height == this.colorBitmap.PixelHeight))
            {
                if (colorFrame.RawColorImageFormat == ColorImageFormat.Bgra)
                {
                    colorFrame.CopyRawFrameDataToBuffer(
                    this.colorBitmap.PixelBuffer);
                }
                else
                {
                    colorFrame.CopyConvertedFrameDataToBuffer(
                    this.colorBitmap.PixelBuffer, ColorImageFormat.Bgra);
                }

                colorFrameProcessed = true;
            }
        }
    }

    if (colorFrameProcessed)
    {
        this.colorBitmap.Invalidate();
    }
}
```

We see in Listing 3-8 that the processes of getting frames from the reader and the data integrity checks are all still the same. Pushing data into a WriteableBitmap for Windows Store Apps consists of calling the CopyRawFrameDataToBuffer method of ColorFrame and specifying the target buffer as the WriteableBitmap's PixelBuffer. We also have the option of converting the raw data to another format on the spot with the CopyConvertedFrameDataToBuffer method, which takes the target buffer and the target format as inputs.

Finally, take note of the colorFrameProcessed Boolean and the colorBitmap.Invalidate(); method statement. We do not have to lock and unlock the WriteableBitmap like in WPF. Invalidate() requests a redraw of the bitmap instead and replaces it with our data, pushing the changes to the UI. Speaking of which, the UI code necessitates virtually no alterations. All we have to do on that end is change the top-level Window tags to Page tags and its Closing attribute to an Unloaded attribute. I suggest you create a new Windows Store App project in Visual Studio to get the markup right.

Before we run our application, there is one last step we must undertake before we can launch. Windows Store Apps require the user to grant permissions, as with iOS or Android apps. You need to declare these capabilities in the app manifest or you will not be able to access the Kinect. To do this, visit the Solutions Explorer and open the `Package.appxmanifest` file. In the *Capabilities* tab, enable the *Microphone* and *Webcam* capabilities, like in Figure 3-2. If you made all the correct changes, your Windows Store App should now stream video exactly like the WPF application did.

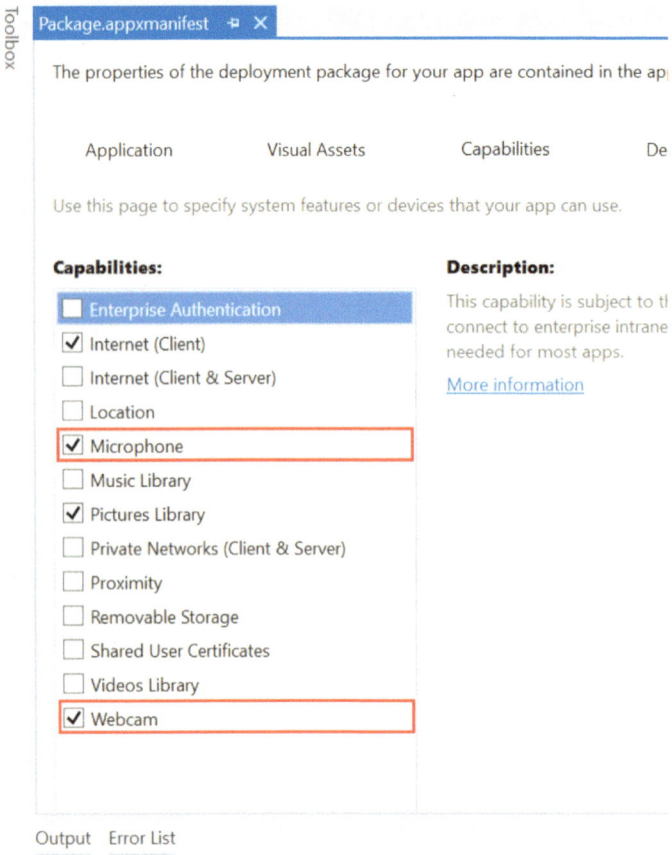

Figure 3-2. *Permissions enabled in the Capabilities tab of the app manifest*

Taking Another Screenshot

Windows Store Apps also have a different set of APIs to handle files and image encoding, so the code for taking screenshots is different as well. We will be saving images to the *My Pictures* folder, so you have to enable the *Pictures Library* permission in the app manifest, as depicted in Figure 3-2. We need to include a few more namespaces, namely `System`, `System.Runtime.InteropServices.WindowsRuntime`, `Windows.Graphics.Imaging`, `Windows.Storage`, and `Windows.Storage.Streams`.

As we did with the WPF example, add a `Button` element to the XAML code and give it an event handler for when it is clicked. You might need to change the color of the button to make it appear, as sometimes it shares the same color as the background. The rest of the code alterations take place in the event-handler body, as shown in Listing 3-9.

Listing 3-9. MainPage.xaml.cs Screenshot Event Handler in a Windows Store App

```
private async void ScreenshotButton_Click(object sender, RoutedEventArgs e)
{

    StorageFolder picFolder = KnownFolders.PicturesLibrary;

    StorageFile picFile = await picFolder.CreateFileAsync("Kinect Screenshot.png",
        CreationCollisionOption.GenerateUniqueName);

    using (IRandomAccessStream stream = await picFile.OpenAsync(FileAccessMode.ReadWrite))
    {
        var encoder = await BitmapEncoder.CreateAsync(BitmapEncoder.PngEncoderId, stream);

        Stream pixelStream = colorBitmap.PixelBuffer.AsStream();
        byte[] pixels = new byte[pixelStream.Length];
        await pixelStream.ReadyAsync(pixels, 0, pixels.Length);

        encoder.SetPixelData(BitmapPixelFormat.Bgra8,
            BitmapAlphaMode.Straight,
            (uint)colorBitmap.PixelWidth,
            (uint)colorBitmap.PixelHeight,
            96.0, 96.0,
            pixels);

        await encoder.FlushAsync();
    }
}
```

We have made a small edit to the event handler method header. We have added the async keyword to modify the method and let the compiler know that there is asynchronous code in its body. Our input and output operations need to be done asynchronously so as to not lock up the UI thread while in operation. Within the method, we first find the *My Pictures* folder and create a PNG image within it. The CreateFileAsync method conveniently handles collisions as we desire, so we tell it to make unique files each time instead of having ourselves add unique timestamps programmatically.

We create a stream with the PNG file and a PNG BitmapEncoder to write to it. We create another stream called pixelStream from our WriteableBitmap's data buffer and use it to write to an array of bytes. Then, the encoder sets the pixel data of our PNG using this array. The BitmapAlphaMode input in the SetPixelData method refers to how we want to save the alpha values to our PNG. Sometimes our final image is in a format that does not measure opacity, so we can choose to ignore or apply opacity to the color channels by multiplying the color values by the alphas. We will simply use the alphas that we already have in hand, thus the Straight option. The two 96.0 values are the horizontal and vertical dpis, respectively. We call the FlushAsync method of encoder to finalize the image operations. After it is called, we can make no more alterations to the image. Your Windows Store App should now be able to take pictures just as well as the WPF application did.

■ **Tip** The await operator forces the code to suspend the execution of a method until it is completed. Meanwhile, control is returned to the caller of the method in a bid to keep the application from blocking.

Casual Image Manipulation

So far we have been copying our Kinect's color data directly to the `WriteableBitmap` construct through its buffer. We are not obliged to simply hand the data off to the `WriteableBitmap` with the Kinect API calls we saw earlier. We can also opt to view the pixel data individually and make alterations. We were able to access the pixel data in the Windows Store App screenshot event handler by making a stream from the `WriteableBitmap`, but the Kinect API has methods to perform this before the data even hits the `WriteableBitmap`.

Why would we want to alter the pixel data? If you have ever used Instagram, Snapchat, or Photoshop (or GIMP!) you have probably applied a filter to one of your grams, snaps, or .psds (or .xcfs!). Seeing as images essentially store all their data as numbers, we can apply mathematical functions to transform this data and ultimately change how the image will look like based on these functions. When some application applies a filter or a digital affect such as red-eye removal, we are applying the necessary mathematical functions to make some data alteration globally in the image or locally on certain pixels.

■ **Tip** How are numerical values used to describe images? Each pixel is described by a handful of bytes describing the color makeup of the pixel. A `WriteableBitmap` image with a BGRA32 format has three color channels and an opacity channel. Each color channel, Blue, Green, and Red, as well as the opacity channel, Alpha, is described with an 8-bit value (8 x 4 = 32, hence BGRA32). The 8-bit value, which goes from 0 to $2^8 - 1$ (255), describes the intensity of the color channel in the pixel. A value of 0 refers to none of the color channel being present in the pixel, a value of 255 means the full intensity is present, and a value in between both extremes means a certain percentage of the color channel is present. The opacity channel describes how transparent the pixel is, with 0 being invisible, 255 being completely opaque, and numbers in between representing various degrees of transparency. The three color channels work like primary colors to make up other colors. Thus, Red 255, Blue 255 and Green 0 can be used to describe purple, and Blue 230, Green 230 and Red 0 can be used to describe a shade of cyan.

We will not go on to apply Instagram-caliber filters on our images right now, as that is beyond the scope of the chapter, but we will explore the procedure to access and manipulate the image data before it is displayed to the user.

Displaying ColorFrameSource Data with a Byte Array

Starting with the Kinect camera-viewing application we just built in the previous section, we will make some adjustments to obtain the image data inside of a byte array instead of a buffer.

Listing 3-10. MainPage.xaml.cs Variable Initialization for Color Data with Use of Byte Array

```
public sealed partial class MainPage : Page
{
    private KinectSensor kinect = null;
    private ColorFrameReader colorFrameReader = null;
    private WriteableBitmap colorBitmap = null;

    private readonly uint bytesPerPixel;
    private byte[] colorPixels = null;
    ...
```

In Listing 3-10, we see the variable initialization stage of our application. `KinectSensor`, `ColorFrameReader`, and `WriteableBitmap` all make a return. The structure of the code is still the same. What *will* change is how `WriteableBitmap` reads the data from our color frames. Previously, we could copy our data directly to the `WriteableBitmap`'s back buffer/pixel buffer, with an `IntPtr` or `IBuffer` provided by the Kinect API. Instead, we will now provide an array of bytes. `colorPixels` will be this array. Earlier, I mentioned that RGBA consisted of 3 bytes of color data + 1 byte of alpha opacity per pixel, for a total of 4 bytes per pixel. `bytesPerPixel` will hold this value as garnered from the `ColorFrameSource`. We could have resorted to using the literal four instead of creating a whole variable, but since the bytes per pixel change based on the color format, we can set it up so that we do not have to manually switch it when we change the format.

Seeing as we introduced two new variables, the bootstrap process is different as well (Listing 3-11).

Listing 3-11. MainPage.xaml.cs Bootstrapping and Variable Assignment for Color Data with Use of Byte Array

```
public MainPage()
{
    this.kinect = KinectSensor.GetDefault();
    this.colorFrameReader = this.kinect.ColorFrameSource.OpenReader();

    this.colorFrameReader.FrameArrived += this.Reader_ColorFrameArrived;

    FrameDescription colorFrameDescription = this.kinect.ColorFrameSource.CreateFrameDescription(ColorImageFormat.Rgba);

    this.bytesPerPixel = colorFrameDescription.BytesPerPixel;

    this.colorPixels = new byte[colorFrameDescription.Width * colorFrameDescription.Height * this.bytesPerPixel];

    this.colorBitmap = new WriteableBitmap(colorFrameDescription.Width, colorFrameDescription.Height);

    this.kinect.Open();

    this.DataContext = this;
    this.InitializeComponent();
}
...
```

In Listing 3-11, we added two new lines preparing the `colorPixels` data array. Since `colorPixels` will hold the bytes of every pixel in our image, we are declaring its total size as the number of bytes per pixel for the relevant format multiplied by the total number of pixels. We get the bytes per pixel directly from the `ColorFrameDescription`, which takes into account the image format that we chose (RGBA in our case).

The rest of the code modification is in the event handler for the color frames (Listing 3-12).

Listing 3-12. MainPage.xaml.cs ColorFrameReader Event Handler for Direct Access to Pixel Data

```
private void Reader_ColorFrameArrived(object sender, ColorFrameArrivedEventArgs e)
{
    using (ColorFrame colorFrame = e.FrameReference.AcquireFrame())
    {
```

```
        if (colorFrame != null)
        {
            FrameDescription colorFrameDescription = colorFrame.FrameDescription;

            if ((colorFrameDescription.Width == this.colorBitmap.PixelWidth) &&
            (colorFrameDescription.Height == this.colorBitmap.PixelHeight))
            {
                if (colorFrame.RawColorImageFormat == ColorImageFormat.Bgra)
                {
                    colorFrame.CopyRawFrameDataToArray(
                    this.colorPixels);
                }
                else
                {
                    colorFrame.CopyConvertedFrameDataToArray(
                    this.colorPixels, ColorImageFormat.Bgra);
                }

                Stream pixelStream = colorBitmap.PixelBuffer.AsStream();
                pixelStream.Seek(0, SeekOrigin.Begin);
                pixelStream.Write(colorPixels, 0, colorPixels.Length);

                colorFrameProcessed = true;
            }
        }
    }

    if (colorFrameProcessed)
    {
        this.colorBitmap.Invalidate();
    }
}
```

Instead of the CopyRawFrameDataToBuffer and CopyConvertedFrameDataToBuffer methods we used earlier, we are using their byte-array equivalents, CopyRawFrameDataToArray and CopyConvertedFrameDataToArray, respectively. Once the pixel data is copied to the arrays, we write to the WriteableBitmap in way a similar to that used when we read from it in the screenshot code: we create a stream from its pixel buffer. The Seek method of our Stream pixelStream is used to set the position of where we will write in the stream. The SeekOrigin.Begin input sets it to the start of the stream, and the first input, 0, is the offset from the beginning position where we want to start. In the Write method of Stream, we specify that we want to write with the contents of the byte array colorPixels for the entire length of the array. Again, the 0 here is also the offset.

For now, the rest of the code is the same. Make sure that the System.IO namespace is included so that we can use the Stream class. Your program should run just as if you had used a buffer to write to WriteableBitmap.

Manipulating Image Pixel Data

We only really benefit from having the image data available as a byte array if we plan to manipulate it. Let us make a very simple alteration. As mentioned earlier, each pixel has three color channels—Blue, Green and Red–with 8 bits of intensity, or 256 different values. We will be making a filter that hides the green and blue channels of the image, meaning we will set their respective intensities to 0.

In our Reader_ColorFrameArrived event handler, our code will find itself in the lines right after ColorFrame data has been copied to an array and right before that array is used by a stream to write to the WriteableBitmap (Listing 3-13).

Listing 3-13. MainPage.xaml.cs ColorFrameReader Event Handler Manipulating Color Data Byte Array Values

```
...
colorFrame.CopyConvertedFrameDataToArray(this.colorPixels, ColorImageFormat.Bgra);
}

for (uint i = 0; i < colorPixels.Length; i += bytesPerPixel)
{
    colorPixels[i] = 0x00;
    colorPixels[i + 1] = 0x00;
}

Stream pixelStream = colorBitmap.PixelBuffer.AsStream();
...
```

In Listing 3-13, the for loop cycles through each pixel in the image and sets the first two channels, Blue and Green, to 0 intensity. We are iterating by the value of bytesPerPixel, which is 4, because the array's indices contain bytes, 4 for each pixel. When we go through the first 4 bytes, we have covered the first pixel and applied the desired operations to its color channels. Since we are manipulating bytes, we set the pixel color channel intensity values using hexadecimal notation. Hence, when we say colorPixels[i + 1] = 0x00, we are saying the green channel (the second channel in BGRA) has an intensity of 0.

Compile the program, and you should see nothing but red throughout the image (Figure 3-3).

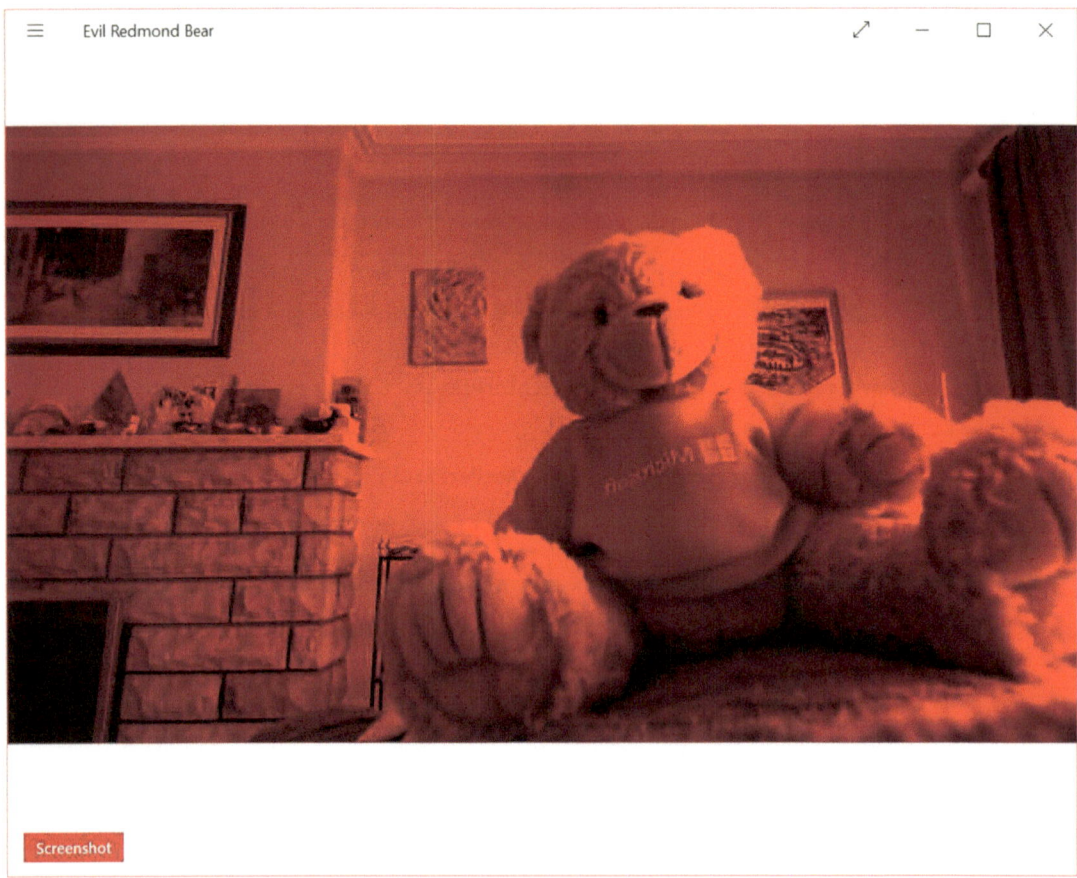

Figure 3-3. *Redmond bear shows his true colors*

On my i5 Surface Pro 2, tinting the picture red resulted in a 7 to 10 point higher reading for CPU usage in Task Manager. Generally, we will want to avoid doing heavy image processing in WPF without C++ interop because of performance issues. After all, we are looping through and applying operations to over 2 million pixels, 30 times a second! You can compile and run your program in a *Release* configuration to improve performance.

InfraredFrameSource: Infrared Image Data

Unlike its predecessor, the Kinect for Windows v2 can give us unfettered access to its infrared camera without having to shut off the color camera. The resolution is much better than in Kinect for Windows v1, and the IR dot speckle pattern no longer clouds the resulting image. Unlike with our (or any average) color camera, our infrared camera ignores ambient lighting conditions (as depicted in Figure 3-4) and can thus see in the dark and through other occlusions, such as smoke. This is particularly helpful in situations where the texture of an object is important, which is why infrared cameras are also used by Windows Hello's face-detection algorithm. All this means more magic™ with which to design and build our NUI applications!

Figure 3-4. *Footage from the Kinect's infrared camera. All three of the lights on the lamp are lit, and the author is holding out a lit candle as well. The lamp light is not picked up, and only the faint outline of the candle flame is visible. In the Kinect for Windows v1, the entire flame would have been picked up.*

In this chapter, we will cover how to extract and display the Kinect's infrared data. This data can be used for various purposes, such as tracking active markers in motion-capture (mocap) systems or green-screening, but that is beyond the scope of this section.

Displaying Infrared Data

As I mentioned a few times before, the pattern of creating a reader to get frames from a data source is the same for all of the Kinect's sensory data, so to see how to write the scaffolding setup for this code, revisit the code for ColorFrameSource, particularly Listings 3-1, 3-2, and 3-3. The only differences in the bootstrapping process are the WriteableBitmap image format, which is Gray32Float, and the names of the data source, reader, frame, and frame description. These are InfraredFrameSource, InfraredFrameReader, InfraredFrame, and InfraredFrameDescription, respectively.

The actual differences in how the data is rendered are entirely located in the InfraredFrameReader event handler and its associated methods. Listing 3-14 shows an example for a WPF .NET application.

Listing 3-14. InfraredFrameReader Event Handler and Helper Method

```
...
//Private variables declared before class constructor
private const float InfrDataScale = 0.75f;
private const float InfrMinVal = 0.01f;
private const float InfrMaxVal = 1.0f;

//Private variable for the InfraredFrameReader
private InfraredFrameReader irFrameReader = null;

/*
Our WriteableBitmap is named infrBitmap and created in a similar manner to the
ColorFrameSource examples, except the PixelFormat's input is PixelFormats.Gray32Float
instead of PixelFormats.Bgr32
*/
private WriteableBitmap infrBitmap = null;
...

//Logic inside the MainWindow() Constructor
//public MainWindow() {
//...
//InfraRed Initialization
this.irFrameReader = this.kinect.InfraredFrameSource.OpenReader();

this.irFrameReader.FrameArrived += Reader_InfraredFrameArrived;

FrameDescription infraRedFrameDescription = this.kinect.InfraredFrameSource.
FrameDescription;

this.infrBitmap = new WriteableBitmap(infraRedFrameDescription.Width,
infraRedFrameDescription.Height, 96.0, 96.0, PixelFormats.Gray32Float, null);

// ... }

private void Reader_InfraredFrameArrived(object sender, InfraredFrameArrivedEventArgs e)
{
    using (InfraredFrame infrFrame = e.FrameReference.AcquireFrame())
    {
        if (infrFrame != null)
        {
            using (KinectBuffer infrBuffer = infrFrame.LockImageBuffer())
            {
                if ((infrFrame.FrameDescription.Width * infrFrame.FrameDescription.Height) ==
                (infrBuffer.Size / infrFrame.FrameDescription.BytesPerPixel))
                    {
                        this.ProcessInfraredFrameData(infrBuffer.UnderlyingBuffer, infrBuffer.
                        Size, infrFrame.FrameDescription.BytesPerPixel);
                    }
            }
        }
    }
}
```

```
private unsafe void ProcessInfraredFrameData(IntPtr infrFrameData, uint infrFrameDataSize,
uint bytesPerPix)
{
    ushort* frameData = (ushort*)infrFrameData;

    this.infrBitmap.Lock();

    float* backBuffer = (float*)this.infrBitmap.BackBuffer;

    for (int i = 0; i < (int)(infrFrameDataSize / bytesPerPix); ++i)
            {
                ushort irValue = frameData[i];

                float irRange = (float)(irValue - ushort.MinValue ) / (float)(ushort.
                MaxValue  - ushort.MinValue );

                float ir_desiredRange = (float)(irRange * (1 - InfrMinVal)) + InfrMinVal;

                float ir_desiredRange_withBrightness = ir_desiredRange * InfrDataScale;

                backBuffer[i] = Math.Min(InfrMaxVal, ir_desiredRange_withBrightness);

            }
    this.infrBitmap.AddDirtyRect(new Int32Rect(0, 0, this.infrBitmap.PixelWidth, this.
    infrBitmap.PixelHeight));

    this.infrBitmap.Unlock();
}
```

Listing 3-14 is based off the *InfraredBasics-WPF* code sample in the SDK Browser. The code is somewhat more involved than the equivalent code for the exploitation of the ColorFrameSource. While we could have gone the simpler route, we opted to make performance optimizations by relying on pointers for the processing of our infrared data.

Since we are using pointers to directly access memory in managed code, we need to add the unsafe keyword to our ProcessInfraredFrameData method header. Additionally, we need to set the /unsafe compiler option as well. To do this, visit the *Project Properties* page under the *Project* tab in Visual Studio. Next, go to the *Build* tab on the Properties page and check the *Allow unsafe code* checkbox, as in Figure 3-5.

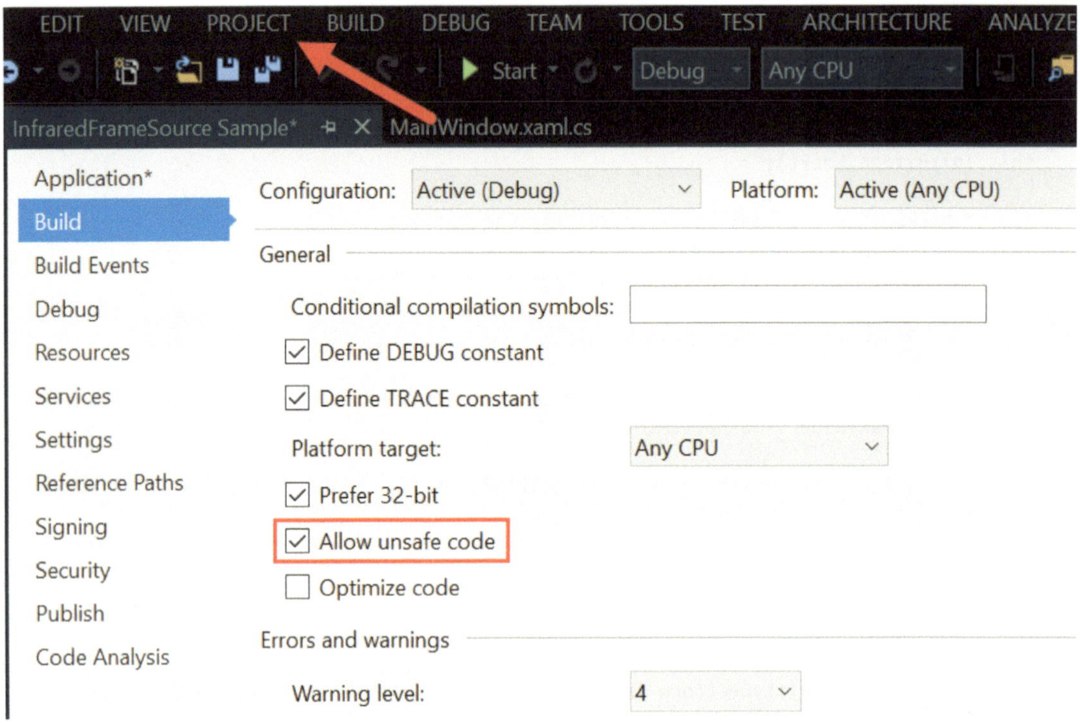

Figure 3-5. *Allowing unsafe code in Visual Studio*

Our frame is acquired in the event handler, and after checking the integrity of our data we pass a reference to the data to our helper method so as to process and display the infrared data. We take an IntPtr to our infrared frame buffer, as well as its size and its number of bytes per pixel as inputs to our helper method. As each pixel of the infrared data is an unsigned 16-bit value, we cast the infrared frame's data buffer to a ushort (unsigned short, 16-bit integer) pointer to access its values. Similarly, we cast the WriteableBitmap's BackBuffer to a float pointer. This is because, unlike our RGBA sample, our grayscale picture is described in fractions of 0 to 1 with Gray32Float. The WriteableBitmap is locked before this to prevent updates to the UI while we work on it.

To transform the 16-bit infrared intensity values from the Kinect into Gray32Float values that can be displayed by the WriteableBitmap, we must normalize them (see the note for further explanation). We loop through all the ushorts in frameData and apply our normalization algorithm to each of them, then write the result in the WriteableBitmap's BackBuffer. Our loop's termination condition is the total byte size of the infrared frame divided by its bytes per pixel value (which is 2). This is because ushort values are 2-byte values; thus, when we iterate through them we go by 2 bytes each time, and consequently the number of ushorts to go through is half the total size of the infrared frame in bytes.

While performing the normalization, we make use of three float variables: InfrMaxVal, InfrMinVal, and InfrDataScale. InfrMaxVal and InfrMinVal represent the maximum and minimum desired values for the gray values displayed on our WriteableBitmap. In Gray32Float, each pixel can be one of over 4 billion shades of gray (beat that, you 6-bit drivel of a romance novel!), and we can thus set a "wall" so that extreme (perhaps erroneous) infrared intensity values picked up by our sensor do not appear lighter or darker than certain desired shades of gray. InfrDataScale is used to scale the gray shades so that they appear lighter or darker. Larger values for this variable will result in a brighter picture, whereas as lower values will result in a darker picture. Numbers between 0.75f and 1f work well for this.

■ **Note**   *Normalization* refers to the adjustment of values from different scales to correspond to a single common scale. In our case, infrared data has 2^{16} (65,536) different values while `WriteableBitmap` supports up to 2^{32} (4,294,967,296) different values. However, our `Gray32` value is actually measured in floats from 0 to 1, as we set the `PixelFormats` to `Gray32Float`. Since we are normalizing on a 0 to 1 scale, the formula we would use is as follows:

$$x_{0\ to\ 1} = \frac{x - x_{min}}{x_{max} - x_{min}}$$

Where x represents the infrared intensity value we want to normalize, x_{min} its lowest possible value, and x_{max} its highest possible value. Our x_{min} and x_{max} values are 0 and 65535 (or `ushort.MaxValue`), respectively. If we ignore the x_{min} (since it is 0), the formula should be familiar to you, as it is the method with which we calculate percentages (as any student who calculates their test scores can attest to). To ensure that our value is normalized within the range of **desired** minimum and maximum of displayable gray intensities and not merely the entire spectrum of 0 to 1, we apply another formula:

$$x_{dMin\ to\ dMax} = x_0 to_1 \times (1 - x_{desiredMin}) + x_{desiredMin}$$

When we multiply by $(1 - x_{desiredMin})$ we make it so that the value can only be within that range, and then we add the $x_{desiredMin}$ to offset the value and make sure it starts at our desired minimum. For example, if our desired minimum is 0.8, we do $1 - 0.8$, which gives us a range of $[0, 0.2]$. If we add this to our desired minimum, which is 0.8, we get a range of $[0.8, 1]$.

While normally this would give us a range between our desired minimum and desired maximum, we have a scaling value to control brightness. Multiplying $x_{dMin\ to\ dMax}$ by this can result in a value larger than the desired maximum, which is why we apply the `Math.Min()` function to make sure that this does not happen.

After normalizing the values, we set the region to be written in the `WriteableBitmap` with the `AddDirtyRect` method and unlock it so that it can update the UI image again, just like in the `ColorFrameSource` example. Compiling and running the code should result in an infrared video feed similar to that seen in Figure 3-4.

DepthFrameSource: Depth Image Data

The Kinect's depth-sensing capability is very possibly its most interesting and useful feature. Many have purchased the Kinect for that functionality alone. Color cameras are common, and gesture, skeletal, and facial recognition can always be implemented through computer vision techniques. "Cheap and easy-to-use depth camera," however, is a category that the Kinect has pioneered and owned.

As you would guess from its name, the Kinect's depth camera gives us details about the depth of objects in its field of view. Relying on the power of math and algorithms, we can use this data to measure objects and implement custom tracking systems. While it is possible to go very in depth (sorry) with this data, in this portion of the book we will focus on techniques to extract it and process it for visualization.

Displaying Depth Data

Like InfraredFrameSource and ColorFrameSource, DepthFrameSource follows the same patterns and utilizes the analogous DepthFrameReader, DepthFrame, and DepthFrameDescription classes. As with the other image Data Sources, we will be creating a WPF application and including the Microsoft.Kinect, System.Windows.Media, and System.Windows.Media.Imaging namespaces in our project. We will still introduce some new variables, however. To deal with the processing of the depth data for visualization we have the code found in Listing 3-15.

Listing 3-15. MainWindow.xaml.cs Variable Initialization and Bootstrapping for Depth Data

```
public partial class MainWindow : Window
{

    ...
    //Other private variables such as KinectSensor and WriteableBitmap go here
    ...

// Declare the DepthFrameReader
    private DepthFrameReader depthFrameReader = null;
    private FrameDescription depthFrameDescription = null;
    private const int MapDepthToByte = 8000 / 256;
    private byte[] depthPixels = null;

    public MainWindow()
    {

    ...
    /*
    Kinect initialization, event handler assignment and etc. code. See listings 3-1, 3-2,
    3-3 for more info.
    */
    ...
     this.depthFrameReader = this.kinect.DepthFrameSource.OpenReader();
     this.depthFrameReader.FrameArrived += Reader_DepthFrameArrived;
     this.depthFrameDescription = this.kinect.DepthFrameSource.FrameDescription;

    this.depthPixels = new byte[this.depthFrameDescription.Width * this.
    depthFrameDescription.Height];
    this.depthBitmap = new WriteableBitmap(this.depthFrameDescription.Width, this.
    depthFrameDescription.Height, 96.0, 96.0, PixelFormats.Gray8, null);

    ...
```

We bring two new variables to the traditional concoction of XFrameSource-related variables. Depth data from the Kinect will be made available to us in 16-bit ushort values representing millimeter distances from the camera's plane. We will be displaying this data in Gray8 format; thus, each pixel of our WriteableBitmap will be one of possible 256 shades of gray. The shades of gray will indicate how far the Kinect is from that position in the image. MapDepthToByte will be used to normalize the millimeter measurements for display in the WriteableBitmap. depthPixels is an array that will hold our normalized depth pixel values for rendering. We declare it with the size of 1 byte per pixel for the area of the image.

Our event handler for the DepthFrameReader and its associated helper will look somewhat like that of the InfraredFrameReader (Listing 3-16).

Listing 3-16. MainWindow.xaml.cs DepthFrameReader Event Handler and Helper Method

```
private void Reader_DepthFrameArrived(object sender, DepthFrameArrivedEventArgs e)
{
    bool depthFrameProcessed = false;

    using (DepthFrame depthFrame = e.FrameReference.AcquireFrame())
    {
        if (depthFrame != null)
        {
            using (KinectBuffer depthBuffer = depthFrame.LockImageBuffer())
            {
                if ((this.depthFrameDescription.Width * this.depthFrameDescription.Height)
                == (depthBuffer.Size / this.depthFrameDescription.BytesPerPixel))
                {

                    ushort maxDepth = ushort.MaxValue;

                    this.ProcessDepthFrameData(depthBuffer.UnderlyingBuffer, depthBuffer.
                    Size, depthFrame.DepthMinReliableDistance, maxDepth);
                    depthFrameProcessed = true;
                }
            }
        }
    }

    if (depthFrameProcessed)
    {

        this.depthBitmap.WritePixels(
            new Int32Rect(0, 0, this.depthBitmap.PixelWidth, this.depthBitmap.PixelHeight),
            this.depthPixels,
            this.depthBitmap.PixelWidth,
            0);
    }
}

private unsafe void ProcessDepthFrameData(IntPtr depthFrameData, uint depthFrameDataSize,
ushort minDepth, ushort maxDepth)
{

    ushort* frameData = (ushort*)depthFrameData;

    for (int i = 0; i < (int)(depthFrameDataSize / this.depthFrameDescription.
    BytesPerPixel); ++i)
    {
```

```
        ushort depth = frameData[i];

        this.depthPixels[i] = (byte)(depth >= minDepth && depth <= maxDepth ?
        (depth / MapDepthToByte) : 0);
    }
}
```

Like the event handler for the `InfraredFrameReader`, we acquire our frame of data and check its integrity before passing it to a helper method for processing. Our helper method, `ProcessDepthFrameData`, also makes use of pointers to access the depth frame data; thus, we must also mark the method header with the unsafe keyword and let the compiler know to allow unsafe code execution (see Figure 3-5).

`ProcessDepthFrameData` takes an `IntPtr` to the depth frame's data buffer, its size, and our definitions for the closest and farthest depths to show. After creating a ushort pointer, we loop through the pixels in the frame and set the equivalent pixel in the `WriteableBitmap` to a normalized value if it is in between our desired maximum or minimum depth values; otherwise, we set it to 0 (or the color black). To normalize our depth data from millimeters to an 8-bit value, we divide the millimeter value by `MapDepthToByte`. If you recall, `MapDepthToByte` is equal to $\dfrac{8000}{256}$, and thus if we have a value such as 8000mm (which is the max measurable distance), we do $8000\ mm * \dfrac{256}{8000\ mm}$, which is equal to 256, or pure white on a Gray8 `WriteableBitmap`. For a more thorough explanation of normalization, visit Listing 3-14 and its associated discussion.

■ **Tip** Confused by the *? x : y* operator? The *ternary*, or *conditional*, *operator* simply returns the first value (x) if the expression it is operating on is true and the second value (y) if it is false. Take a look at the following expression:

```
string wealth = (name == "Bill Gates") ? "billionaire" : "not a billionaire";
```

If the string name is equal to "Bill Gates", the string wealth will be set to "billionaire"; otherwise, it will be set to "not a billionaire". While this should be an element of any experienced developer's arsenal, as programming becomes more accessible to the general population and younger audiences, the if/else-statement equivalent is often preferred for readability, and thus some readers may have not encountered the use of this operator previously.

After our entire `WriteableBitmap` is filled with depth data, we return control to `Reader_DepthFrameArrived` and finish the rendering process of the `WriteableBitmap`. If all goes well, you should get an image similar to Figure 3-6. The lightest shades of gray represent the farthest distances and vice versa. Any black pixels represent locations where the Kinect could not determine the depth at all. The black jitter is a result the uncertainty of the Kinect in determining depth distances at certain locations and ranges. This can be reduced with smoothing techniques and was not implemented by the Kinect development team, as different applications have different requirements for performance and jitter filtering.

Figure 3-6. *The author's living room visualized in a depth image. Can you spot Redmond bear?*

Visualization Techniques

Although the grayscale representation of our depth data is cool to look at, it is not very pretty, nor is it very useful. It is simply a proof of concept; we can view the depth data through other visualizations to appreciate and utilize it.

Color Gradient Depth Feed

We are not obliged to use a mundane gray to display our depth feed. On the Internet or in demo apps such as Kinect Evolution you will often see the depth feed shaded with various colors. We can choose to do that too.

A common depth feed coloring technique is the color gradient. Right now, our depth feed has a gradient going from black to white. We can change this so that it goes from two or more different colors to white. The first step to achieving this is to change our color format from Gray8 to a more appropriate format. We can choose anything, but Bgra32 is a good bet for any basic color work.

We start by changing the format of our `WriteableBitmap` and the size of the byte array that will render data to it (Listing 3-17).

Listing 3-17. Variable Declaration for Colorized Depth Feed

```
//Inside the MainWindow() Constructor

public MainWindow() {
// ...

this.depthPixels = new byte[this.depthFrameDescription.Width * this.depthFrameDescription.
Height * 4];
this.depthBitmap = new WriteableBitmap(this.depthFrameDesription.Width, this.
depthFrameDescription.Height, 96.0, 96.0, PixelFormats.Bgra32, null);

// ...
```

The depthPixels array's size must be increased to accommodate the bytes from the additional color channels it will now hold for each pixel. We multiply the total number of pixels by 4, as we have 4 bytes of color data per pixel. Remember, Bgra32 is 32 bits of color data, which is equivalent to 4 bytes.

Inside our ProcessDepthFrameData method from earlier, we will have to change how we put data into our depthPixels array, as it is now four times the size (Listing 3-18).

Listing 3-18. Processing Depth Data for Bgra32 WriteableBitmaps

```
private unsafe void ProcessDepthFrameData(IntPtr depthFrameData, uint depthFrameDataSize,
ushort minDepth, ushort maxDepth)
{

    ushort* frameData = (ushort*)depthFrameData;
    int colorByteIndex = 0;

    for (int i = 0; i < (int)(depthFrameDataSize / this.depthFrameDescription.
    BytesPerPixel); ++i)
    {

        ushort depth = frameData[i];

          // Set Blue Channel
        this.depthPixels[colorByteIndex++] = (byte)(depth >= minDepth && depth <= maxDepth ?
        depth : 0);

          // Set Green Channel
        this.depthPixels[colorByteIndex++] = (byte)(depth >= minDepth && depth <= maxDepth ?
        depth : 0);

          // Set Red Channel
        this.depthPixels[colorByteIndex++] = 0;

          // Set Alpha /transparent Channel
        this.depthPixels[colorByteIndex++] = 255;

    }
}
```

The same line is repeated twice—is that a printing error? Not quite. Since we have three color channels, we have to set the color value for each of them every time we iterate a single pixel. colorByteIndex tracks which channel we are inspecting in the depthPixels array. The fourth channel, the one for alpha, is always set to 255, the highest 8-bit value, because we always want full opacity. Before we try to get a color gradient between multiple colors for our image, let us get it working for one color. We set the red channel to 0 so that we can get different shades of cyan as the color of our depth feed.

The one last change needed in order to get our depth feed displaying data through Bgra32 must be made in the method so as to render the data to the WriteableBitmap (Listing 3-19).

Listing 3-19. WriteableBitmap's WritePixel Method for Bgra32

```
// Inside the Reader_DepthFrameArrived() {
// ...

this.depthBitmap.WritePixels(
    New Int32Rect(0, 0, this.depthBitmap.PixelWidth, this.depthBitmap.PixelHeight),
        this.depthPixels,
        this.depthBitmap.PixelWidth * 4,
        0);

// ...
```

The penultimate input for this particular configuration of the WritePixels method is the image's stride. We must multiply by 4 since we now have 4 bytes per pixel worth of data. Note that we use the depthBitmap's PixelWidth property and not its Width property, as the latter does not give us the width of the picture in pixels.

■ **Tip** What's a stride? A stride is the width of a single row of pixels in the image, rounded up to a 4-byte boundary. An easy way to calculate the stride for any image in WPF is to multiply the width of the image in pixels by the number of bytes per pixel. Thus, previously when our depthBitmap displayed the depth feed using the Gray8 format, our stride was simply the width of the WriteableBitmap, as Gray8 pixels are measured using 1-byte.

Compile and run the program when you have finished, and you should get something like Figure 3-7, but in cyan instead of magenta.

Figure 3-7. *The author's living room in a colorized depth feed. Magenta was obtained instead of cyan by setting the first and third color channels (Blue and Red, respectively) of each pixel with the depth value and setting the second color channel (Green) as 0.*

Wait, what is this? What are the black bands repeated throughout the image? Keen readers will have noticed that I omitted the MapDepthToByte divisor from the depth-intensity calculation. Since our ushort depth value is not normalized, the color intensity goes from black to magenta every 0mm to 255mm and then restarts. This is because when the ushort value is converted to a byte value, we truncate it and take the last 8 bits of its 16-bit value. While this can make clustered objects harder to see, when compared to our 8-bit grayscale image in Figure 3-6, it definitely makes the shape of the room easier to visualize. The back wall is clearer, along with the curtains and the rear sofa. Additionally, we can make out the contours of some of the stickers on the HP printer box at the bottom left of the image. Unfortunately, it does give a bit more of a damned vibe to my living room.

Let us now apply the algorithm to get our depth feed mapped to a color gradient (Listing 3-20).

Listing 3-20. Depth Data Processed with a Color Gradient

```
private unsafe void ProcessDepthFrameData(IntPtr depthFrameData, uint depthFrameDataSize,
ushort minDepth, ushort maxDepth)
{

    ushort* frameData = (ushort*)depthFrameData;
    int colorByteIndex = 0;
```

```
for (int i = 0; i < (int)(depthFrameDataSize / this.depthFrameDescription.
BytesPerPixel); ++i)
{

    ushort depth = frameData[i];
    float depthPercentage = (depth / 8000f);

    this.depthPixels[colorByteIndex++] = 0;
    this.depthPixels[colorByteIndex++] = (byte)(depth >= minDepth && depth <= maxDepth ?
    (depthPercentage) * 255 + (1 - depthPercentage) * 0 : 0);
    this.depthPixels[colorByteIndex++] = (byte)(depth >= minDepth && depth <= maxDepth ?
    (depthPercentage) * 0 + (1 - depthPercentage) * 255 : 0);
    this.depthPixels[colorByteIndex++] = 255;

}
}
```

The algorithm to apply a color gradient to our depth feed is not particularly difficult. We are trying to go from one color to another color, so as the depth increases, we use more of the color for the maximum depth and a smaller percentage of the color we started from. Since our colors are measured with three color channels, we do this for the intensity of each channel using the RGB component of the desired colors. This can be described by the following formula (using the Green channel as an example):

$$Green = G \text{ component of } Color1 \times Depth\ Percentage + G \text{ component of } Color2 \times (1 - Depth\ Percentage)$$

In Listing 3-20, the colors I went for were Green and Red, whose BGR components are 0, 255, 0 and 0, 0, 255, respectively. The Blue channel is not used, so I set it to 0 from the get-go. When depthPercentage, which is a measure of how close we are to 8m (the official limit of the Kinect's depth-sensing abilities), reaches 1, we want the depth to be colored green. Looking at the formula for the Green channel, we thus have 1 x 255 + (1 – 1) x 0 = 255. If we were at 4m, we would expect 50 percent green (and 50 percent red), so the formula for green would result in 0.5 x 255 + (1 – 0.5) x 0 = 127.5, which is rounded to 127. The resulting color gradient depth feed should look like that seen in Figure 3-8.

Figure 3-8. *The depth feed under a Red-Green color gradient*

We use colors in other ways to help us visualize the depth of a scene. Instead of having a linear gradient where we go from one color to another gradually, we can choose to have a handful of colors represent different ranges of depth. We could choose yellow for anything between 1 and 2 meters, green for anything between 2 and 3 meters, and so on. This would provide a legend as to where objects are positioned in a scene, rather than saying "it is kind of green and kind of red." Fortunately, this is not difficult at all. It is simply a matter of making a handful of `if` statements for each color region (Listing 3-21).

Listing 3-21. Visualizing Depth Data with Color Regions

```
// Inside the ProcessDepthFrameData() {
    ...
// Replace all the logic inside the method with the following:
    ushort* frameData = (ushort*)depthFrameData;

    int colorByteIndex = 0;
    Color color = new Color();
```

```
    for (int i = 0; i < (int)(depthFrameDataSize / this.depthFrameDescription.
    BytesPerPixel); ++i)
    {
        ushort depth = frameData[i];
        float depthPercentage = (depth / 8000f);

        if (depthPercentage <= 0) {
            color.B = 0;
            color.G = 0;
            color.R = 0;
        }
        if (depthPercentage > 0f & depthPercentage <= 0.2f) {
            color.B = 255;
            color.G = 0;
            color.R = 0;
        }
        if (depthPercentage > 0.2f & depthPercentage <= 0.4f) {
            color.B = 0;
            color.G = 255;
            color.R = 0;
        }
        if (depthPercentage > 0.4f & depthPercentage <= 0.6f) {
            color.B = 0;
            color.G = 0;
            color.R = 255;
        }
        if (depthPercentage >= 0.6f) {
            color.B = 0;
            color.G = 255;
            color.R = 255;
        }

        this.depthPixels[colorByteIndex++] = color.B;
        this.depthPixels[colorByteIndex++] = color.G;
        this.depthPixels[colorByteIndex++] = color.R;
        this.depthPixels[colorByteIndex++] = 255;
    }
// ...
```

In Listing 3-21, we made use of WPF's *Color* structure to keep track of our desired colors for the various regions. Each Color object has properties for Blue, Green, Red, and Alpha values that we can set. Our series of conditional statements simply determines which depth region a pixel falls in and then assigns its color. For example, in Figure 3-9, if our depth value for a pixel is between 20 percent and 40 percent of the maximum possible value of 8000mm, we set it as green. Looking at the image, we can instantly determine that any area colored green is between 1.6m and 3.2m. Of course, we could have picked smaller ranges and added more colors to make it even more practical. If you are measuring anything with this method, remember to take into account the fact that distances are measured from the camera's plane. This means measuring by holding a ruler or tape perpendicular to the Kinect's face. In Figure 3-10, you can see this imaginary plane, which is created by the x and y axes.

Figure 3-9. *The author's living room in bands of 1.6m (yellow is anything farther than 4.8m)*

Figure 3-10. *The camera measures depth perpendicular from the x–y plane, on the z-axis*

Determining the Depth of a Pixel

You might be wondering at this point why, if we can get numerical depth measurements, are we wasting our time with colors? Fair enough. I do have to meet a page quota to have this book published, so bear with me. Let us make a tool that will allow us to determine the depth in millimeters of a particular pixel in an image.

Before I give you the code, let us consider the fact that the Kinect camera's field of view, like most other cameras, is shaped like a pyramid, as in Figure 3-11. This means that if the Kinect sees a surface near it, the depth points represented in 3D space by the pixels are more packed together than if the Kinect sees a surface farther away. The x and y coordinates of our image therefore do not correspond to heights and widths in the real world.

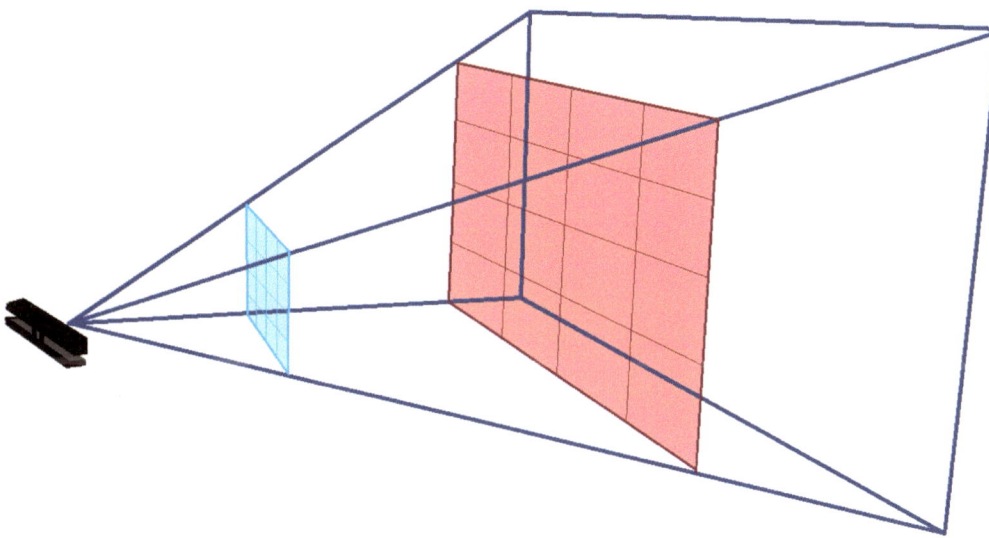

Figure 3-11. *The Kinect's field of view. The blue 4 × 4 pixel grid closer to the Kinect is more densely packed than the 4 × 4 pixel grid farther from the Kinect. In reality, the Kinect has way more pixels, but a 4 × 4 grid was arbitrarily picked for demonstration purposes.*

Since we want to access depth data outside of the DepthFrameReader event handler on a mouse click, we will have to store the depth data in some type of data structure. We could take the data from the byte array depthPixels that we used to write to our WriteableBitmap, but depth data is 16-bits, so we need something larger. We could use a Gray16 image format, or hold it in 2 of the 4 bytes in each RGBA color channel, but we will not bother with the hassle and instead will create another array to hold the depth data (Listing 3-22).

Listing 3-22. Getting Depth Distance for a Pixel on Mouse Click

```
...
//among other private variables such as WriteableBitmap depthBitmap
private int[] depthData = null;
...
```

```
public MainWindow()
{
    ...
    //other variables assigned here
    this.depthData = new int[this.depthFrameDescription.Width * this.depthFrameDescription.
    Height];
    ...
}
private unsafe void ProcessDepthFrameData(IntPtr depthFrameData, uint depthFrameDataSize,
ushort minDepth, ushort maxDepth)
{
    ushort* frameData = (ushort*)depthFrameData;

    int index = 0;

    for (int i = 0; i <(int)(depthFrameDataSize / this.depthFrameDescription.BytesPerPixel); ++i)
    {
        ushort depth = frameData[i];

        this.depthData[i] = depth;

        this.depthPixels[index++] = (byte)(depth >= minDepth && depth <= maxDepth ? depth : 0);
        this.depthPixels[index++] = (byte)(depth >= minDepth && depth <= maxDepth ? depth /
        MapDepthToByte : 0);
        this.depthPixels[index++] = (byte)(depth >= minDepth && depth <= maxDepth ? depth /
        MapDepthToByte : 0);

        this.depthPixels[index++] = 255;
    }
}
private void Image_MouseDown(object sender, System.Windows.Input.MouseEventArgs e)
{
    Point p = e.GetPosition(DepthImage);

    int index = ((int)p.X + ((int)p.Y * depthBitmap.PixelWidth));
    int depth = depthData[index];

    ToolTip toolTip = new ToolTip { Content = depth + "mm", FontSize = 36, IsOpen = true,
    StaysOpen = false };
    DepthImage.ToolTip = toolTip;

}
```

We added an array of integers, depthData, to cache the depth for each pixel in the image every time we have a new frame. The depth data processing code was barely modified to accommodate this. We simply added a line, this.depthData[i] = depth;, to have it collect the depth every time we process a new pixel. An event handler called Image_MouseDown was added to be called whenever a pixel in our image is clicked. It gets the x and y coordinates of the pixel that was clicked, and we use it to find the corresponding pixel in our integer array. Our integer array is one-dimensional, whereas the picture coordinates are two-dimensional. The integer array stores each row of the image one after another, so we can multiply the y coordinate by the width of the image in pixels to find out how far into the integer array to go to get the desired depth data. To better imagine this, pretend our image is a 10-pixel by 10-pixel bitmap and we want to find the depth of the pixel at column 7 and row 5. The size of our integer array is 100. Counting from the top left and **starting at 0**, we get the pixel number as 5 x 10 + 6, or 56. The corresponding depth value in our integer array would be at index 56 as well. We start at 0, of course, as C# arrays start their indexes at 0.

To display the data, we create a tooltip UI element that pops up wherever on the image we clicked.

On the front end, we need to modify our Image tag to support the mouse-clicked event (Listing 3-23).

Listing 3-23. Image XAML for Getting Depth Distance of a Pixel on Mouse Click

```
<Image Name="DepthImage" Source="{Binding ImageSource}" Stretch="UniformToFill"
MouseDown="Image_MouseDown" />
```

In addition to adding the MouseDown event-handler tag for when the mouse is clicked, we also added the name DepthImage so that our event handler knows to which image to add its tooltip. After compiling the program, click on any pixel on the image to find out how far it is from the Kinect's plane, as demonstrated in Figure 3-12.

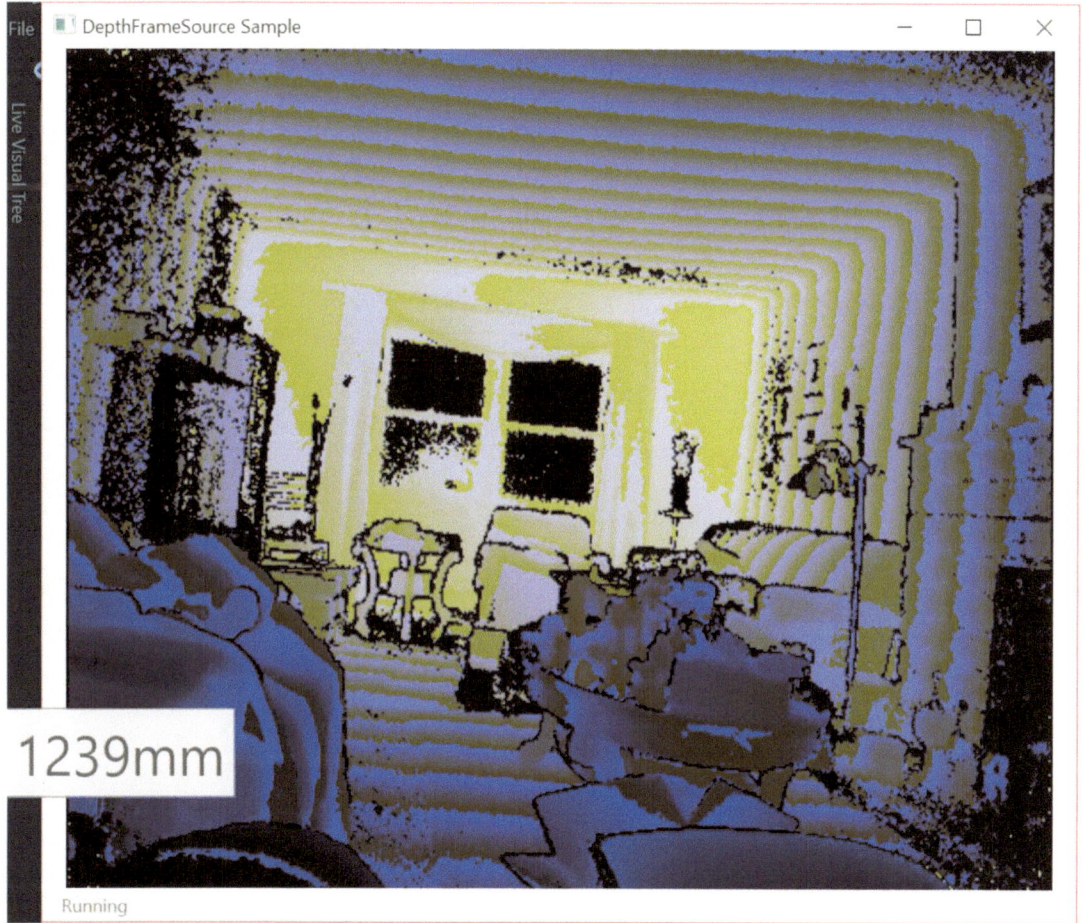

Figure 3-12. *Distance from the Kinect to the couch. The pixel that was clicked is located at the top-right corner of the tooltip.*

Keen observers will again have noted why the image gets progressively yellower further in. In the code to set the intensity for each color channel, we divided the depth values for the Green and Red channels by the MapDepthToByte value, which we did not do for the Blue channel. Thus, the Blue channel has the wrapping effect throughout the image, whereas the Green and Red do not. If you have memorized some RGB values by now, you will know that Red + Green = Yellow.

Summary

In this chapter, we discovered that the Kinect's color, infrared, and depth data can be conveyed by various means. Although there are convenient API methods to display the data to the UI, we can choose to inspect and manipulate the data to further enhance its use for our application's users. The data is malleable and can be saved to the disk as images or cached in the application. It is only by taking advantage of all these capabilities that we can hope to build the best experience for our users.

In the next chapter, we will take advantage of the Kinect's audio sensors and create applications by parsing vocal input.

CHAPTER 4

■ ■ ■

Audio & Speech

In the previous chapter, we covered the Kinect's depth camera. In conjunction with its gesture-recognition capabilities, the depth camera tends to be the Kinect's most touted feature. The Kinect's audio and speech abilities, on the other hand, are typically overlooked. This is partly to do with marketability. From a video-gaming perspective, these features just do not help sell the Kinect nearly as much as the more immersive aspects do, such as gestures. The other important reason, though, is that the paradigm for audio input is somewhat misunderstood. We have all witnessed the "Xbox. . ." commands used to manipulate our video game consoles, but let us be completely honest: it is often easier to rely on the controller. To best realize the potential of the Kinect's audio capabilities, we have to be sincere with ourselves. Audio is not a replacement for hand input, whether that be with a mouse, an Xbox One controller, or a touchpad. It only takes a discreet 2mm translation of our thumb to confirm a selection using a gamepad. We have to clear our voice and awkwardly talk to our device and wait for some latency in the voice-recognition technology to do the same with a microphone.

Thus, most applications that introduce voice commands for the Kinect need to create novel ways for the user to communicate their intentions. There is only so much data bandwidth we can convey to the user using a Graphical User Interface and so much we can take in from hand peripherals. What audio does is provide us with another channel by which to convey information between the user and the software. Take, for example, how what we can do in a video game, in contrast to what the developers and the players would *like* to do, is often limited by how many keys are found on the controller and how many digits a pair of human hands have to physically key them (which are usually limited to the thumb and index fingers anyway). We have some clever ways of trying to overcome this, such as button combinations for special moves in fighting games. This is not ideal for other genres, however.

Let us consider an example from a first-person shooter. Say we were trying to introduce an airstrike radio call capability in an FPS game such as *Battlefield*, where a user could call for a fighter jet to drop munitions on a specific location (yes, I know, that would be OP). We would currently have to take the user out of the game context and have them manipulate a crosshair on a top-down map to select where to drop their ordinance (think *Call of Duty* Predator missile). In the heat of battle, which is when such an airstrike could come most in handy, this would prove very distracting, taking the soldier off the front line and hindering her from aiding her teammates. This might also make the capability unacceptably unrealistic by allowing the player to target areas of the map where they have no situational awareness (e.g., the home-base spawn point). If we incorporated the use of voice in this hypothetical scenario, the player could call the airstrike based on the description of their local environment while still being able to maneuver on the battlefield. They could say, "Calling firing mission behind the tree line at Radio Tower, approaching from the east, danger close." Forgiving my poor knowledge of military jargon, you can see how a user could conveniently convey much more information this way while at the same time giving themselves the ability to do other things with their hands.

© Mansib Rahman 2017
M. Rahman, *Beginning Microsoft Kinect for Windows SDK 2.0*, DOI 10.1007/978-1-4842-2316-1_4

In addition to giving average users the power to do more, let us not forget that audio also helps us bridge the accessibility gap for visually impaired individuals as well as those who have difficulty using hand-input devices because of physical constraints (paralysis, injury, age, etc.). Although having to confirm selections by voice might be tedious for some, for others it could be an indispensable blessing. Thus, even if it is potentially redundant for your application's typical users, you should still strive to make voice equivalents for gesture or peripheral commands. The World Health Organization has defined *disability* as "a mismatch in interaction between the features of a person's body and the features of the environment in which they live." Directional audio enhanced with speech recognition in union with the Kinect's other capabilities grants us an unprecedented opportunity to create venues in which disabled individuals can interact with the environment.

Working with Kinect's Audio Features

The Kinect has a directional microphone array, meaning your application can figure out where in the room sound is coming from. The manner in which this works is analogous to how a telescope is used. Normally, our eyes have a combined horizontal vision range of about 200 degrees. If we wanted to focus on a 1-degree strip from this range, we could use a telescope so that only photons within its conical field of view would be seen by our eyes. Likewise, the Kinect's microphone array has a horizontal range of about 180 degrees, and we can find out where audio is coming from within 5-degree increments of this range; these increments are called *audio beams*. The Kinect also features ambient noise cancellation, which tries to completely ignore sounds coming from the rear and single out voices coming from the front.

Recording Audio

Let us start off by developing a very barebones microphone application. We will take in audio from the Kinect when a Record button is clicked and stop recording when the Stop button is clicked. The resulting audio will be saved to disk in a `.wav` file.

As we have done a handful of times before, create a new WPF project in Visual Studio. Make sure that when you do that, .NET Framework 4.5 is selected in the New Project dialog (see Figure 4-1).

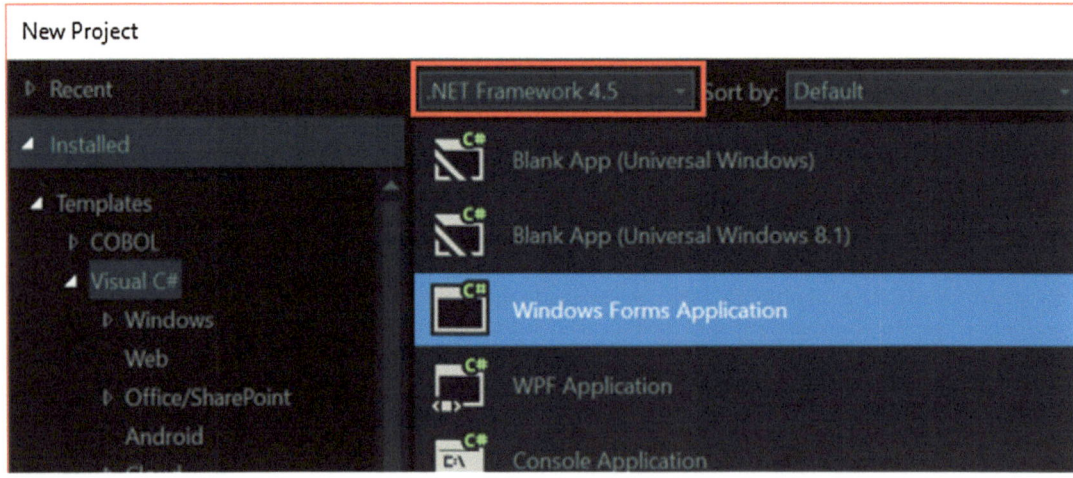

Figure 4-1. *The .NET Framework selector in the New Project dialog. Those are . . . uh, my sister's COBOL templates.*

The namespaces and reference we will be using this time around are System.Windows, System.IO, System.Collections.Generic, System.Text, System, and, as always, Microsoft.Kinect. Many of these will be included by default in any new WPF project. AudioSource is the one Data Source whose data is not gleaned visually (i.e., with a camera). Consequently, the audio data is extracted in a slightly different manner from the other Data Sources. Normally, we have a reader that presents us with a series of frames from which we can access individual data buffers. With audio, we receive frames with references to AudioBeamsFrameLists. These in turn contain AudioBeamSubFrames, from which we gather our raw audio buffer data as well as beam metadata. Our application will progressively copy the audio buffer data to a .wav file if the user has chosen to record. See Listing 4-1.

Listing 4-1. MainWindow.xaml.cs Variable Initialization for Audio Recording

```
public partial class MainWindow : Window
{

    private KinectSensor kinect = null;
    private AudioBeamFrameReader audioBeamFrameReader = null;

    private readonly byte[] audioBuffer = null;

    private bool isRecording = false;

    private int size = 0;
    private FileStream fileStream;
    ...
```

As with every Data Source, we start off with the KinectSensor and reader variables. An array of bytes is created to hold our raw audio data. The isRecording Boolean will be toggled by our Recording/Stop button to keep track of the button state. size will keep track of how many bytes of audio data are going into the .wav file, which we will later include as a part of the file's header data. Filestream is an I/O object we saw earlier in the screenshot example in Chapter 3. We will write our .wav file with it. See Listing 4-2.

Listing 4-2. MainWindow.xaml.cs Variable Assignment and Data Source Bootstrapping for Audio Recording

```
public MainWindow()
{

    this.kinect = KinectSensor.GetDefault();
    this.kinect.Open();

    AudioSource audioSource = this.kinect.AudioSource;

    this.audioBuffer = new byte[audioSource.SubFrameLengthInBytes];

    this.audioBeamFrameReader = audioSource.OpenReader();
    this.audioBeamFrameReader.FrameArrived += Reader_FrameArrived;

    this.audioBeamFrameReader.IsPaused = true;

    InitializeComponent();
}
```

79

The constructor in Listing 4-2 is pretty average looking. The audio buffer array is given a size of 1024 bytes, which is exactly enough to hold the data from each `AudioBeamSubFrame`. Each sub-frame is 16 milliseconds long and is sampled at 16 KHz; thus, we get 256 samples. Each sample is 4 bytes, which is how we get to the figure of 1024 bytes.

The reader is initialized in a paused state, and we re-enable it whenever the user starts recording. There are other ways to control the sensor's recording process, such as adding and subtracting `FrameArrived` event handlers or creating and disposing of readers, but I chose to pause the reader instead as it is remarkably convenient for our simple application. Other methods may be more suitable for more complex, multipart applications.

▪ **Tip**   When we refer to *sampling rate*, i.e., sampled at 16 KHz, we refer to how often an electronic audio signal is measured in a given interval. Normally, sound waves consist of continuous signals, but as with everything else digital, we must convert them to a discrete signal so that the computer can interpret them with bits. When we get 256 samples, we are saying we measured the amplitude of the sound wave 256 times within the 16-millisecond timeframe. This is later reconstructed as a continuous signal for analog circuits in your speakers to amplify and play back. The discrete digital signal can be processed in a way similar to what we did with images to change the character of the sounds.

Let us explore the Record button code before we check the `FrameArrived` event handler and other ancillary code. The program's primary logic starts and ends here, so it will be easier to understand the rest after looking here first.

Listing 4-3. MainWindow.xaml.cs button_Click Event Handler for Audio Recording

```
private void button_Click(object sender, RoutedEventArgs e)
{

    //If currently recording, then stop recording
    if (isRecording == true)
    {

        button.IsEnabled = false;
        button.Content = "Record";
        this.isRecording = false;
        audioBeamFrameReader.IsPaused = true;

        fileStream.Seek(0, SeekOrigin.Begin);
        WriteWavHeader(fileStream, size);
        fileStream.Seek(0, SeekOrigin.End);
        fileStream.Flush();
        fileStream.Dispose();
        size = 0;

        button.IsEnabled = true;
    }
```

```
        //If not recording, start recording
    else if (isRecording == false)
    {

        button.IsEnabled - false;
        button.Content = "Stop";
        this.isRecording = true;
        audioBeamFrameReader.IsPaused = false;

        string time = DateTime.Now.ToString("d MMM yyyy hh-mm-ss");
        string myMusic = Environment.GetFolderPath(Environment.SpecialFolder.MyMusic);
        string filename = System.IO.Path.Combine(myMusic, "Kinect Audio-" + time + ".wav");
        fileStream = new FileStream(filename, FileMode.Create);
        WriteWavHeader(fileStream, size);

        button.IsEnabled = true;
    }
}
```

The button event handler in Listing 4-3 basically toggles whether the application is recording or not. In either case—stop recording and start recording—we commence by disabling the button and finish by re-enabling it, so that the application completes the file-encoding process before permitting further action. This will likely never be a problem, as the process is fairly quick. We then toggle whether AudioBeamFrameReader is recording by setting its IsPaused property.

If we are ending the recording process, we move to the start of the soon-to-be .wav file's FileStream with its Seek(long offset, SeekOrigin origin) method (as seen in the screenshot examples in Chapter 3) and append a header with the WriteWavHeader method, which we will be creating shortly. Once the header is written, we move back to wherever we were in the stream and write all the buffered data to the file with Flush(). The FileStream is disposed of to free up resources and so that it will be ready to be reused for future recording operations.

To start the recording process, we create a new file in the *My Music* folder with a unique timestamp and then write a placeholder header in the file with WriteWavHeader. We do this to ensure that the start of the file's data is not occupied by sound data before we include the proper header during completion.

In C#, there is no built-in way to write .wav files, as we saw for images with classes like PngEncoder and BitmapEncoder. Thus, we must create our own metadata and write it to the start of the file as its header. Following convention, we use a custom C# adoption of the *WAVEFORMATEX* structure found in the DirectShow API (which is a part of the Windows SDK) used to define our data as waveform-audio. The WriteWavHeader method is used to work with the structure.

Listing 4-4. MainWindow.xaml.cs WriteWavHeader and Associated Code

```
private static void WriteWavHeader(FileStream fileStream, int size)
{

    using (MemoryStream memStream = new MemoryStream(64))
    {

        int cbFormat = 18;
        WAVEFORMATEX format = new WAVEFORMATEX()
        {
```

```
                wFormatTag = 3,
                nChannels = 1,
                nSamplesPerSec = 16000,
                nAvgBytesPerSec = 64000,
                nBlockAlign = 4,
                wBitsPerSample = 32,
                cbSize = 0

        };

        using (var bw = new BinaryWriter(memStream))
        {

            WriteString(memStream, "RIFF");
            bw.Write(size + cbFormat + 4);
            WriteString(memStream, "WAVE");
            WriteString(memStream, "fmt ");
            bw.Write(cbFormat);
            bw.Write(format.wFormatTag);
            bw.Write(format.nChannels);
            bw.Write(format.nSamplesPerSec);
            bw.Write(format.nAvgBytesPerSec);
            bw.Write(format.nBlockAlign);
            bw.Write(format.wBitsPerSample);
            bw.Write(format.cbSize);
            WriteString(memStream, "data");
            bw.Write(size);
            memStream.WriteTo(fileStream);
        }
    }
}

private static void WriteString(Stream stream, string s)
{

    byte[] bytes = Encoding.ASCII.GetBytes(s);
    stream.Write(bytes, 0, bytes.Length);
}

struct WAVEFORMATEX
{

    public ushort wFormatTag;
    public ushort nChannels;
    public uint nSamplesPerSec;
    public uint nAvgBytesPerSec;
    public ushort nBlockAlign;
    public ushort wBitsPerSample;
    public ushort cbSize;
}
```

I will not cover the code in Listing 4-4 in too much detail, as it is beyond the scope of this book. What is important to know is that the `WriteWavHeader` method writes the WAVEFORMATEX structure to the header of the .wav file. We have a `WriteString` helper method that writes certain strings to the memory stream using bytes. These strings demarcate the different parts of the WAVEFORMATEX construct in the .wav file. WAVEFORMATEX consists mainly of data that helps an audio player like VLC or Windows Media Player interpret the data into sound. This is where knowing the Kinect's audio specs is critical; for example, setting the nSamplesPerSec value to anything but 16,000 would result in unintelligible audio. You might be able to tamper with the values I have used to get a more enhanced sound, but, in my experience, this is the best configuration.

It is noteworthy that we write the size of all the audio buffer data into the memory stream with the `bw.Write(size);` statement. This size value is gathered progressively as we get more frames and the length of the audio file grows.

■ **Tip** WAVEFORMATEX is essentially an extension of the basic 16-byte *WAVEFORMAT* structure with the addition of a 1-byte `cbSize` value to keep track of metadata size and additional bytes for the metadata itself. All additional data would be written after `cbSize` is written to the memory stream. For a cool write-up of the history of WAVEFORMAT[EX], check out this informative MSDN post: `https://blogs.msdn.microsoft.com/larryosterman/2007/10/18/the-evolution-of-a-data-structure-the-waveformat/`

The final piece of the code-behind to cover is the `FrameArrived` event handler. Its only purpose is to chunk audio data into the .wav file and keep track of the buffer's size.

Listing 4-5. MainWindow.xaml.cs Reader_FrameArrived Event Handler for Audio Recording

```
private void Reader_FrameArrived(object sender, AudioBeamFrameArrivedEventArgs e)
{

    AudioBeamFrameReference frameReference = e.FrameReference;
    AudioBeamFrameList frameList = frameReference.AcquireBeamFrames();

    if (frameList != null)
    {

        using (frameList) {

            IReadOnlyList<AudioBeamSubFrame> subFrameList = frameList[0].SubFrames;

            foreach (AudioBeamSubFrame subFrame in subFrameList)
            {

                subFrame.CopyFrameDataToArray(this.audioBuffer);
                if (fileStream.CanWrite == true)
                {
                    fileStream.Write(audioBuffer, 0, audioBuffer.Length);
                    size += audioBuffer.Length;
                }
            }
        }
    }
}
```

In Listing 4-5, after having extracted the AudioBeamFrameList, we acquire the AudioBeamSubFrame with the IReadOnlyList<AudioBeamSubFrame> subFrameList = frameList[0].SubFrames statement. The AudioBeamFrameList only ever contains one beam, which is why we access the zeroth index instead of iterating through the list. We then loop through all the sub-frames, which will typically number from 1 to 3, but sometimes up to 8, and copy their data to the Filestream if it is open. We also add to the total size counter.

The front end is probably the simplest you will encounter in this book. I have made uglier UIs, so try not to cringe too much (Listing 4-6).

Listing 4-6. MainWindow.xaml Audio Recording UI

```
//Use rest of <Window/> scaffolding code from your own project
<Window Height="75" Width="100" ... >
    <Grid>
        <Button x:Name="button" Content="Record" HorizontalAlignment="Left"
        Margin="10, 10, 10, 10" VerticalAlignment="Top" Width="75" Click="button_Click" />
    </Grid>
</Window>
```

The finished result will look like Figure 4-2. Once you click Record, simply click Stop when you have recorded what you need. The resulting .wav recording of the Kinect's audio will be saved to the *My Music* folder. It can be played back with pretty much any modern music player. When playing the recorded audio, you will notice that when there was no prominent noise the Kinect ignored any minor noises. This might include low-volume music, whispering, shuffling, and so on. During a loud song, it might cut out when the song features an instrumental or a decrease in volume on a certain beat and then pick up again momentarily when the volume or human voice is restored. This is the Kinect filtering out ambient noise on its own accord. While the Kinect for Windows SDK v1 had ways to configure this, the SDK v2 does not.

Figure 4-2. *The Kinect audio recording application. The Record button becomes the Stop button once started.*

Recording Audio in Windows Store Apps

The process of recording audio in a Windows Store app is very similar to doing so in a WPF app. I will not cover it in detail, as there are only minor differences in the Windows Store APIs. The source code to do so is included in the book samples, however. Before you proceed to copy it, I encourage you to attempt to convert the WPF code yourself.

The main difference you have to keep in mind is how I/O is done. There is no Filestream class in Windows Store apps. Instead, as you saw in the screenshot example in Chapter 3, we rely on the StorageFile class. This can open a stream that can be written to. Remember to take into account that much of the code will be async.

Audio Beam Properties

What really differentiates the Kinect's microphone array from any run-of-the-mill laptop microphone is its ability to determine from which direction a sound is coming. As mentioned earlier, the Kinect achieves this with the help of audio beams. Audio beams can be configured in two ways. We can either set it to *Automatic* mode, in which the Kinect tries to find the most prominent voice in the area, or *Manual* mode, where you choose an angle at which to focus and the Kinect favors sounds coming from that direction.

I say audio beams, but I should really say audio beam, because the Kinect will only ever return one audio beam to you. This audio beam is a weighted sum of all the audio beams it detects internally. So if, for example, two humans are talking near the Kinect, one louder than the other, the Kinect will process two separate beams internally and return one `AudioBeamSubFrame` to you with the `BeamAngle` aimed toward the louder speaker. `BeamAngle` is the property that gives you the angle of the sound source from the center.

■ **Note**   While it is not possible to get the multiple internal audio beams that are summed to give the resulting audio beam in the API, it may be possible to access them by using the *Windows Audio Session API* (WASAPI). It is unlikely that Microsoft will implement this functionality in the foreseeable future.

For a great visualization of how audio beams work, you should check out the *AudioBasics-WPF* (or D2D) sample in SDK Browser v2.0. As shown in Figure 4-3, it features a needle that points to the loudest audio source in front of the Kinect, as well as an energy bar representing the intensity of that audio.

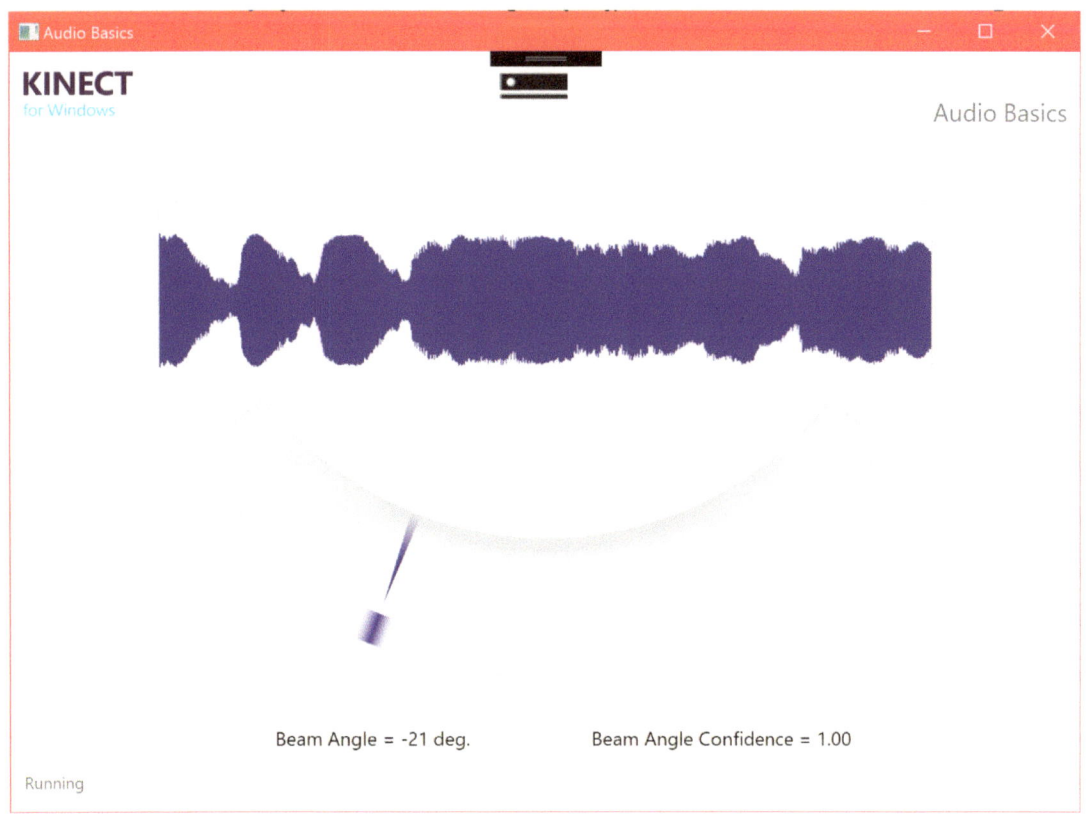

Figure 4-3. *Audio Basics sample picking up my voice*

If you tried to register sounds from both sides of the Kinect, you will notice that the beam angle only ever goes ± 50 degrees from the center (negative being left of the Kinect and vice versa). I told a small lie earlier when I said the microphone detects anything in a 180-degree range. While it does technically do this, the Kinect team has limited the value that the API returns to within ± 0.87 radians, or ± 50 degrees from the center. Even if a sound is originated at 51 degrees right of the center, the Kinect will not inform you of this. In manual mode, you can set the beam angle to anything within $\pm\frac{\pi}{2}$ radians or ± 90 degrees from the center, but you will not receive any audio from values approximately larger than ± 0.915 radians or ± 53 degrees from the center.

Beam Angle Confidence refers to how confident the Kinect is that the loudest sound came from the direction of the beam angle. Its value ranges from 0 to 1.0, where 0 refers to no sound being present and 1.0 refers to total confidence of sound being present. In Manual mode, the value can be either 1 or 0, whereas in Automatic mode it can be anything in between.

By default, Automatic mode is configured for use by the Kinect. Taking a look at the *AudioBasics-WPF* sample, you will notice the following lines commented out in the constructor of MainPage:

```
audioSource.AudioBeams[0].AudioBeamMode = AudioBeamMode.Manual;
audioSource.AudioBeams[0].BeamAngle = 0;
```

As we saw earlier, there is only one AudioBeam available for use, so we can simply access it through the first index of the AudioBeams property of the Kinect's AudioSource. Setting the AudioBeamMode back to automatic is simply a matter of writing audioSource.AudioBeams[0].AudioBeamMode = AudioBeamMode. Automatic;.

To access the AudioBeam values returned by the Kinect during operation, we must grab them from the AudioBeamSubFrame. We can easily access its BeamAngle or BeamAngleConfidence properties this way:

```
foreach (AudioBeamSubFrame subFrame in subFrameList)
{
    // Check if beam angle and/or confidence have changed
    bool updateBeam = false;

    if (subFrame.BeamAngle != this.beamAngle)
    {
        this.beamAngle = subFrame.BeamAngle;
        updateBeam = true;
    }

    if (subFrame.BeamAngleConfidence != this.beamAngleConfidence)
    {
        this.beamAngleConfidence = subFrame.BeamAngleConfidence;
        updateBeam = true;
    }
...
}
```

■ **Note** If AudioBeamMode is Manual, the application must be actively in the foreground for beamforming to work. The code relevant to manual beamforming cannot run in a background thread either.

The AudioBeamSubFrame has an additional property called AudioBodyCorrelations. In conjunction with the skeletal data received from the BodyFrameSource (which will be explored in Chapter 5), it enables us to figure out who is currently speaking. It only tells us the ID of the most prominent voice speaking at the time. To get this ID, access the BodyTrackingId of the property's first index, as follows:

```
if (subFrame.AudioBodyCorrelations.Count != 0) {
    AudioTrackingId = subFrame.AudioBodyCorrelations[0].BodyTrackingId;
}
```

■ **Note** The reason AudioBodyCorrelations is a list and not a single value is because there was the potential to support the detection of multiple voice sources simultaneously, but seeing as multiple audio beams were not incorporated, this never came about.

Using Kinect for Speech Recognition

Developers will most probably use the Kinect's microphone array for speech recognition above all else. Other than some *language models* that were custom-prepared by Microsoft, there is not much to developing speech recognition that is inherent to the Kinect. The extent of the Kinect's built-in audio capabilities was covered in the first half of this chapter. The rest must be developed and integrated by us. Fortunately, this is not as torturous a task as it might sound at first. Microsoft has a separate Speech SDK that we can use in our application. You are not obliged to use this SDK; you can use the Bing Speech API (not the same as the Microsoft Speech SDK we are using) or probably even the recently announced (this section was written in 2016) Google Speech API, among others. We will be using Microsoft Speech because, as mentioned earlier, Microsoft has prepared Microsoft Speech language models specifically for the Kinect. These will have better recognition capabilities than other speech recognizers when used by the Kinect. This is because they take into account that users will be far away from the microphone array, whereas other language packs assume the user's microphone is within arm's width.

■ **Tip** *Language models* represent the ways in which the words of a language are combined. Along with *acoustic models* (speech patterns) and *lexicons* (word lists and their pronunciations), they make up *language packs*, which are used by the Microsoft Speech runtime to interpret audio into a specific language. There are a handful of language packs prepared specifically for the Kinect. If your language is represented by one of them, you should definitely use them, barring other constraints. Else, you can use one of the generic language packs that are available for a greater variety of languages. The Kinect-specific language packs are available for Canadian English, American English, British English, Australian English, New Zealand English, Irish English, Canadian French, France French, Spain Spanish, Mexican Spanish, German, Italian, and Japanese.

Installing Microsoft Speech

Before we get started, you need to ensure that the Microsoft Speech SDK 11.0 is installed, along with the associated runtime and language packs. You can grab these in the Kinect SDK 2.0 Browser, or you can download them at the following links. The Kinect for Windows team recommends that you get the x86 version of the SDK and runtime.

- **Microsoft Speech Platform SDK Version 11.0** – `https://www.microsoft.com/en-us/download/details.aspx?id=27226`

- **Microsoft Speech Platform Runtime Version 11.0** – `https://www.microsoft.com/en-us/download/details.aspx?id=27225`

- **Kinect for Windows SDK 2.0 Language Packs** – `https://www.microsoft.com/en-us/download/details.aspx?id=43662`

Many developers have had trouble getting speech recognition to work because of configuration issues with the SDK. For this reason, it is generally recommended that you install the necessary components in a specific order:

1. Kinect for Windows SDK

2. Microsoft Speech Platform SDK

3. Microsoft Speech Platform Runtime

4. Language Packs

The language packs do not seem to give a confirmation once they have finished installing. You can check whether they have been properly installed by looking for their entries on Windows' *Programs and Features* page. You might have to restart your machine to complete the installation. I recommend you try out the *SpeechBasics-WPF* sample to verify that everything has been installed properly.

Integrating Microsoft Speech in an Application

This time around, our little demo project will be a speech-driven Tic Tac Toe game. It will be played against a dumb AI, and the available voice commands will be "Start," "Restart," and the names of the various Tic Tac Toe boxes (e.g., "Top Left" or "Center"). When the user says a command, a circle will be drawn on the box of the user's choice, and the AI will then take a turn. The dependencies we will be using are `System`, `System.Collections.Generic`, `System.Windows`, `System.Windows.Controls`, `Microsoft.Kinect`, `Microsoft.Speech.Recogniton`, and `Microsoft.Speech.AudioFormat`. The two `Microsoft.Speech.*` dependencies need to be added in the Reference Manager. You might not see it listed as an option, so you may need to click "Browse…" and grab it from `C:\Program Files (x86)\Microsoft SDKs\Speech\v11.0\Assembly`.

We will need to create a `KinectAudioStream` class that wraps around the `Stream` class. This is because the Kinect's audio data comes in 32-bit IEEE float format, whereas Microsoft Speech uses signed 16-bit integers to represent its audio data. To add a new class, press **Shift** + **Alt** + **C**. Name the class `KinectAudioStream` and confirm (Listing 4-7).

Listing 4-7. KinectAudioStream.cs

```
namespace KinectTicTacToe
{
    using System;
    using System.IO;

    internal class KinectAudioStream : Stream
    {
        private Stream kinect32BitStream;
```

```csharp
public KinectAudioStream(Stream input)
{
    this.kinect32BitStream = input;
}

public bool SpeechActive { get; set; }

public override bool CanRead
{
    get { return true; }
}

public override bool CanWrite
{
    get { return false; }
}

public override bool CanSeek
{
    get { return false; }
}

public override long Position
{
    get { return 0; }
    set { throw new NotImplementedException(); }
}

public override long Length
{
    get { throw new NotImplementedException(); }
}

public override void Flush()
{
    throw new NotImplementedException();
}

public override long Seek(long offset, SeekOrigin origin)
{
    return 0;
}

public override void SetLength(long value)
{
    throw new NotImplementedException();
}

public override void Write(byte[] buffer, int offset, int count)
{
    throw new NotImplementedException();
}
```

```csharp
public override int Read(byte[] buffer, int offset, int count)
{
    const int SampleSizeRatio = sizeof(float) / sizeof(short);
    const int SleepDuration = 50;

    int readcount = count * SampleSizeRatio;
    byte[] kinectBuffer = new byte[readcount];

    int bytesremaining = readcount;

    while (bytesremaining > 0)
    {
        if (!this.SpeechActive)
        {
            return 0;
        }

        int result = this.kinect32BitStream.Read(kinectBuffer, readcount
        - bytesremaining, bytesremaining);
        bytesremaining -= result;

        if (bytesremaining > 0)
        {
            System.Threading.Thread.Sleep(SleepDuration);
        }
    }

    for (int i = 0; i < count / sizeof(short); i++)
    {
        float sample = BitConverter.ToSingle(kinectBuffer, i * sizeof(float));

        if (sample > 1.0f)
        {
            sample = 1.0f;
        }
        else if (sample < -1.0f)
        {
            sample = -1.0f;
        }

        short convertedSample = Convert.ToInt16(sample * short.MaxValue);

        byte[] local = BitConverter.GetBytes(convertedSample);
        System.Buffer.BlockCopy(local, 0, buffer, offset + (i * sizeof(short)),
        sizeof(short));
    }

    return count;
}
```

You might be puzzled as to why we have so much dead code in Listing 4-7. We are deriving the Stream class, so we have to inherit all its abstracted members. Almost none of it will be used by the Microsoft Speech API. Only Read(byte[] buffer, int offset, int count) and a few helper members are really necessary for the code to work. This is because Microsoft Speech will be doing nothing else but reading audio data from the stream to get voice commands. It has no need to write to it or do other operations. There are still some expected return values for the inherited methods, however, so just plastering ThrowNotImplementedException() in them will not cut it. Understanding the Read method is not too important. As mentioned earlier, it reads 32-bit float samples from the stream and translates them to 16-bit signed integers. At first, it keeps writing data into the kinectBuffer byte array within the while loop until it has no more data to stream. It takes a 50-millisecond break on each iteration because the speech engine can read data faster than the Kinect's microphones can provide it. It then extracts the 32-bit values from kinectBuffer, converts them, and then puts in the buffer to be used by Microsoft Speech.

Listing 4-8. MainWindow.xaml.cs Private Variables for Kinect Tic Tac Toe

```
public partial class MainWindow : Window
{

    private bool inGame = false;
    private TextBlock[,] textBlockGrid = new TextBlock[3, 3];

    private KinectSensor kinect = null;
    private KinectAudioStream kinectAudioStream = null;

    private SpeechRecognitionEngine speechEngine = null;
    ...
```

In Listing 4-8, we see that all our game-state logic is contained in the inGame and textBlockGrid variables. We could have opted for an internal representation of the game grid, but for this simple example, all the code will operate based on what characters (i.e., "X" or "O") show up on the UI. We initialize an instance of the KinectAudioStream class that we just created. Our SpeechRecognitionEngine speechEngine is then declared. It works similarly to one of the Kinect's Data Sources. After being configured, it has an event handler that gets called every time a new voice command has been received.

Listing 4-9. MainWindow.xaml.cs Constructor for Kinect Tic Tac Toe Part 1: Kinect Setup

```
public MainWindow()
{
    kinect = KinectSensor.GetDefault();
    kinect.Open();

    IReadOnlyList<AudioBeam> audioBeamList = this.kinect.AudioSource.AudioBeams;
    System.IO.Stream audioStream = audioBeamList[0].OpenInputStream();

    this.kinectAudioStream = new KinectAudioStream(audioStream);
    ...
```

In Listing 4-9, instead of having an event handler for audio frames, we open a direct stream and wrap the KinectAudioStream class around it. This converted stream will be piped into the speech engine.

Listing 4-10. MainWindow.xaml.cs Constructor for Kinect Tic Tac Toe Part 2: Speech Engine Vocabulary Creation

```
...
RecognizerInfo ri = TryGetKinectRecognizer();

if (null != ri)
{
    this.speechEngine = new SpeechRecognitionEngine(ri.Id);

    Choices commands = new Choices();
    commands.Add(new SemanticResultValue("start", "START"));
    commands.Add(new SemanticResultValue("restart", "RESTART"));

    commands.Add(new SemanticResultValue("top left", "TOPLEFT"));
    commands.Add(new SemanticResultValue("top left corner", "TOPLEFT"));
    commands.Add(new SemanticResultValue("upper left corner", "TOPLEFT"));
    commands.Add(new SemanticResultValue("top", "TOP"));
    commands.Add(new SemanticResultValue("top center", "TOP"));
    commands.Add(new SemanticResultValue("top right", "TOPRIGHT"));
    commands.Add(new SemanticResultValue("top right corner", "TOPRIGHT"));
    commands.Add(new SemanticResultValue("upper right corner", "TOPRIGHT"));

    commands.Add(new SemanticResultValue("left", "LEFT"));
    commands.Add(new SemanticResultValue("center left", "LEFT"));
    commands.Add(new SemanticResultValue("center", "CENTER"));
    commands.Add(new SemanticResultValue("middle", "CENTER"));
    commands.Add(new SemanticResultValue("right", "RIGHT"));
    commands.Add(new SemanticResultValue("center right", "RIGHT"));

    commands.Add(new SemanticResultValue("bottom left", "BOTTOMLEFT"));
    commands.Add(new SemanticResultValue("bottom left corner", "BOTTOMLEFT"));
    commands.Add(new SemanticResultValue("lower left corner", "BOTTOMLEFT"));
    commands.Add(new SemanticResultValue("bottom", "BOTTOM"));
    commands.Add(new SemanticResultValue("bottom center", "BOTTOM"));
    commands.Add(new SemanticResultValue("bottom right", "BOTTOMRIGHT"));
    commands.Add(new SemanticResultValue("bottom right corner", "BOTTOMRIGHT"));
    commands.Add(new SemanticResultValue("lower right corner", "BOTTOMRIGHT"));
    ...
```

The first thing we do in Listing 4-10 is look for installed language packs on the PC. Our TryGetKinectRecognizer() method, which we will write later, will take care of this. If we find a language pack, we proceed to add all the different voice commands that can be accepted by our application. The first input of SemanticResultValue(string phrase, object value) is the spoken phrase that will trigger a speech-recognized event, and the second input is the value that will be provided by the event handler for us to work with. A value can have more than one phrase that triggers it, but you should not try to trigger more than one value with a single phrase. We are not obliged to add all the voice commands programmatically. You can also create an XML file, parse it, and pass the resulting string into the Grammar object, which is then loaded onto the speech engine.

Listing 4-11. MainWindow.xaml.cs Constructor for Kinect Tic Tac Toe Part 3: Speech Engine Configuration

```
...
var gb = new GrammarBuilder { Culture = ri.Culture };
gb.Append(commands);

var g = new Grammar(gb);
this.speechEngine.LoadGrammar(g);

this.speechEngine.SpeechRecognized += this.SpeechRecognized;

this.kinectAudioStream.SpeechActive = true;
this.speechEngine.SetInputToAudioStream(
this.kinectAudioStream, new SpeechAudioFormatInfo(EncodingFormat.Pcm, 16000,
16, 1, 32000, 2, null));
this.speechEngine.RecognizeAsync(RecognizeMode.Multiple);

}
else
{
    Application.Current.Shutdown();
}
...
```

In Listing 4-11, the GrammarBuilder object takes our voice command list as well as our language pack and creates a Grammar object. This is loaded into the speech engine, after which we assign it the SpeechRecognized event handler. We input the KinectAudioStream into it with the format of the audio data. You will notice that it is a bit different from the values we used in WAVEFORMATEX. This is because we changed the sample value to 16-bit. The RecognizeAsync(RecognizeMode mode) method then starts the engine asynchronously. The RecognizeMode can either be either Single or Multiple. Single stops the recognition after the first phrase is completed. Multiple does not.

Listing 4-12. MainWindow.xaml.cs Constructor for Kinect Tic Tac Toe Part 4: Gameboard Setup

```
...
InitializeComponent();

textBlockGrid[0, 0] = topLeft;
textBlockGrid[0, 1] = top;
textBlockGrid[0, 2] = topRight;

textBlockGrid[1, 0] = left;
textBlockGrid[1, 1] = center;
textBlockGrid[1, 2] = right;

textBlockGrid[2, 0] = bottomLeft;
textBlockGrid[2, 1] = bottom;
textBlockGrid[2, 2] = bottomRight;
}
```

In Listing 4-12, we fill a grid with the location of each square on the Tic Tac Toe board for more convenient access. topLeft, top, topRight, and so on are all TextBlock controls that are declared in the XAML. They will display the "X" or "O" value on the UI in the correct location. InitializeComponent() must be called first so that they can be created before we do anything with them.

Listing 4-13. MainWindow.xaml.cs TryGetKinectRecognier Method

```
private static RecognizerInfo TryGetKinectRecognizer()
{
    IEnumerable<RecognizerInfo> recognizers;

    try
    {
        recognizers = SpeechRecognitionEngine.InstalledRecognizers();
    }
    catch (System.Runtime.InteropServices.COMException)
    {
        return null;
    }

    foreach (RecognizerInfo recognizer in recognizers)
    {
        string value;
        recognizer.AdditionalInfo.TryGetValue("Kinect", out value);
        if ("True".Equals(value, StringComparison.OrdinalIgnoreCase) && "en-US".
        Equals(recognizer.Culture.Name, StringComparison.OrdinalIgnoreCase))
        {
            return recognizer;
        }
    }

    return null;
}
```

In Listing 4-13, TryGetKinectRecognizer() first checks to see what language packs are installed on our computer. The gist of its logic is found in the statement recognizers = SpeechRecognitionEngine. InstalledRecognizers();. We have to wrap it around a try block to deal with the possibility that no language packs or runtime are installed on the machine. We then loop through the installed recognizers looking for one with "Kinect" in its title as well as "en-US". If you have another language pack that you want to use, you can ignore the "Kinect" requirement and/or change the "en-US" string to one that reflects your chosen language pack.

■ **Note**   Microsoft has some server-side speech recognizers in addition to client-side ones. These will not work with the Kinect. InstalledRecognizers() will find the server one by its registry entry, which is very similar to the client recognizer's.

Listing 4-14. MainWindow.xaml.cs SpeechRecognized Event Handler

```
private void SpeechRecognized(object sender, SpeechRecognizedEventArgs e)
{
    if (e.Result.Confidence >= 0.35)
    {
        switch (e.Result.Semantics.Value.ToString())
        {
            case "START":
                if (inGame == false)
                {
                    for (int i = 0; i < 3; i++)
                    {
                        for (int j = 0; j < 3; j++)
                        {
                            textBlockGrid[i, j].Text = "";
                            textBlockGrid[i, j].Foreground = System.Windows.Media.Brushes.
                            Black;
                        }
                    }
                    inGame = true;
                    resultbox.Text = "Playing... Say 'Restart' to Restart";
                    AITurn();
                }
                break;
            case "RESTART":
                if (e.Result.Confidence >= 0.55)
                {
                    if (inGame == true)
                    {
                        for (int i = 0; i < 3; i++)
                        {
                            for (int j = 0; j < 3; j++)
                            {

                                textBlockGrid[i, j].Text = "";
                                textBlockGrid[i, j].Foreground = System.Windows.Media.
                                Brushes.Black;
                            }
                        }
                        AITurn();
                    }
                }
                break;
            case "TOPLEFT":
                if (inGame == true)
                    HumanTurn(0, 0);
                break;
            case "TOP":
                if (inGame == true)
                    HumanTurn(0, 1);
                break;
```

95

```
        //Omitted for redundancy, have a case for each value
        [ ... ]
        case "BOTTOMRIGHT":
            if (inGame == true)
                HumanTurn(2, 2);
            break;
        }
    }
}
```

In Listing 4-14, the first thing we do after receiving a voice command is check the confidence values with e.Result.confidence. This value ranges from 0 to 1 for low to high confidence, respectively. They do not indicate absolutely whether an input was recognized. Rather, they are relative values used to compare the likeliness of being recognized. For example, a phrase recognized with 0.1223 confidence is a lot less likely to be the correct match for the desired input than a phrase with 0.7625 confidence. It does not mean that they have a 12.23 percent or 76.25 percent chance of matching, respectively.

We then extract the value derived from the recognized phrase with e.Result.Semantics.Value. ToString(). This is put through a switch statement so that we can decide what to do for each situation. If the input is the command to draw a circle in one of the boxes, we attempt to put a circle in the relevant box with the HumanTurn() method. Otherwise, if the commands are "Start" or "Restart," it resets the board accordingly and lets "X" (the AI) play a turn. For brevity, most of the possible cases for values were omitted from Listing 4-14. You can probably figure them out on your own, but they are available to be copied from the code samples included with the book should you need them.

Pay heed to the fact that there was a further confidence value check for the "RESTART" command. We want it to be harder for a user to restart a game so that they do not accidentally restart while saying another command. Ideally, we would have done the same for the commands to draw a circle on the corner boxes. "Upper left corner" and "lower left corner" can be easily misinterpreted by the speech engine, as two-thirds of each phrase is the same as the other phrase.

■ **Note**   Although we received a value from the speech engine in response to a voice command, we could have also directly accessed the phrase the user uttered with their voice. e.Results, which consists of an object named RecognizedPhrase, has a member called Words that contains an array of the words in the inputted phrase. It also has a Text property that returns the input normalized into a display form. For example, if you had said "two kilo apple," e.Results.Words[1] would return the string "kilo", and e.Results.Text would return the string "2kg apple". For more information on the results returned by the speech engine, check out the MSDN documentation for RecognizedPhrase at https://msdn.microsoft.com/en-us/library/system.speech. recognition.recognizedphrase(v=vs.110).aspx

---

Listing 4-15. MainWindow.xaml.cs Game-Logic Methods

```
//AI plays a turn
void AITurn()
{
    Random r = new Random();
    while (true)
    {
        int row = r.Next(0, 3);
        int col = r.Next(0, 3);
```

```
            if (textBlockGrid[row, col].Text == "")
            {
                textBlockGrid[row, col].Text = "X";
                break;
            }
        }
    if (CheckIfWin("X") == true)
    {
        inGame = false;
        resultbox.Text = "Oh no! You lost :( Say 'Start' to play again";
        return;
    }
    if (CheckIfTie() == true)
    {
        inGame = false;
        resultbox.Text = "Game is a TIE :/, say 'Start' to play another round";
        return;
    }
}

//Human plays a turn
void HumanTurn(int row, int col)
{

    if (textBlockGrid[row, col].Text == "")
    {
        textBlockGrid[row, col].Text = "O";
        if (CheckIfWin("O") == true)
        {
            inGame = false;
            resultbox.Text = "Congrats! You won! Say 'Start' to play again";
            return;
        }
        if (CheckIfTie() == true)
        {
            inGame = false;
            resultbox.Text = "Game is a TIE :/, say 'Start' to play another round";
            return;
        }
        AITurn();
    }
}

//Check if player won
private bool CheckIfWin(string v)
{
    if (textBlockGrid[0, 0].Text == v && textBlockGrid[0, 1].Text == v && textBlockGrid
    [0, 2].Text == v)
```

```
        {
            textBlockGrid[0, 0].Foreground = System.Windows.Media.Brushes.Red;
            textBlockGrid[0, 1].Foreground = System.Windows.Media.Brushes.Red;
            textBlockGrid[0, 2].Foreground = System.Windows.Media.Brushes.Red;
            return true;
        }
        else if (textBlockGrid[1, 0].Text == v && textBlockGrid[1, 1].Text == v &&
        textBlockGrid[1, 2].Text == v)
        {
            textBlockGrid[1, 0].Foreground = System.Windows.Media.Brushes.Red;
            textBlockGrid[1, 1].Foreground = System.Windows.Media.Brushes.Red;
            textBlockGrid[1, 2].Foreground = System.Windows.Media.Brushes.Red;
            return true;
        }
        //Omitted for redundancy
        [...]
        return false;
}


//Check if there is a tie
bool CheckIfTie()
{
    for (int i = 0; i < 3; i++)
    {
        for (int j = 0; j < 3; j++)
        {
            if (textBlockGrid[i, j].Text == "")
            {
                return false;
            }
        }
    }
    return true;
}
```

There is nothing special about the game-logic code in Listing 4-15. I could have gone with *Minimax* or *alpha-beta pruning*, but instead I let the AI literally be a pseudo-random number generator that places Xs on the board at random. Seeing as this makes it fallible, it is actually somewhat more fun to play (and it still beat me half the time. . .). CheckIfWin() goes through one of the eight possible solutions and checks whether there is a line of three TextBlocks with characters that match. Again, I omitted six of the solutions for brevity, but you can easily divine what they are or copy them from the provided samples.

Listing 4-16. MainWindow.xaml Kinect Tic Tac Toe XAML Front End

```
<Window x:Class="KinectTicTacToe.MainWindow"
        xmlns="http://schemas.microsoft.com/winfx/2006/xaml/presentation"
        xmlns:x="http://schemas.microsoft.com/winfx/2006/xaml"
        xmlns:d="http://schemas.microsoft.com/expression/blend/2008"
        xmlns:mc="http://schemas.openxmlformats.org/markup-compatibility/2006"
        xmlns:local="clr-namespace:KinectTicTacToe"
        mc:Ignorable="d"
```

```
        Title="Kinect Tic Tac Toe" Height="375" Width="330">
    <Grid Margin="0,0,0,5.5">
      <Grid HorizontalAlignment="Left" Height="300" Margin="10,10,0,0" VerticalAlignment="Top"
      Width="300" Grid.Column="3">
        <Grid.RowDefinitions>
          <RowDefinition Height="100"/>
          <RowDefinition Height="100"/>
          <RowDefinition Height="100"/>
        </Grid.RowDefinitions>
        <Grid.ColumnDefinitions>
          <ColumnDefinition Width="100"/>
          <ColumnDefinition Width="100"/>
          <ColumnDefinition Width="100"/>
        </Grid.ColumnDefinitions>
        <Border BorderBrush="Black" BorderThickness="0,0,2,2">
          <TextBlock x:Name="topLeft" HorizontalAlignment="Left" TextWrapping="Wrap"
          Text="" VerticalAlignment="Top" FontSize="48" Height="100" Width="100"
          TextAlignment="Center"/>
        </Border>
        <Border BorderBrush="Black" BorderThickness="2,0,2,2" Grid.Column="1">
          <TextBlock x:Name="top" HorizontalAlignment="Left" TextWrapping="Wrap"
          Text="" VerticalAlignment="Top" FontSize="48" Height="100" Width="100"
          TextAlignment="Center"/>
        </Border>
        <!--Omitted for redundancy-->
       [...]
      </Grid>
      <TextBox x:Name="resultbox" HorizontalAlignment="Left" Height="22" Margin="10,316,0,0"
      TextWrapping="Wrap" Text="Say 'Start' to commence a new round" VerticalAlignment="Top"
      Width="305"/>
    </Grid>
</Window>
```

Listing 4-16 features a Grid element with TextBlocks in each square. The Border element draws the lines of the Tic Tac Toe grid. There is a TextBlock on the bottom that informs the user of the game's state. Compile the application and run it. The final result should look like Figure 4-4.

Figure 4-4. *Kinect Tic Tac Toe game window. Getting whooped by the "dumb" AI.*

The quality of the speech recognition depends on your local environment, accent, and the integrity of the Kinect's microphone array, but at least for all the occasions I have seen it used, it has been remarkably accurate. Of the two dozen rounds of Tic Tac Toe that I have played with the Kinect, my command was only misrecognized once, when I said "lower right corner" and the speech engine detected "upper right corner."

While the form of speech recognition we just implemented is suitable for a variety of applications, keen readers will realize that it has its limitations. The Microsoft Speech SDK is appropriate for short voice commands, but not for dictation. It is no Cortana. This is because Microsoft Speech is supposed to be a server-based speech engine. It is suited to dealing with grainy, telephone-quality audio with latency. Thus, the *Battlefield* airstrike example we saw at the start of the chapter is not as practical, unless all the commands follow a strict pattern.

There is another speech engine that is included with .NET that you can make use of. The `System.Speech` library is an API that on its face is similar to `Microsoft.Speech`. It enables dictation and has better recognition than `Microsoft.Speech`. Then, why are we not using it? It does not have Kinect language packs, so the recognition ultimately suffers. Despite this, `System.Speech` can be adapted for use with the Kinect. The code is largely the same, though it has some extra classes so as to work with dictation.

Speech Recognition in Windows Store Apps

It is sad to say, but the previously discussed code is totally unusable for Windows Store apps, as Microsoft Speech is not available for use with it. Instead, Microsoft recommends that you use the Bing Speech API. This is a paid API that you need to have a Microsoft Azure account to use. You can make 500,000 free calls to it each month, so for personal projects, it is not too bad.

The interesting thing about using the Bing Speech API with the Kinect is that you do not reference the Kinect in any way within the code. The API just opens your default microphone and works with what it's given. Given the circumstances, exploring the API is out of the scope of this book. If you do need to use its services, however, you should check out Microsoft's official tutorial at `http://kinect.github.io/tutorial/lab13/index.html`.

Human Interface Guidelines for Voice Input

When it comes to voice input, the possibilities for how the user can interact with the Kinect are endless. That is not necessarily a good thing, however. An ambiguous user experience can hinder the user from being able to appreciate your application. For this reason, Microsoft has drafted a set of *Human Interface Guidelines* to help developers design their projects in a manner that appeals to a potential user. These guidelines are based on insights gathered from the Kinect v1 years, as well as on the feedback received during the beta phase of the Kinect v2. Adhering to these suggestions can help build a positive and coherent experience for your users.

Environment

Unless your workplace promotes the dreaded "open plan" office, you are likely developing your Kinect applications in a relatively quiet place. The Tic Tac Toe speech-recognition sample works great in my basement, but if I were to bring it to an elementary school during recess, I would experience the Kinect equivalent of *Twitch Plays Pokémon*. The Kinect's ambient noise-cancellation works for noises up to 26 decibels. An average human conversation is 50 decibels. In a loud conference, the ambient noise is 60–70 decibels, and speech recognition may no longer be feasible. You should try out your application in its target environment before putting it into production to mitigate these types of issues. How close the user is to the Kinect also affects the recognition quality. If the user is eight meters away and is shouting to speak the voice command, the speech engine might have difficulty parsing the user's words. If you cannot control these environmental constraints, you can try to hack together homegrown noise-cancellation algorithms or use third-party APIs. There is no magic solution to pick out a voice from a crowd of a thousand, but you might be able to cancel out certain songs or sounds if you know they will be present at the time of the application's use.

Not only can the environment interfere with the Kinect, but the Kinect can interfere with the environment as well. Voice inputs are not always appropriate. In a quiet office space or library, raised voices might cause distractions and draw weird looks from bystanders, making your user feel uneasy. Even when using voice commands might be appropriate, the exact phrases being used might not. For example, having a Kinect game where the user has to loudly and repeatedly exclaim "Faster!" inside of a department store might prevent the user from wanting to try it out.

Confidence Levels

The acceptable range of possible confidence-level values that do not cause either incessant false positives or exasperation for the user from having to strain their voice can be very small. The environment can affect the appropriateness of the confidence-level requirements as well. Developers should strive to make them accommodating of the different situations in which the applications might be used.

As we saw earlier with the "restart" and "upper/lower left/right corner" voice commands, not every voice command needs to have the same confidence-level requirement. The situations in which this is particularly important are when the user's resources (e.g., time, money, files) are being expended or when they are being removed. A command to purchase or delete something should always require a lot of confidence in the user's voice input. Some other situations, such as adding a friend on a social network, can also fall in that category. Sometimes it might even be worth it to ask the user for a confirmation.

Word Choice

Voice commands that sound the same can be difficult for the speech engine to distinguish. Generally, these include words that rhyme or that for the most part have the same letters. This applies to phrases as well. Take, for example, a hypothetical Kinect dating app called "Date or No Date." You can probably already guess how the app works. The user is presented with a potential partner, and they must say "Date" or "No Date." These are voice commands that are easily misinterpreted by the recognition engine. Not only do they mostly have the same letters, but they also have the same words! "Yes" or "No" and "Love" or "Ignore" are better voice commands that can help users avoid any awkward encounters.

The voice commands should be as short as possible while still describing the action. "I Agree" can be shortened to "Agree," and "Clear the board" can be replaced with "Clear" or "Restart." That being said, if you are having issues with false positives, longer phrases can be a solution. "Clear the board" can be a lot less ambiguous for the speech engine than "Clear." You have to try out the different variations and see whether your speech engine or users have issues with them, then balance the concerns. The voice command should also be straightforward and intuitive. With this aim, you should try to limit the number of available voice commands the user has at their disposal. If there are a lot of options from which a user can pick on a menu, it could be better to have numbers associated with each option instead of a phrase (e.g., Option 6).

What might be intuitive for a certain demographic of users might not be for others. A Kinect application that allows a user to order from McDonald's could have the voice command "supersize" in the United States and in Canada, but other English-speaking countries have different vernaculars and another saying, such as "extra-large," might be more appropriate for them.

Interaction Design

The way the user interacts with the application should always be constrained and consist of a catered experience. This is not limiting the user's freedom. Quite the contrary, this provides them with guidelines on how they can most efficiently take advantage of the application, thus empowering them to focus on the actual task at hand. This art of setting clear paths for the user to follow is called *interaction design*.

A prime example of this is the "Xbox. . ." command on the Xbox. Since the speech engine is only listening for that one word, it eliminates false positives arising from chatter in the area. The word is rather unique as well, so it is not likely to be confused by the speech engine.

More important, it defines guidelines as to how a user should interact with the application. As soon as you say "Xbox," a screen with available commands is presented. This prevents a list of possible commands from occluding the view permanently, which is normally detrimental to experienced users. The user knows that if they want to interact with the application, they have to follow this procedure. The alternative is to have them talk freely, and although certain commands would work out fine, others might never be discovered by the user.

In general, some form of trigger activation will usually be desired. When the speech engine is always listening for a set of words, it tends to pick up many more false positives. If a button or voice command is used to signify that the user wants to communicate with the app, you can be certain that the subsequent inputs picked up by the microphone array are not erroneous. If the microphone is always listening, it may be wise to indicate this with an icon or other sign.

Often, when you type a query into Google, you can envision the result you are searching for in your head, but you do not know the words to describe it exactly. After making a query, Google might ask you "Did you mean X" to launch another query and guide you to a better search term. The effectiveness of the suggestion is even more pronounced to those who are new to computers and are not accustomed to writing search queries on Google. Doing something similar with the voice commands of your application when the user's commands are unclear or have low confidence readings can be similarly helpful. Your Kinect File Explorer app might have a "delete" command to delete files, but a user with a weaker mastery of the English language might say "destroy" instead. Just like a search engine, your application can suggest the proper command for the user to say.

Summary

The Kinect's audio and speech capabilities are a powerful and underused asset in the toolbox of any developer. They empower users by granting them additional ways to communicate with the application. There is no standard .NET library to record audio, so we must encode it ourselves. With the help of its directional capabilities, we build experiences where the Kinect knows the source of the audio. The Microsoft Speech Platform can further enhance these experiences by providing a framework in which to interpret the audio data into instructions for our applications to act upon. Taking into account the Human Interface Guidelines, we can couple the technology with user experience design to grant the users the additional communication venues that we so desire.

Let us venture on. In the next chapter, we will learn how to detect bodies and use their data in our applications.

CHAPTER 5

■ ■ ■

Body & Face Tracking

At the heart of a user's desire to interact with the Kinect is the ability to physically manipulate a digital reality. It is an experience that is nigh on magical for most people. Just imagine if you were to obscure a Kinect from someone's view and enable them to wield the environment about them. Perhaps in a kitchen setting, a person could stand in a spot and control all the appliances surrounding them solely with hand gestures. Or in a department store, as soon as a person walked in front of a two-way mirror, clothing that the store was retailing could be superimposed on the person's body on a screen behind the mirror, and the user could use gestures to have the screen change which clothes were appearing on them. There are not many technologies that can so viscerally interact with humans. While you might not be able to build these projects within a few hours of having read this book, the Kinect's body-tracking features are unparalleled by anything else that is generally available on the market, and the possibilities for you to pursue with it are essentially limitless.

Before we can look at capturing complex gestures, we ought to understand how the Kinect tracks one's body and face. By doing so, we can later build heuristics and analytical systems that are capable of processing this data. Body and face tracking in the Kinect SDK is divided into four Data Sources: BodyFrameSource, BodyIndexFrameSource, FaceFrameSource, and HighDefinitionFrameSource. We will explore all of their components and see how to best exploit them.

BodyFrameSource: Skeletal Data

Almost all of the Kinect's raw skeletal and body data is embodied in BodyFrameSource. As with Depth, Infrared, and ColorFrameSource, BodyFrameSource gives data frames roughly in sync with the other Data Sources, each one detailing positional and directional data about the body joints observed by the Kinect. In Chapter 2, Table 2-1 mentioned that the Kinect for Windows v2 tracks 25 joints, 5 more than the Kinect for Windows v1 did. These 25 joints are depicted in Figure 5-1 and include the **Neck**, **Left Hand Tip**, **Right Hand Tip**, **Left Thumb**, and **Right Thumb**. Overall, the skeletal-tracking accuracy has been drastically improved since the previous generation. This enables a much more fluid experience for a user using gesture-based applications.

© Mansib Rahman 2017
M. Rahman, *Beginning Microsoft Kinect for Windows SDK 2.0*, DOI 10.1007/978-1-4842-2316-1_5

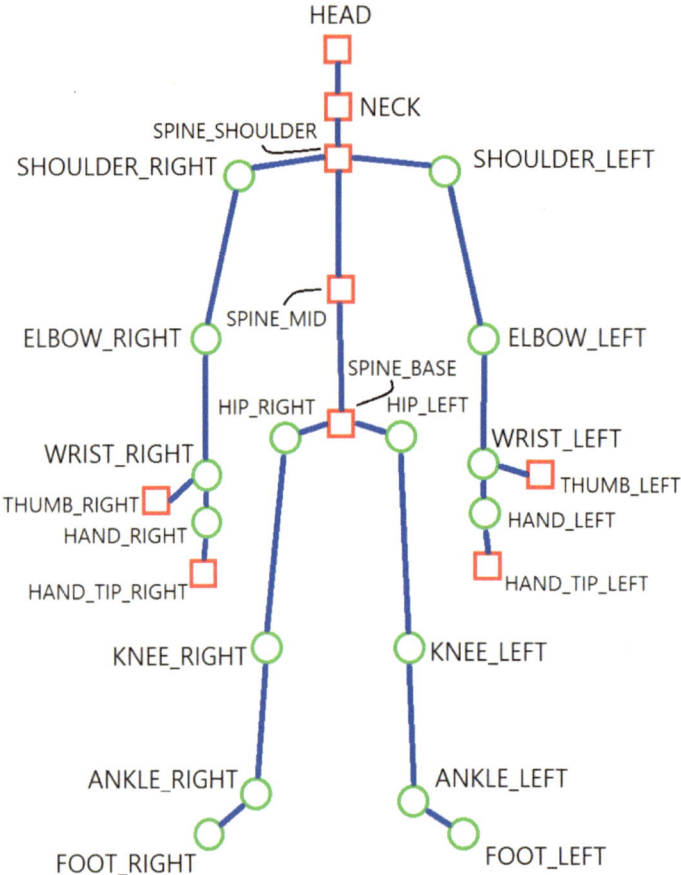

Figure 5-1. *The various joints represented in the Kinect for Windows v2 API. Square joints represent joints that have been altered from or were unavailable in the v1 SDK.*

The process of grabbing body frames is a bit different than that for the more visual frames. There is a special `BodyFrame` method called `GetAndRefreshBodyData(IList<Body> bodies)` that we have to call in every `FrameArrived` event handler call in order to get our body data. To reduce memory consumption, the SDK directly updates the necessary parts of the Body array that is inputted into this method. This array should be maintained throughout the lifetime of the `BodyFrameReader`'s operation.

Drawing Skeletons

The quintessential demonstration of the Kinect's skeletal-tracking abilities has always been the superimposition of a stick figure or dots representing the joints and bones of a tracked user on a video feed. It can be amusing to watch people unfamiliar with the Kinect prance around trying to see if the Kinect can keep up through its body tracking. They are always quite surprised to see their skeletal reflections rendered onto a screen in real-time, never mind that a comparatively low-tech solution, putting stickers on yourself and looking at a mirror, can yield much better results. There is something about computers slowly but surely adopting human-like perception that is entertaining to watch.

Drawing Joints

We will get started with a simple dots-for-joints demonstration. We will draw purple dots on the color camera feed for the location of tracked joints and gray dots for the location of inferred joints, which are joints that the Kinect cannot see at the moment but assumes are present. Start a new C# WPF project and include the standard assortment of namespaces: System, System.Collections.Generic, System.Linq, System.Windows, System.Windows.Media, System.Windows.Media.Imaging, and Microsoft Kinect. Some extra namespaces that also need to be included this time around are System.Windows.Shapes and System. Windows.Controls so that we can draw joints as circles on a Canvas control.

Listing 5-1. MainWindow.xaml.cs Private Variable Declaration for Displaying Skeletal Joints

```
public partial class MainWindow : Window
{

    private KinectSensor kinect;
    private MultiSourceFrameReader multiSourceFrameReader;
    //private BodyFrameReader bodyFrameReader;
    private CoordinateMapper coordinateMapper;

    FrameDescription colorFrameDescription;
    private WriteableBitmap colorBitmap;

    private bool dataReceived;
    private Body[] bodies;
    ...
}
```

In Listing 5-1, we opted to create a MultiSourceFrameReader instead of a BodyFrameReader. This will allow us to read frames from multiple Data Sources relatively synchronously without having to open multiple readers. You should keep in mind that there is no Data Source called MultiFrameSource. It is mostly an abstraction of convenience, and we could have called the readers for the color and skeletal Data Sources individually instead.

By this point in the book, you might be wondering about how the depth camera and color camera have different images that by default do not align. Quite simply, if you look at the front of your Kinect, the cameras are not in the same spot, not to mention have different resolutions. As a consequence, we have to align the images. We can do this mathematically, but fortunately the Kinect has already done this for us. In Listing 5-1, we declare an instance of the CoordinateMapper utility class to let us align points in both images along with the positions of skeletal joints. It is not a static class and is a property of each individual KinectSensor object.

Culminating the private variables, we have the dataReceived boolean, which will toggle whether we draw our joints based on the availability of new data, and the bodies array, which will hold the skeletal data collected by the Kinect.

Listing 5-2. MainWindow.xaml.cs Constructor and Variable Assignment for Displaying Skeletal Joints

```
public MainWindow()
{

    kinect = KinectSensor.GetDefault();

    multiSourceFrameReader = kinect.OpenMultiSourceFrameReader(FrameSourceTypes.Color |
    FrameSourceTypes.Body);
```

```
multiSourceFrameReader.MultiSourceFrameArrived += MultiSourceFrameReader_
MultiSourceFrameArrived;

coordinateMapper = kinect.CoordinateMapper;
colorFrameDescription = kinect.ColorFrameSource.CreateFrameDescription
(ColorImageFormat.Bgra);

colorBitmap = new WriteableBitmap(colorFrameDescription.Width, colorFrameDescription.
Height, 96.0, 96.0, PixelFormats.Bgra32, null);

kinect.Open();

DataContext = this;
InitializeComponent();

}


public ImageSource ImageSource
{
    get
    {
        return colorBitmap;
    }
}
```

In Listing 5-2, we initialize our reader and the other variables necessary for skeletal tracking. Note that the method signature for creating MultiSourceFrameReaders is different from that for the other Data Sources. Instead of calling KinectSensor.XFrameSource.OpenReader() as with the others, we called the KinectSensor's OpenMultiSourceFrameReader(FrameSourceTypes enabledFrameSourceTypes) method. As mentioned previously, there is no MultiFrameSource Data Source property in the KinectSensor object to access directly. FrameSourceTypes indicates which Data Sources we want our reader to offer frames from. It is an enum that can be delineated by the overloaded | operator (which acts as a bitmask). Like the other readers, we can assign it event handlers to be triggered when frames arrive. We also assign CoordinateMapper and WriteableBitmap in the constructor.

Listing 5-3. MainWindow.xaml.cs Event Handler for Displaying Skeletal Joints

```
private void MultiSourceFrameReader_MultiSourceFrameArrived(object sender,
MultiSourceFrameArrivedEventArgs e)
{

    bool dataReceived = false;
    MultiSourceFrame multiSourceFrame = e.FrameReference.AcquireFrame();

    using (ColorFrame colorFrame = multiSourceFrame.ColorFrameReference.AcquireFrame())
    {
        if (colorFrame != null)
        {
            using (KinectBuffer colorBuffer = colorFrame.LockRawImageBuffer())
            {
                colorBitmap.Lock();
```

```
            if ((colorFrameDescription.Width == colorBitmap.PixelWidth) &&
            (colorFrameDescription.Height == colorBitmap.PixelHeight))
            {
                colorFrame.CopyConvertedFrameDataToIntPtr(
                    colorBitmap.BackBuffer,
                    (uint)(colorFrameDescription.Width *
                    colorFrameDescription.Height * 4),
                    ColorImageFormat.Bgra);

                colorBitmap.AddDirtyRect(new Int32Rect(0, 0, colorBitmap.PixelWidth,
                colorBitmap.PixelHeight));
            }

            colorBitmap.Unlock();
        }
    }
}

using (BodyFrame bodyFrame = multiSourceFrame.BodyFrameReference.AcquireFrame())
{
    if (bodyFrame != null)
    {
        if (bodies == null)
        {
            bodies = new Body[bodyFrame.BodyCount];
        }
        bodyFrame.GetAndRefreshBodyData(bodies);
        dataReceived = true;
    }
}

if (dataReceived)
{
    canvas.Children.Clear();

    foreach (Body body in bodies.Where(b => b.IsTracked))
    {
        foreach (var joint in body.Joints)
        {
            CameraSpacePoint position = joint.Value.Position;
            if (position.Z < 0)
            {
                position.Z = 0.1f;
            }

            ColorSpacePoint colorSpacePoint = coordinateMapper.MapCameraPointToColorSpace
            (position);

            if (joint.Value.TrackingState == TrackingState.Tracked)
            {
                DrawJoint(new Point(colorSpacePoint.X, colorSpacePoint.Y), new
                SolidColorBrush(Colors.Purple));
            }
```

```
            if (joint.Value.TrackingState == TrackingState.Inferred)
            {
                DrawJoint(new Point(colorSpacePoint.X, colorSpacePoint.Y), new
                SolidColorBrush(Colors.LightGray));
            }
        }
    }
}
}
```

In Listing 5-3, we have a large event handler method that encapsulates the code for both displaying color images and grabbing skeletal frames. In a larger application, we would probably split this up by passing the `MultiSourceFrameArrivedEventArgs` e or the frames to separate methods. For simplicity's sake, I put them together here. It is noteworthy that there is a bit of *Inception* going on with the frame references for `MultiSourceFrames`. With the other Data Sources, the process of grabbing data can be summated with xFrameArrivedEventArgs ➤ xFrameReference ➤ xFrame ➤ *Underlying Data Buffer*. With multi-source data, the process is MultiSourceFrameEventArgs ➤ MultiSourceFrameReference ➤ MultiSourceFrame ➤ xFrameReference ➤ xFrame ➤ *Underlying Data Buffer*. The important thing to know is that the `MultiSourceFrames` contain the references for the other frames, and the rest of the paradigm works as you would expect.

There is nothing interesting about our code to acquire and present color data; it is the same as in Chapter 3. In the `using (BodyFrame ...)` portion of the code, we create a new array of bodies if none exists and then update it with the `bodyFrame.GetAndRefreshBodyData(bodies)` statement. We set the dataReceived flag to true so that we know to execute the joint-drawing portion of the code. This is reset to false every time the event handler is called.

We could have chosen to draw the joints directly onto the Image WPF control, writing over the relevant color pixels in our `colorBitmap`, but instead we will draw the joints on a *Canvas* control that will be overlaid on top of our color image. This is mainly a matter of preference, though there are some trade-offs. For example, if we wanted to take a screenshot, we would not be able to use the code from Chapter 3 to do so if we were using the Canvas control. We would instead have to rely on the `RenderTargetToBitmap` class to take a snapshot of the Image and Canvas controls simultaneously. We opted for Canvas out of convenience.

We start off the drawing portion of our code by calling `canvas.Children.Clear()`, which deletes all the existing joint dots on the canvas. These were created as child WPF controls of our Canvas (called canvas).

■ **Reminder**   *WPF controls* are simply the XAML UI elements with which we design the front end of our application. Think of them as HTML tags, such as `<Button />`.

Using LINQ, we loop through each body in our bodies array that is being tracked to apply our drawing code. The `Joints` property of Body contains a dictionary of key-value pairs for each joint in the tracked body. If we wanted to access a specific joint directly, we could query the `JointType` in the dictionary as such:

```
Joint j = joints[JointType.ElbowRight];
```

Note that the identifiers for the JointTypes are all written as camel-cased renditions of those listed in Figure 2-1 (i.e., Elbow_Right is written as ElbowRight in this context).

■ **Tip** *LINQ*, or *Language Integrated Query*, is a .NET component that allows us to make queries on lists, arrays, and other data structures. It has statements such as `Select` or `Where` that help us map, filter, extract, and process data to get the interesting bits. It helps us eliminate superfluous loops and `if` statements. It has even been ported to other languages such as Java. To learn more, visit `https://msdn.microsoft.com/en-us/library/mt693024.aspx`.

Looping through the joints, we start off by extracting the `Position` property of each joint as seen from the Depth camera, which is defined as a `CameraSpacePoint`. Next, we clamp its depth value (`position.Z`), ensuring that it is above 0, to prevent the `CoordinateMapper` from returning negative infinity values when we convert it to a `ColorSpacePoint`. These depth values can be negative as a result of algorithms used by the Kinect SDK to determine the position of inferred joints. It might seem odd that the depth value is relevant at all in the conversion from a 3D coordinate system to a 2D coordinate system, but it makes sense. If you think about it, the smaller the distance between the camera and a person, the larger their skeleton (and their joints) should appear on the screen. Thus, the relationship between the depth value and the size of the joints in the color space is inversely proportional.

■ **Tip** `CameraSpacePoints` are one of the three types of coordinate space points in the Kinect 2 SDK. It refers to the 3D coordinate system used by the Kinect. Its origin is located at the center of the depth camera, and each unit of measure is equivalent to one meter. `ColorSpacePoint` marks the location of coordinates on a 2D color image (as garnered from a `ColorFrame`). Likewise, `DepthSpacePoint` marks the location of coordinates on a 2D depth image.

After the depth value is clamped, we need to convert the depth positions to the color camera view positions. Thus, we call the `MapCameraPointToColorSpace(CameraSpacePoint cameraPoint)` method of `CoordinateMapper` to find the location of the joint on our color image. Checking the `TrackingState` property of each joint, we determine whether to color it purple for `Tracked` or light gray for `Inferred`. We give the coordinates of the `ColorSpacePoint` as a `Point` object to our custom `DrawJoint(Point jointCoord, SolidColorBrush s)` method, along with the desired color of the joint on the canvas.

Listing 5-4. MainWindow.xaml.cs Drawing Method for Displaying Skeletal Joints

```
private void DrawJoint(Point jointCoord, SolidColorBrush s)
{
    if (jointCoord.X < 0 || jointCoord.Y < 0)
        return;

    Ellipse ellipse = new Ellipse()
    {
        Width = 10,
        Height = 10,
        Fill = s
    };
```

```
Canvas.SetLeft(ellipse, (jointCoord.X/colorFrameDescription.Width) * canvas.ActualWidth
- ellipse.Width / 2);
Canvas.SetTop(ellipse, (jointCoord.Y/colorFrameDescription.Height) * canvas.ActualHeight
- ellipse.Height / 2);
canvas.Children.Add(ellipse);
}
```

In Listing 5-4, we have the method that is responsible for drawing the joints on our canvas. We first confirm that the joint coordinates on the color image have positive values. We create a new ellipse object for the joint and give it a 10-pixel diameter and our chosen brush as the fill color. We then set the x coordinate and y coordinate of the ellipse on the Canvas with the Canvas class' SetLeft(UIElement element, double length) and SetTop(UIElement element, double length) methods, respectively. The (0, 0) coordinate for the Canvas is the top-left pixel, as with images. There is an interesting bit of calculating we must do to determine where we put the circle on the Canvas:

```
Canvas.SetLeft(ellipse, (jointCoord.X/colorFrameDescription.Width) * canvas.ActualWidth
- ellipse.Width / 2);
```

The preceding calculation also applies for the y coordinate. Normally, ColorSpacePoints refer to Cartesian coordinates on the color camera image. This image is normally 1920 x 1080 pixels, but for whatever reason, it might be displayed in a different resolution. In our case, we set the image to 1024 x 576 in our application so that it can fit better in our monitor. Thus, we must scale down the coordinate value so that it fits in this lower-resolution image. In the example of the x coordinate, we divide its value by the width of the original image and then multiply it by the width of our Canvas. SetLeft(...) and SetTop(...) set elements on the canvas by their top-left corners, so we subtract the radius of the ellipse from the length to ensure it is set to the desired location by its center.

Listing 5-5. MainWindow.xaml Front End for Displaying Skeletal Joints

```
<Window [...]
        Title="Skeletal Joints" Height="604" Width="1028" ResizeMode="NoResize">
    <Grid>
        <Image Source="{Binding ImageSource}" Width="1024" Height="576" />
        <Canvas Name="canvas" Width="1024" Height="576" />
    </Grid>
</Window>
```

In the front end (Listing 5-5), we add the Canvas element right after the Image element so that it is displayed in front of the image and not behind. The rest of the XAML is otherwise uninteresting. Compile and run the application, then stand a short distance away from the Kinect. If you can see the joint dots on your screen like as in Figure 5-2, then you can try sitting down or appearing in other positions to see that it is working as anticipated.

Figure 5-2. *Skeletal joints displayed on a user. Notice the light gray dots under the waist, which represent the inferred positions of the leg joints. The torso and waist joints are being partially shifted as a result of the table's obstructing the lower right side of the user's abdomen.*

▪ **Note**   Are your joint dots flashing erratically and ruining your aesthetic? This is happening because we are calling the `canvas.Children.Clear()` method several times per second, thus causing all the dots to disappear on the screen for brief instants before they are redrawn. To mitigate this issue, you can keep track of all the canvas joint ellipses with a data structure and then loop through them in each frame, changing their locations on the canvas systematically instead of clearing them. Alternatively, you can draw on the image directly with the use of the `DrawingContext` class. Refer to the *BodyBasics-WPF* sample to see how. The exact implementation of the solution is left to the reader as an exercise.

Drawing Bones

There is no concept of "bones" in the Kinect SDK. We have to enumerate the various connections between the provided joints ourselves before we draw them. The process of drawing them is similar to doing so for the joints. The primary difference is that instead of drawing ellipses, we will be drawing lines.

Listing 5-6. MainWindow.xaml.cs Additional Private Variables to Display Skeletons

```
...
private Body[] bodies; //previously declared private variables

private List<Tuple<JointType, JointType>> bones;
private List<SolidColorBrush> bodyColors;
...
```

In Listing 5-6, we make a list of tuples to represent the various joint relationships in the body that constitute bones. Additionally, we make a list for the colors we will use to paint each body.

Listing 5-7. MainWindow.xaml.cs Additional Constructor Code to Display Skeletons

```
public MainWindow()
{
    bones = new List<Tuple<JointType, JointType>>();

    //Torso
    bones.Add(new Tuple<JointType, JointType>(JointType.Head, JointType.Neck));
    bones.Add(new Tuple<JointType, JointType>(JointType.Neck, JointType.SpineShoulder));
    bones.Add(new Tuple<JointType, JointType>(JointType.SpineShoulder, JointType.SpineMid));
    bones.Add(new Tuple<JointType, JointType>(JointType.SpineMid, JointType.SpineBase));
    bones.Add(new Tuple<JointType, JointType>(JointType.SpineShoulder, JointType.ShoulderRight));
    bones.Add(new Tuple<JointType, JointType>(JointType.SpineShoulder, JointType.ShoulderLeft));
    bones.Add(new Tuple<JointType, JointType>(JointType.SpineBase, JointType.HipRight));
    bones.Add(new Tuple<JointType, JointType>(JointType.SpineBase, JointType.HipLeft));

    //Left Arm
    bones.Add(new Tuple<JointType, JointType>(JointType.ShoulderLeft, JointType.ElbowLeft));
    bones.Add(new Tuple<JointType, JointType>(JointType.ElbowLeft, JointType.WristLeft));
    bones.Add(new Tuple<JointType, JointType>(JointType.WristLeft, JointType.HandLeft));
    bones.Add(new Tuple<JointType, JointType>(JointType.HandLeft, JointType.HandTipLeft));
    bones.Add(new Tuple<JointType, JointType>(JointType.WristLeft, JointType.ThumbLeft));

    //Right Arm
    bones.Add(new Tuple<JointType, JointType>(JointType.ShoulderRight, JointType.ElbowRight));
    bones.Add(new Tuple<JointType, JointType>(JointType.ElbowRight, JointType.WristRight));
    bones.Add(new Tuple<JointType, JointType>(JointType.WristRight, JointType.HandRight));
    bones.Add(new Tuple<JointType, JointType>(JointType.HandRight, JointType.HandTipRight));
    bones.Add(new Tuple<JointType, JointType>(JointType.WristRight, JointType.ThumbRight));

    //Left Leg
    bones.Add(new Tuple<JointType, JointType>(JointType.HipLeft, JointType.KneeLeft));
    bones.Add(new Tuple<JointType, JointType>(JointType.KneeLeft, JointType.AnkleLeft));
    bones.Add(new Tuple<JointType, JointType>(JointType.AnkleLeft, JointType.FootLeft));

    //Right Leg
    bones.Add(new Tuple<JointType, JointType>(JointType.HipRight, JointType.KneeRight));
    bones.Add(new Tuple<JointType, JointType>(JointType.KneeRight, JointType.AnkleRight));
    bones.Add(new Tuple<JointType, JointType>(JointType.AnkleRight, JointType.FootRight));
```

```
bodyColors = new List<SolidColorBrush>();

bodyColors.Add(new SolidColorBrush(Colors.Red));
bodyColors.Add(new SolidColorBrush(Colors.Green));
bodyColors.Add(new SolidColorBrush(Colors.Orange));
bodyColors.Add(new SolidColorBrush(Colors.Blue));
bodyColors.Add(new SolidColorBrush(Colors.Yellow));
bodyColors.Add(new SolidColorBrush(Colors.Pink));

//remaining code in constructor
kinect = KinectSensor.GetDefault();
...
```

In the `MainWindow` constructor shown in Listing 5-7, we add a bunch of "bones" to the list, which are supposed to be reasonable-looking connections between the body joints that the SDK gives us access to. Like the joints, they do not necessarily depict our anatomy exactly as in a textbook, but instead have some abstractions applied to facilitate working with them through software. For example, there are two bones between our wrist and elbow joint, the radius and the ulna, but in our demonstration code we have only one bone, described as `Tuple<JointType, JointType>(JointType.ElbowLeft, JointType.WristLeft)`. In addition to the bones, we add six colors to the bodyColors list for the six different skeletons.

Listing 5-8. MainWindow.xaml.cs Modified FrameArrived Event Handler

```
...
//color and body frames grabbed and processed before this
if (dataReceived)
{
    canvas.Children.Clear();

    //Add this line
    int colorIndex = 0;

    foreach (Body body in bodies.Where(b => b.IsTracked))
    {
        //Add the following lines
        SolidColorBrush colorBrush = bodyColors[colorIndex++];
        Dictionary<JointType, Point> jointColorPoints = new Dictionary<JointType, Point>();

        foreach (var joint in body.Joints)
        {
            //[...] depth clamping omitted for brevity

            ColorSpacePoint colorSpacePoint = coordinateMapper.MapCameraPointToColorSpace(position);

            //Add this line
            jointColorPoints[joint.Key] = new Point(colorSpacePoint.X, colorSpacePoint.Y);

            //[...] DrawJoint calls omitted for brevity

        }
```

```
        //Add this foreach loop
    foreach (var bone in bones)
    {
        DrawBone(body.Joints, jointColorPoints, bone.Item1, bone.Item2, colorBrush);
    }
  }
}
```

Listing 5-8 contains the skeleton- and joint-drawing portions of the MultiSourceFrameReader_
MultiSourceFrameArrived event handler. We keep track of body colors with colorIndex so that we
can maintain a unique color for each body in the frame with the SolidColorBrush colorBrush =
bodyColors[colorIndex++]; statement. A dictionary is created to keep track of the ColorSpacePoints
being mapped for each joint in the body. Once we have this data, we loop through all the joint-joint tuples
in our bones list and draw them. We have a new DrawBone(...) method for this; we will not be modifying
the DrawJoint(...) method from earlier, which is still needed for the joints.

Listing 5-9. MainWindow.xaml.cs Drawing Method for Bones

```
private void DrawBone(IReadOnlyDictionary<JointType, Joint> joints, IDictionary<JointType,
Point> jointColorPoints, JointType jointType0, JointType jointType1, SolidColorBrush color)
{
    Joint joint0 = joints[jointType0];
    Joint joint1 = joints[jointType1];

    if (joint0.TrackingState == TrackingState.NotTracked || joint1.TrackingState ==
    TrackingState.NotTracked)
        return;

    if (jointColorPoints[jointType0].X < 0 || jointColorPoints[jointType0].Y < 0 ||
    jointColorPoints[jointType1].X < 0 || jointColorPoints[jointType0].Y < 0)
        return;

    Line line = new Line()
    {
        X1 = (jointColorPoints[jointType0].X / colorFrameDescription.Width) * canvas.ActualWidth,
        Y1 = (jointColorPoints[jointType0].Y / colorFrameDescription.Height) * canvas.ActualHeight,
        X2 = (jointColorPoints[jointType1].X / colorFrameDescription.Width) * canvas.ActualWidth,
        Y2 = (jointColorPoints[jointType1].Y / colorFrameDescription.Height) * canvas.ActualHeight,
        StrokeThickness = 5,
        Stroke = color
    };
    canvas.Children.Add(line);
}
```

In the DrawBone(...) method featured in Listing 5-9, we begin by refusing to draw any bones for situations where the necessary data is not present. This is namely when any of the joints are not tracked, or when their color-image coordinates are not positive values. Next, we create a line with the coordinates of the joints as its start and end points, along with the desired color inputted from bodyColor[colorIndex++]. Notice that the joint coordinates fed into the line are scaled as we did with the joints. Finally, we added the line to the canvas. Compiling the application should result in a scene like that in Figure 5-3.

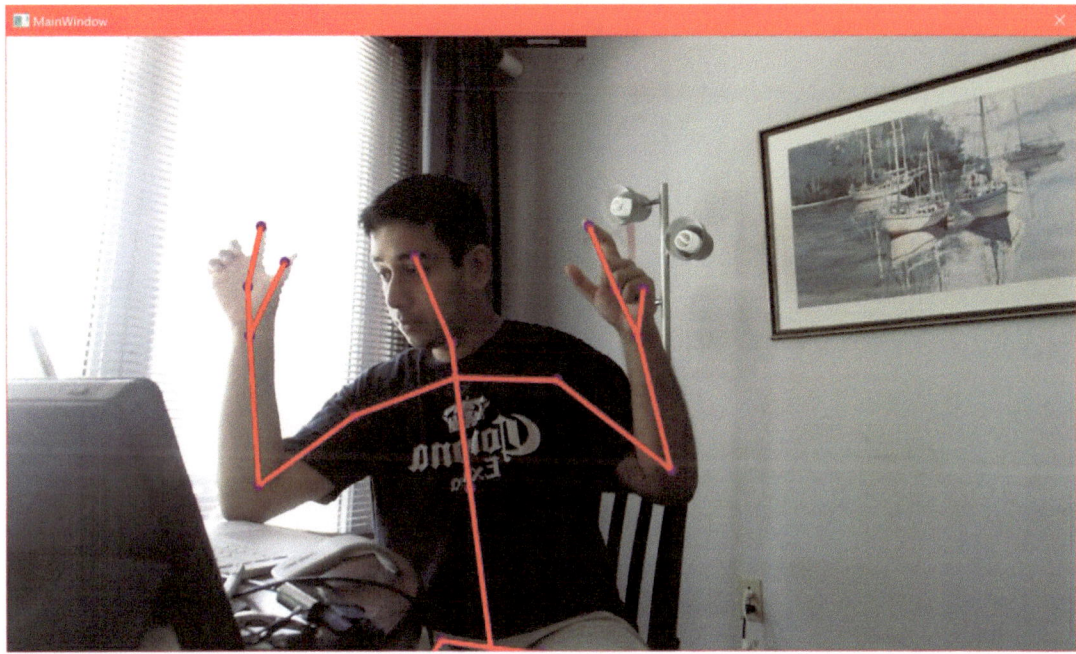

Figure 5-3. *Skeleton displayed on a user*

Showing Clipped Edges

More often than not, users will not fully position their body within the view range of the Kinect. The Kinect SDK has an easy way for us to find out if such is the case. Each Body object has a ClippedEdge property that tells us on which edges of the image the user's skeleton is being clipped on. This can be useful in many ways, such as knowing when to tell the user to move back into the Kinect's view range, as in Figure 5-4.

Figure 5-4. *Clipped edges shown as red lines on the sides. The left edge is being shown as clipped because the person's right leg joints are inferred as being past that edge on the frame.*

Drawing clipped edges in our existing application only requires minor edits. We will include a method to draw lines on the sides of the image representing the clipped edges and call it from the event handler for MultiSourceFrame.

Listing 5-10. MainWindow.xaml.cs Method for Drawing Clipped Edges

```
private void DrawClippedEdges(Body body)
{
    FrameEdges clippedEdges = body.ClippedEdges;

    if (clippedEdges.HasFlag(FrameEdges.Bottom))
    {
        Line edge = new Line()
        {
            X1 = 0,
            Y1 = canvas.ActualHeight - 9,
            X2 = canvas.ActualWidth,
            Y2 = canvas.ActualHeight - 9,
            StrokeThickness = 20,
            Stroke = new SolidColorBrush(Colors.Red)
        };
        canvas.Children.Add(edge);
    }
```

```
    if (clippedEdges.HasFlag(FrameEdges.Top))
    {
        Line edge = new Line()
        {
            X1 = 0,
            Y1 = 0,
            X2 = canvas.ActualWidth,
            Y2 = 0,
            StrokeThickness = 20,
            Stroke = new SolidColorBrush(Colors.Red)
        };
        canvas.Children.Add(edge);
    }

    if (clippedEdges.HasFlag(FrameEdges.Left))
    {
        Line edge = new Line()
        {
            X1 = 0,
            Y1 = 0,
            X2 = 0,
            Y2 = canvas.ActualHeight,
            StrokeThickness = 20,
            Stroke = new SolidColorBrush(Colors.Red)
        };
        canvas.Children.Add(edge);
    }

    if (clippedEdges.HasFlag(FrameEdges.Right))
    {
        Line edge = new Line()
        {
            X1 = canvas.ActualWidth - 9,
            Y1 = 0,
            X2 = canvas.ActualWidth - 9,
            Y2 = canvas.ActualHeight,
            StrokeThickness = 20,
            Stroke = new SolidColorBrush(Colors.Red)
        };
        canvas.Children.Add(edge);
    }
}
```

In Listing 5-10, we can see that checking whether an edge is clipped is a matter of checking which FrameEdges are in body.ClippedEdges. Once we have this, we can draw a line on the canvas for each flagged frame edge. Finally, the DrawClippedEdges(Body body) method needs to be called in the loop in which we go through all the bodies and draw their joints and bones, as follows:

```
foreach (Body body in bodies.Where(b => b.IsTracked))
{
    //[...] Rest of code to gather and draw joints
```

```
foreach (var bone in bones)
{
    DrawBone(body.Joints, jointColorPoints, bone.Item1, bone.Item2, colorBrush);
}

//Add this line to call DrawClippedEdges
DrawClippedEdges(body);
```

}

Showing Hand States

While the Kinect supports more advanced gesture-recognition capabilities, being able to near-instantaneously determine whether the hands are clenched (closed), open, or in a lasso state is very convenient for smaller projects where we do not want to spend time defining custom gestures.

As with the clipped edges, showing hand states will not require a significant alteration of our existing application. We will add a method to paint a circle around the hand depending on the hand state. We will have Red represent a closed first, Green an open one, and Blue for one in the lasso form.

Listing 5-11. MainWindow.xaml.cs Method to Show Hand State with Colored Circles

```
private void DrawHandStates(HandState handState, Point handCoord)
    {
        switch (handState)
        {
            case HandState.Closed:
                Ellipse closedEllipse = new Ellipse()
                {
                    Width = 100,
                    Height = 100,
                    Fill = new SolidColorBrush(Color.FromArgb(128, 255, 0, 0))
                };
                Canvas.SetLeft(closedEllipse, (handCoord.X / colorFrameDescription.
                Width) * canvas.ActualWidth - closedEllipse.Width / 2);
                Canvas.SetTop(closedEllipse, (handCoord.Y / colorFrameDescription.
                Height) * canvas.ActualHeight - closedEllipse.Width / 2);
                canvas.Children.Add(closedEllipse);
                break;

            case HandState.Open:
                Ellipse openEllipse = new Ellipse()
                {
                    Width = 100,
                    Height = 100,
                    Fill = new SolidColorBrush(Color.FromArgb(128, 0, 255, 0))
                };
                Canvas.SetLeft(openEllipse, (handCoord.X / colorFrameDescription.Width)
                * canvas.ActualWidth - openEllipse.Width / 2);
```

```
                Canvas.SetTop(openEllipse, (handCoord.Y / colorFrameDescription.Height)
                * canvas.ActualHeight - openEllipse.Width / 2);
                canvas.Children.Add(openEllipse);
                break;

            case HandState.Lasso:
                Ellipse lassoEllipse = new Ellipse()
                {
                    Width = 100,
                    Height = 100,
                    Fill = new SolidColorBrush(Color.FromArgb(128, 0, 0, 255))
                };
                Canvas.SetLeft(lassoEllipse, (handCoord.X / colorFrameDescription.Width)
                * canvas.ActualWidth - lassoEllipse.Width / 2);
                Canvas.SetTop(lassoEllipse, (handCoord.Y / colorFrameDescription.Height)
                * canvas.ActualHeight - lassoEllipse.Width / 2);
                canvas.Children.Add(lassoEllipse);
                break;
        }
    }
```

In Listing 5-11, we see the method that is used to paint circles representing hand states onto the canvas. HandState is an enum with five possible values, the aforementioned *Open*, *Closed*, and *Lasso*, as well as *Unknown* and *NotTracked*. The switch-case statement finds out what state the hand is in and then basically draws the circle in a manner similar to the DrawJoint method using the ColorSpacePoint coordinates of the hand. The method is called twice per hand, as follows:

```
...
//[...] Rest of code to gather and draw joints and bones
DrawClippedEdges(body);

//Add these lines to MultiSourceFrameReader_MultiSourceFrameArrived
DrawHandStates(body.HandRightState, jointColorPoints[JointType.HandRight]);
DrawHandStates(body.HandLeftState, jointColorPoints[JointType.HandLeft]);
...
```

This method is called in the MultiSourceFrame event handler right after (or before) the DrawClippedEdges(Body body) method. The HandState enumeration can be obtained from the HandRightState and HandLeftState properties of Body. The color space coordinates of the joints that we derived earlier are used here to tell the method where to draw the hand-state circles. Compiling and running the application at this point should result in a scene similar to that in Figure 5-5.

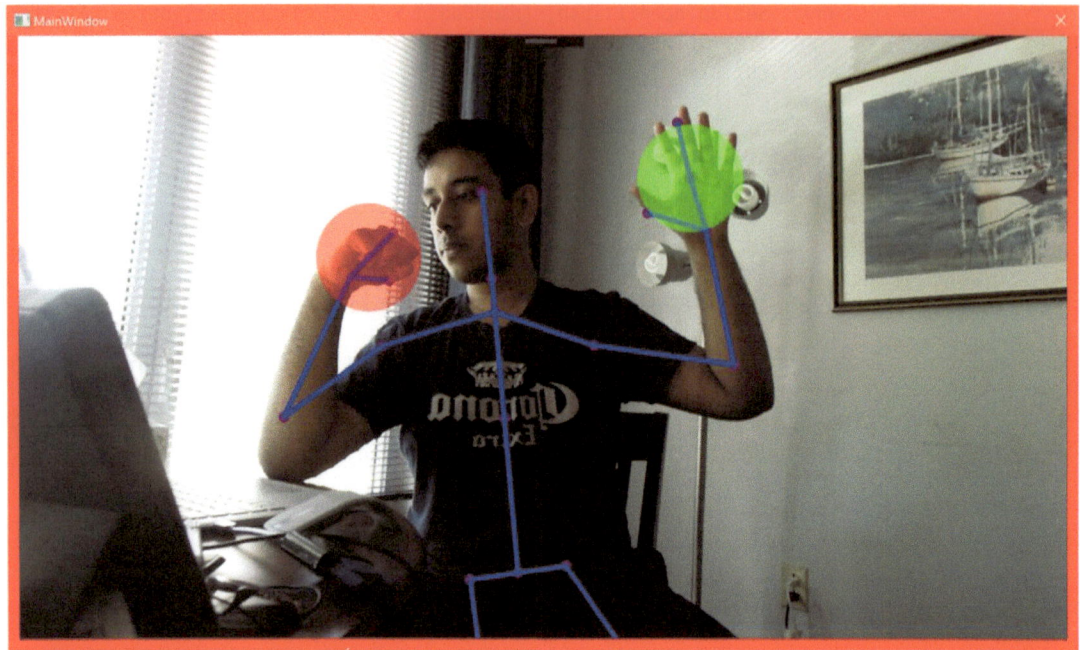

Figure 5-5. *Hand states depicted with colored circles*

It is worth knowing that Body has `HandLeftConfidence` and `HandRightConfidence` properties that return a `TrackingConfidence` value for the left and right hand states, respectively. `TrackingConfidence` is nothing but an enum with two values, *High* and *Low*, whose meanings are self-described.

Understanding Joint Orientations

In the "Drawing Skeletons" section, we saw how `CameraSpacePoint` coordinates could be used to reconstruct a user's skeleton. There is an alternative way to achieve this, which is through *joint orientations*. Joint orientation can be one of the more difficult concepts to grasp in the Kinect SDK, but is especially useful in applications such as avateering, where the proportions of an animated character might not correspond to a user's actual body. Just think of the calamity that would ensue if we tried to rig a LeBron James model in a Kinect basketball game with the joint coordinates of an eight-year-old kid! Each playable athlete in the game would have different arm and leg proportions, and they would all be different from the wide range of possible users playing the game. Or think about if we tried to avateer a T-Rex, which are notorious for their short, stubby arms. It's easy to see how absolute metric coordinates are not as helpful in such cases.

Joint orientations sidestep this issue by ignoring bone length and instead permitting us to rebuild the skeleton with knowledge of the joints' local *quaternion* orientation values. These quaternion values indicate the rotation of a joint about the bone originating from its parent joint. Each joint has its place in a hierarchy, as depicted in Figure 5-6, which can be traversed to reconstruct a skeleton.

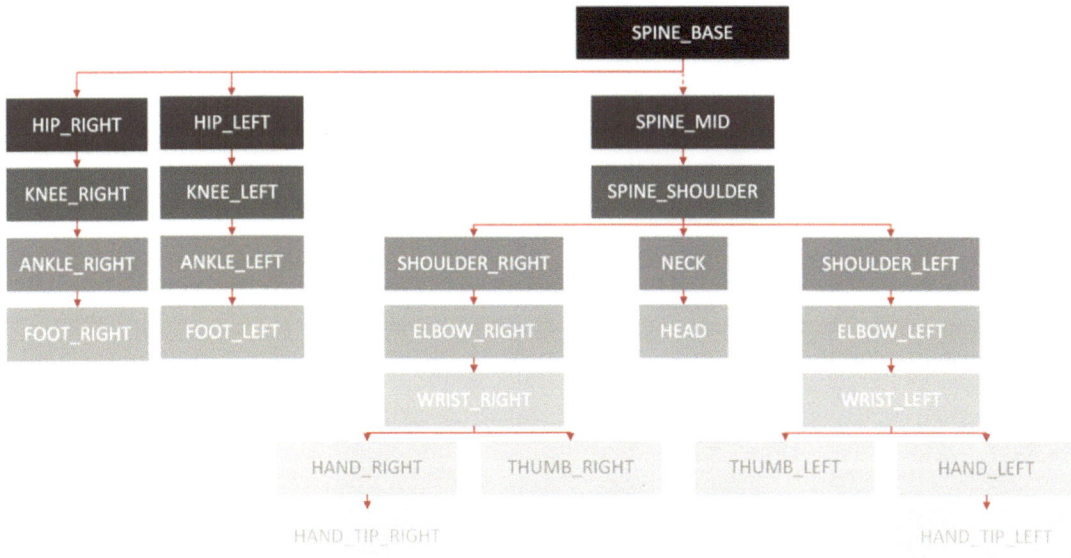

Figure 5-6. *Joint hierarchy in the Kinect v2 SDK*

■ **Tip** Rotations in 3D-coordinate systems are often described with either *Euler angles* or *quaternions*. There are trade-offs to choosing either; Euler angles are conceptually simpler to understand, whereas quaternions avoid a critical issue with Euler angles known as *gimbal lock* and provide better results for certain computational operations. Typically, your graphical programming environment, whether it be something closer to metal such as DirectX or a full-fledged game engine such as Unity, will have utilities to abstract the complexities of working in between both. A more thorough discussion of how to work with both rotational description systems falls outside the purview of this book and could easily take up a few college lectures. A brief but succinct tutorial on rotations can be found at `http://www.opengl-tutorial.org/intermediate-tutorials/tutorial-17-quaternions/`.

Normally, I would prefer to build a sample from scratch for us to explore, but considering the amount of code necessary to get a project working, we will inspect the excellent **Kinect Evolution** sample instead. We tinkered with it a bit in Chapter 1, but now we will go more in depth, specifically with its "Block(wo) man" demo (Figure 5-7). This is the one time that we will touch C++ in this book, but it will be brief and relatively painless. To find the sample, go to the SDK Browser v2.0 and select *Install from Web* for the **Kinect Evolution-XAML** entry. Navigate to the folder you installed it in and open the solution file.

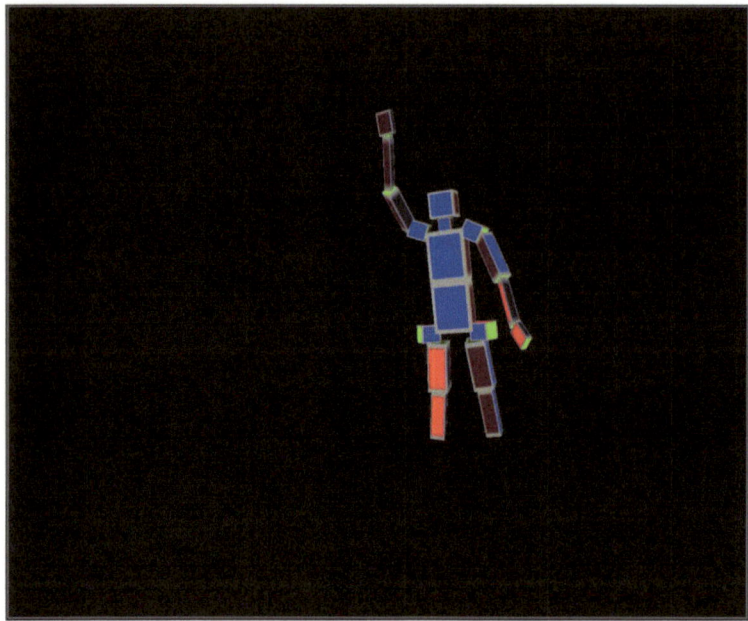

Figure 5-7. *The Block(wo)man demo in the Kinect Evolution sample*

The code to get the whole avatar fleshed out in DirectX is somewhat involved, so we will focus on a top-level overview of the code that extracts and processes joint orientations. This portion is found in `BlockManPanel.cpp`, which is situated in the `KinectEvolution.Xaml.Controls` project within the solution.

Listing 5-12. BlockManPanel.cpp OnBodyFrame Method Part 1: Obtaining Joint Orientations

```
[...]
//Body was obtained from body frame before this point
IMapView<JointType, Joint>^ joints = body->Joints;
IMapView<JointType, JointOrientation>^ jointOrientations = body->JointOrientations;
{
    for (UINT i = 0; i < joints->Size; i++)
    {
        JointType jt = static_cast<JointType>(i);

        //this bit is unimportant
        if (jt == JointType::Head)
        {
            continue;
        }

        JointOrientation jo = jointOrientations->Lookup(jt);
        _blockMen[iBody]._JointOrientations[i] = SmoothQuaternion(_blockMen[iBody]._
        JointOrientations[i], XMLoadFloat4((XMFLOAT4*) &jo.Orientation));
    }
    ...
```

In Listing 5-12, we have the code right after the start of the OnBodyFrame(_In_ BodyFrame^ bodyFrame) event handler, which is the C++ equivalent of an event handler for BodyFrameReader's FrameArrived event in C#. The body was already obtained, and now we proceed to extract the joints and joint orientations. We then apply a function to each joint orientation, SmoothQuaternion(...), that, in short, normalizes the quaternion and interpolates between its prior and new states using spherical linear interpolation. Do not worry if that bit made no sense. The point is to smooth the motion of the quaternion between frames to reduce noise and jerkiness. SmoothQuaternion(...) is made up of DirectX quaternion helper methods.

Listing 5-13. BlockManPanel.cpp OnBodyFrame Method Part 2: Inspecting Skeleton Blocks

```
...
for (int blockIndex = 0; blockIndex < BLOCK_COUNT; ++blockIndex)
{
    JointType jointIndex = g_SkeletonBlocks[blockIndex].SkeletonJoint;
    int parentIndex = g_SkeletonBlocks[blockIndex].ParentBlockIndex;

    XMVECTOR translateFromParent = g_SkeletonBlocks[blockIndex].CenterFromParentCenter;
    XMMATRIX parentTransform = (parentIndex > -1) ? _blockMen[iBody]._
    BlockTransforms[parentIndex] : XMMatrixIdentity();
    XMVECTOR position = XMVector3Transform(translateFromParent, parentTransform);
    ...
```

In the header file, BlockManPanel.h, we defined a SkeletonBlock structure for each joint pair that is used by DirectX to construct the 3D blocks used to make our Block(wo)man. In Listing 5-13, we are simply calculating each SkeletonBlock's position and base rotation based off its parent SkeletonBlock's properties.

Listing 5-14. BlockManPanel.cpp OnBodyFrame Method Part 3: Determining Joint Rotation

```
...
XMVECTOR rotationQuat = XMVectorZero();

if (XMVector4Equal(_blockMen[iBody]._JointOrientations[(int)jointIndex], XMVectorZero()))
{
    if (parentIndex > -1)
    {
        if (XMVector4Equal(_blockMen[iBody]._JointOrientations[parentIndex],
        XMVectorZero()))
        {
            rotationQuat = XMQuaternionIdentity();
        }
        else
        {
            rotationQuat = _blockMen[iBody]._JointOrientations[parentIndex];
        }
    }
}
else
{
    rotationQuat = _blockMen[iBody]._JointOrientations[(int) jointIndex];
}
...
```

In Listing 5-14, we get the rotation quaternions from the joints based off their orientation. In certain cases, the joint's orientation is not available, so we use the parent joint's orientation. If this is not available, we assume no further orientation.

Listing 5-15. BlockManPanel.cpp OnBodyFrame Method Part 4: Creating Model Matrix

```
    ...
    XMMATRIX translation = XMMatrixTranslationFromVector(position);
    XMMATRIX scale = XMMatrixScalingFromVector(g_SkeletonBlocks[blockIndex].Scale);
    XMMATRIX rotation = XMMatrixRotationQuaternion(rotationQuat);

    _blockMen[iBody]._BlockTransforms[blockIndex] = scale * rotation * translation;
    }

    Joint joint = joints->Lookup(JointType::SpineBase);

    _blockMen[iBody]._position = XMLoadFloat4((XMFLOAT4*) &(joint.Position));
}
```

In Listing 5-15, we bring everything together and calculate the translation, scale, and rotation matrices for the SkeletonBlock. These are multiplied by each other to determine the SkeletonBlock's overall transformation matrix, which will essentially decide how the block will appear on screen. At the completion of the iteration through the SkeletonBlocks, we set the SpineBase's position as the position of the skeleton, as it is the root joint of the joint hierarchy.

The important thing to remember throughout is that the orientation of a joint is based off its parent joint's orientation. If a body's thumb joint rotates by 20 degrees on a certain axis, it is not rotating 20 degrees from the global Cartesian space, but rather 20 degrees from the hand joint.

Determining Body Lean

The Kinect SDK has a convenient way to determine whether a user is leaning or not. This capability is limited and does not provide us with the exact characteristics of the lean, such as its angular direction or how pronounced it is, but it is still useful if you only need to know roughly whether a person is leaning forward or backward and/or right or left. This can be a quick way to implement a "dodge" feature in a game so that the player can avoid projectiles, or a way to see if the user is engaged (i.e., leaning forward).

Determining the lean value is simply a matter of accessing the Lean property of Body (Listing 5-16).

Listing 5-16. Determining the Lean Value of a Body

```
foreach (Body body in bodies.Where(b => b.IsTracked))
{
    PointF lean = body.Lean;
    float xLean = lean.X;
    float yLean = lean.Y;

    DrawLean(xLean, yLean, body.LeanTrackingState);
 [...]
}

private void DrawLean(float xLean, float yLean, TrackingState leanTrackingState)
{
    var drawLean = false;
```

```
//The following shows the different lean states
switch (leanTrackingState)
{
    case TrackingState.Inferred:
        //Do nothing
        break;
    case TrackingState.NotTracked:
        //Do nothing
        break;
    case TrackingState.Tracked:
        drawLean = true;
        break;
}

//We'll only draw this if the user leaning state is tracked
if (drawLean)
{
    //Draw a Left/Right Lean Meter at top of screen
    Rectangle mainHorizontalMeter = new Rectangle()
    {
        Width = canvas.ActualWidth - 50,
        Height = 50,
        Fill = new SolidColorBrush(Color.FromArgb(100, 128, 100, 200))
    };

    //Draw a Forward/Back Lean Meter at Right side of screen
    Rectangle mainVerticalMeter = new Rectangle()
    {
        Width = 50,
        Height = canvas.ActualHeight - 50,
        Fill = new SolidColorBrush(Color.FromArgb(100, 50, 250, 250))
    };

    //Set Position and Add the meter to the drawing canvas
    Canvas.SetLeft(mainHorizontalMeter, 0);
    Canvas.SetTop(mainHorizontalMeter, 0);
    canvas.Children.Add(mainHorizontalMeter);

    //Set Position and Add the meter to the drawing canvas
    Canvas.SetLeft(mainVerticalMeter, canvas.ActualWidth - 60);
    Canvas.SetTop(mainVerticalMeter, 50);
    canvas.Children.Add(mainVerticalMeter);

    //Draw out measurement rectangle
    Rectangle leftRightLean = new Rectangle()
    {
        Width = 10,
        Height = 38,
        Fill = new SolidColorBrush(Color.FromArgb(150, 0, 0, 0))
    };
```

127

```
//Draw a Forward/Back Lean Meter at Right side of screen
Rectangle forwardBackLean = new Rectangle()
{
    Width = 38,
    Height = 10,
    Fill = new SolidColorBrush(Color.FromArgb(150, 0, 0, 0))
};

//Figure out how much to draw based on xLean and yLean values
//xLean and yLean are normalized between -1 and 1, so we will
//use those values to be width values of our meter
//0 ... 1 - 1/2 width to full width of meter
//-1 ... 0 - 0 width to 1/2 width of meter
var halfWidth = mainHorizontalMeter.Width/2;
var horizontalMeter = ( (xLean > 0) ? xLean * (mainHorizontalMeter.Width  ) : Math.
Abs(xLean * (halfWidth)) );
var halfHeight = mainVerticalMeter.Height/2;
var verticalMeter =( (yLean > 0) ? yLean*(mainVerticalMeter.Height ): Math.Abs(yLean
* (halfHeight)) )  + 50;

if (xLean < 0)
{
    Canvas.SetLeft(leftRightLean, (canvas.ActualWidth / 2) - horizontalMeter);
    Canvas.SetTop(leftRightLean, 5);
    leftRightLean.Width = horizontalMeter;
}
else
{
    Canvas.SetLeft(leftRightLean, (canvas.ActualWidth / 2));
    Canvas.SetTop(leftRightLean, 5);
    leftRightLean.Width = horizontalMeter;

}
canvas.Children.Add(leftRightLean);

if (yLean > 0)
{
    Canvas.SetLeft(forwardBackLean, (canvas.ActualWidth - 55));
    Canvas.SetTop(forwardBackLean, halfHeight - verticalMeter);
    forwardBackLean.Height = verticalMeter;

}
else
{
    Canvas.SetLeft(forwardBackLean, canvas.ActualWidth - 55);
    Canvas.SetTop(forwardBackLean, canvas.ActualHeight / 2);
    forwardBackLean.Height = verticalMeter;

}
canvas.Children.Add(forwardBackLean);
```

```
//Now we'll add some simple text labels in the meter
TextBlock labelLeft = new TextBlock() { Text="<< --- Leaning Left", FontSize=16.0,
Foreground = new SolidColorBrush(Color.FromArgb(150, 0, 0, 0)) };

Canvas.SetLeft(labelLeft, (canvas.ActualWidth *.25 ) );
Canvas.SetTop(labelLeft, 10);
canvas.Children.Add(labelLeft);

TextBlock labelRight = new TextBlock() { Text = "Leaning Right --- >>>", FontSize =
16.0, Foreground = new SolidColorBrush(Color.FromArgb(150, 0, 0, 0)) };

Canvas.SetLeft(labelRight, canvas.ActualWidth * .75 );
Canvas.SetTop(labelRight, 10);
canvas.Children.Add(labelRight);

TextBlock labelForward = new TextBlock() { Text = "Forward", FontSize = 8.0,
Foreground = new SolidColorBrush(Color.FromArgb(150, 0, 0, 0)) };

Canvas.SetLeft(labelForward, canvas.ActualWidth - 55);
Canvas.SetTop(labelForward, canvas.ActualHeight * .25);

canvas.Children.Add(labelForward);

TextBlock labelBackward = new TextBlock() { Text = "Backward", FontSize = 8.0,
Foreground = new SolidColorBrush(Color.FromArgb(150, 0, 0, 0)) };

Canvas.SetLeft(labelBackward, canvas.ActualWidth - 55);
Canvas.SetTop(labelBackward, canvas.ActualHeight * .75);
canvas.Children.Add(labelBackward);
    }
}
```

The x and y values of lean are normalized values that refer to the confidence of the lean. The values range from –1 to 1, which is left to right on the x-axis and back to front on the y-axis, respectively. So, a lean value of x: –0.17, y: 0.94 would mean a possible but unlikely lean to the left and a probable lean forward.

As with hand states, there is a TrackingState property for lean called LeanTrackingState. This can be used to determine whether a body's lean itself is being tracked.

BodyIndexFrameSource: Body Index Data

There are times when we need to determine whether a pixel in a depth or color image belongs to a player. We might want to draw an outline around a player to indicate that it is their turn, or we could cut out their body from the image and use it in a green screen application. BodyIndexFrameSource is the Data Source that enables us to do so. It lets us know whether pixels in a depth image are a part of a specific player or the background. Using CoordinateMapper, we can extract a player's image from a color image as well.

Displaying BodyIndex Data

Working with BodyIndexFrameSource is very much like working with color, infrared, or depth data. It is an image from which we can copy, process, and present image data. It features the familiar CopyFrameDataToArray and CopyFrameDataToIntPtr methods, as well as LockImageBuffer. The resolution is naturally the same as that of the depth and infrared frames (as skeletal data is being processed internally using those frames), and each pixel is represented by an 8-bit unsigned integer. Each pixel in the array will have a value from 0 to 5 that corresponds to the index of the Body array provided by BodyFrameSource; a pixel with the value 2 represents a depth pixel occupied by the player found in bodies[2].

I will only show the FrameArrived event handler, as the rest of the code can be inferred from the color or depth image samples. If need be, the source code for the full application can be found in the book samples.

Listing 5-17. BodyIndexFrameReader Event Handler

```
private void Reader_BodyIndexFrameArrived(object sender, BodyIndexFrameArrivedEventArgs e)
{
    using (BodyIndexFrame bodyIndexFrame = e.FrameReference.AcquireFrame())
    {
        if (bodyIndexFrame != null)
        {
            bodyIndexFrame.CopyFrameDataToArray(bodyIndexPixels);
        }

        for (int i = 0; i < bodyIndexPixels.Length; ++i)
        {
            if (bodyIndexPixels[i] != 255)
            {
                //player exists in space, draw their color
                var color = bodyIndexColors[bodyIndexPixels[i]];
                bitmapPixels[i * 4 + 0] = color.B;
                bitmapPixels[i * 4 + 1] = color.G;
                bitmapPixels[i * 4 + 2] = color.R;
                bitmapPixels[i * 4 + 3] = 255;
            }
            else
            {
                //no player found, write pixels as black
                bitmapPixels[i * 4 + 0] = 0;
                bitmapPixels[i * 4 + 1] = 0;
                bitmapPixels[i * 4 + 2] = 0;
                bitmapPixels[i * 4 + 3] = 255;
            }
        }

        bodyIndexBitmap.WritePixels(new Int32Rect(0, 0,
            bodyIndexBitmap.PixelWidth,
            bodyIndexBitmap.PixelHeight),
            bitmapPixels,
            bodyIndexBitmap.PixelWidth * 4, 0);
    }
}
```

In Listing 5-17, we transcribe the body index data from the BodyIndexFrame into the bodyIndexPixels byte array and then process its data for display in the bitmapPixels byte array, which represents RGBA data to be displayed through the bodyIndexBitmap WriteableBitmap. To make it clear which pixels are occupied by a player and which are not, we show an image that is black for the background and colored for any pixels with players in them. Going through each pixel of bodyIndexPixels, if it does not contain a player, it will have an unsigned integer value of 255, and thus we display an RGBA value of black for it. In cases where the value is **not** 255, we can assume that there is a player occupying the pixel. We have an array of six `System.Windows.Media.Colors` called `bodyIndexColors`, each of which coincides with a body the Kinect recognizes: the player index. We set its RGBA components as the corresponding four bytes in `bitmapPixels`. The result is then written to a `WriteableBitmap` to be used as an `ImageSource`.

In Figure 5-8, we have a screenshot of the application in operation. Two users are present in front of the Kinect. The general outlines of both can be clearly seen, but the edges are not well defined in certain areas. For the blue player, the bottom portion of one leg is obscured, so the Kinect attempts to resolve pixels near that leg through approximation, which is why there are blocky specks of blue. For the red player, the Kinect erroneously assumes that the area between the legs is part of the body. Errors like this are not uncommon, though overall the Kinect manages to do a decent job.

Figure 5-8. *BodyIndex data displayed as an image*

Rudimentary Green Screen App

The most evident use case for `BodyIndexFrameSource` is to create green screen apps with the Kinect. It is not real green-screening (otherwise known as Chroma-keying), as we are not subtracting the green color channel from the camera feed received by the Kinect. Instead, the Kinect is using machine-learning algorithms to find humans in the vicinity of the camera, and we can extract those directly. The final result is not as robust as what would have been obtained through Chroma-keying, but is a lot simpler to set up and has the added bonus of not removing any colors from the scene.

The process of extracting human images from the color feed is based off coordinate mapping the BodyIndexFrameSource data and keeping the corresponding pixels in the color feed while rendering the others transparent. We will inspect the event handler for the Data Sources directly without reviewing the variable declaration and constructor section of the app, as it features code we have gone over countless times before. As always, the full source code is included in the book samples.

Listing 5-18. MultiSourceFrameReader Event Handler for Green Screen Application Part 1

```
private void MultiSourceFrameReader_MultiSourceFrameArrived(object sender,
MultiSourceFrameArrivedEventArgs e)
{
    System.Array.Clear(bitmapPixels, 0, bitmapPixels.Length);
    MultiSourceFrame multiSourceFrame = e.FrameReference.AcquireFrame();

    using (DepthFrame depthFrame = multiSourceFrame.DepthFrameReference.AcquireFrame())
    {
        if (depthFrame != null)
        {
            depthFrame.CopyFrameDataToArray(depthPixels);
            coordinateMapper.MapDepthFrameToColorSpace(depthPixels, colorSpacePoints);
        }
    }

    using (BodyIndexFrame bodyIndexFrame = multiSourceFrame.BodyIndexFrameReference.
    AcquireFrame())
    {
        if (bodyIndexFrame != null)
        {
            bodyIndexFrame.CopyFrameDataToArray(bodyIndexPixels);
        }
    }
```

In Listing 5-18, we have half of the MultiSourceFrameReader event handler for our green screen application. As expected, we make use of body index data, but what you might not have foreseen is that we need to use depth data too. Technically speaking, we do not need the depth data itself, but there is no built-in way to map body index coordinates to color space. Building our method to do so would not be tremendously difficult, but seeing as depth pixels correspond directly to body index pixels, we might as well rely on what already exists. In addition to grabbing depth and body index data, we also clear the bitmapPixels array. This is the array that holds the final processed color data to be displayed in the UI. We have to reset it each time, as we are not overwriting every pixel like when we normally display an image. We do this because otherwise there would be artefacts from a person's body appearing on areas of the image where the user is no longer located, as they have not been cleared.

Listing 5-19. MultiSourceFrameReader Event Handler for Green Screen Application Part 2

```
using (ColorFrame colorFrame = multiSourceFrame.ColorFrameReference.AcquireFrame())
{
    if (colorFrame != null)
    {
        colorFrame.CopyConvertedFrameDataToArray(colorPixels, ColorImageFormat.Bgra);
        for (int y = 0; y < depthFrameDescription.Height; y++)
        {
```

```
        for (int x = 0; x < depthFrameDescription.Width; x++)
        {
            int depthIndex = (y * depthFrameDescription.Width) + x;

            byte player = bodyIndexPixels[depthIndex];

            if (player != 255)
            {
                ColorSpacePoint colorPoint = colorSpacePoints[depthIndex];

                int colorX = (int)System.Math.Floor(colorPoint.X);
                int colorY = (int)System.Math.Floor(colorPoint.Y);

                if ((colorX >= 0) && (colorX < colorFrameDescription.Width) &&
                (colorY >= 0) && (colorY < colorFrameDescription.Height))
                {
                    int colorIndex = ((colorY * colorFrameDescription.Width) + colorX) * 4;
                    int displayIndex = depthIndex * 4;

                    bitmapPixels[displayIndex + 0] = colorPixels[colorIndex];
                    bitmapPixels[displayIndex + 1] = colorPixels[colorIndex + 1];
                    bitmapPixels[displayIndex + 2] = colorPixels[colorIndex + 2];
                    bitmapPixels[displayIndex + 3] = 255;
                }
            }
        }
    }
    colorBitmap.WritePixels(new Int32Rect(0, 0,
        depthFrameDescription.Width,
        depthFrameDescription.Height),
        bitmapPixels,
        depthFrameDescription.Width * 4, 0);
    }
  }
}
```

In Listing 5-19, after ensuring that color data is available for use, we iterate through the body index data and check if the pixel is not equal to 255, and thus is indicative of a human user. Normally, we can go through the image arrays with one for loop, but we opted to use two here so that we can track the x and y coordinates of the pixel we are currently inspecting in the image. This is used to get its equivalent color pixel's ColorSpacePoint. After verifying that it is usable, we find its color pixel counterpart in colorPixels and copy the data over to the bitmapPixels array, which is used to display our final image.

⬛ **Tip** Copying the color data into colorPixels and then further copying it into bitmapPixels is a redundant process. Through the use of LockRawImageBuffer() the underlying color data could have been accessed at will and copied for the pixels where it was needed. The implementation of this is left to the reader as an exercise. For help on how to make use of Kinect data buffers, visit the examples for depth processing in Chapter 3.

You might have noticed that when we called `WritePixels(...)` on our `colorBitmap` we passed in a rectangle with the width and height of our depth image instead of our color image. Since we are coordinate-mapping a 512 x 424 body index data image to a 1920 x 1080 color image, we must reduce the resolution of the color image to accommodate the lower-resolution source image. Failure to do so results in a green screen image with a larger size, but with low color density/transparent bits all over. The alternative to this would have been to dilate the range of included pixels in the color image based on the body index data. For example, if a certain body index pixel included a player and returned a `ColorSpacePoint` with the coordinates (100, 140), we could include the color pixels in ranges $[100 \pm t, 140 \pm t]$, where t is the number of pixels on one side of the `ColorSpacePoint` to include. This method would reduce the accuracy of the image's included pixels and increase overheard, but would better preserve the resolution.

Listing 5-20. XAML Code for Green Screen Application

```
<Window [...]
        Title="MainWindow" Height="927.5" Width="1092">
    <Grid>
        <Image Source="Images/background.jpg" Stretch="UniformToFill" />
        <Image Source="{Binding ImageSource}" Stretch="UniformToFill" />
    </Grid>
</Window>
```

Listing 5-20 details the front-end code for our app. In the XAML, we must include a background picture in order to create a true green screen application. This image can be included in the project files through File Explorer, but it might be more convenient to simply include it through Visual Studio's Solution Explorer, as shown in Figure 5-9. Right-click the project in Solution Explorer, which is `GreenScreenSample` in our example, and click *Add*, followed by *Add New Folder*, naming it *Images*. Repeat the same steps, but this time click *Add Existing Items. . .* and then add the image you want to use as your background. The name must be the same as indicated in the XAML code (`background.jpg` in the sample). After running the application, the results will be similar to Figure 5-10, minus the banana.

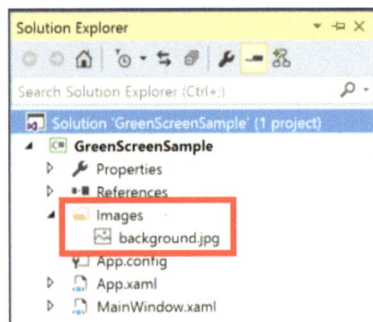

Figure 5-9. *The background image visible in Solution Explorer*

Figure 5-10. *Marisa holding a banana at Microsoft's TechReady conference*

FaceFrameSource & HighDefinitionFaceFrameSource: Facial Structure Data

Normally, discussions of the Kinect's tracking abilities revolve around its more macroscopic skeletal-capture features. The Kinect can also focus specifically on the face of a user and conduct tracking and assaying of its structures and anatomy. It can do this for six people simultaneously, though certain facial-analysis features are computationally intensive and should be applied to one person at a time.

The Kinect's face-tracking APIs are in a different namespace called `Microsoft.Kinect.Face` and are divided into two subsets. There is Face, which provides a general 2D overview of the face's features, and HD Face, which offers a more comprehensive 3D model of the face's structure. An application that merely needs limited or brief tracking of facial features, such as detecting whether a user is looking at the screen, should rely on the Face API. An application that wants to recreate the user's face on an avatar or track minute differences on the face should make use of HD Face.

■ **Note**   If your application uses Face or HD Face, it makes use of `Microsoft.Kinect.Face.dll`. Along with the DLL, you must include the `NuiDatabase` folder that appears in the same folder. For example, if you use the DLL from `C:\Program Files\Microsoft SDKs\Kinect\v2.0_1409\Redist\Face\x64\Microsoft.Kinect.Face.dll`, the `NuiDatabase` folder from within must also be used. If you encounter a *"Failed to Load NuiDatabase"* error during compilation, verify that the `NuiDatabase` folder is present in the configurations found within your project's bin, as in Figure 5-11. An easy way to solve this issue is by copying the DLL and `NuiDatabase` folder from the bin in the *FaceBasics-WPF* sample project from the SDK Browser. Alternatively, handle your references through NuGet.

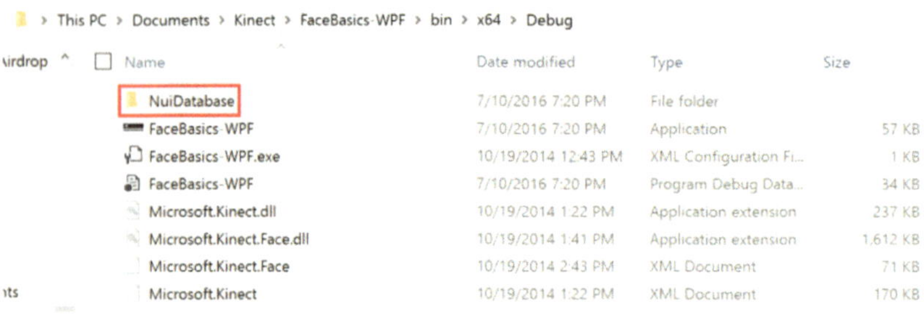

Figure 5-11.   *The NuiDatabase folder in the Visual Studio project's bin folder*

Additionally, you will need to set the CPU architecture in Visual Studio's Configuration Manager. To do so, click the *Any CPU* drop-down on Visual Studio's toolbar and select *Configuration Manager*. In the *Project contexts*, click on the *Any CPU* drop-down under *Platform* and click *x64* (use *x86* for these instructions if needed for certain DLLs in your project) if already present, else click *<New...>*. In the *New Project Platform* window that pops up, choose x64 for *New Platform* and select *Any CPU* from Copy settings from. Keep the *Create new solution platforms* box checked. Finally, click *OK* and close the Configuration Manager. Make sure that x64 is now picked in the configuration drop-down in the toolbar. The entire procedure is depicted in Figure 5-12. The `Microsoft.Kinect.Face` DLL and `NuiDatabase` folder must match the chosen platform architecture.

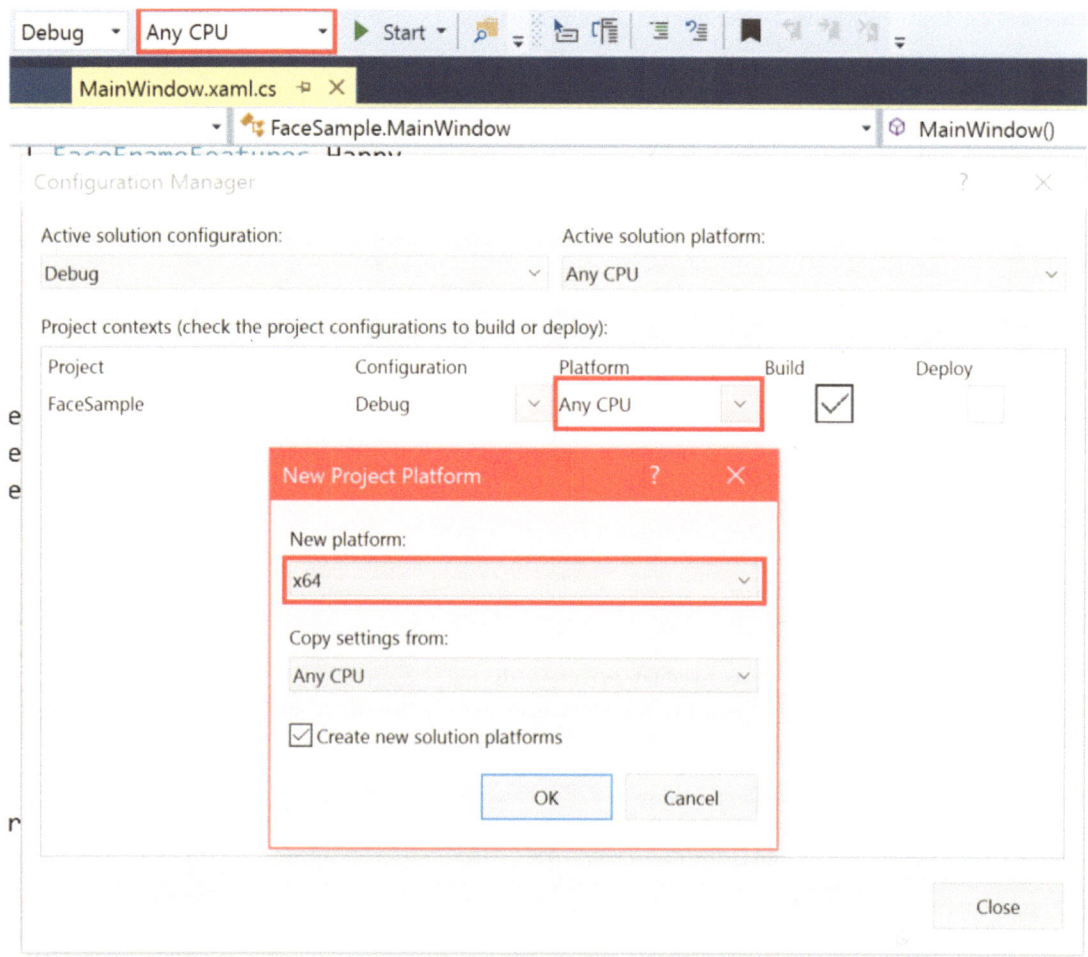

Figure 5-12. *Setting the CPU architecture of the project in Configuration Manager*

Face

There are four key components to the Face API: *detection, alignment, orientation,* and *expressions.* These are not hard and fast Kinect terms that must be memorized, but they are a useful way to describe the API's features with single words. They can all be accessed by opening readers on the FaceFrameSource Data Source.

Obtaining Face Data

Data from the Face API is obtained in a manner similar to doing so for the other Data Sources. The primary differences are that we must designate a separate FaceFrameSource and FaceFrameReader for each face we are looking to pick out and that we have to extract a FaceFrameResult, which is what contains the recognized facial features and data, from each FaceFrame.

Listing 5-21. Private Variables Declaration to Obtain Face Data

```
KinectSensor kinect;

const float InfraredSourceValueMaximum = (float)ushort.MaxValue;
const float InfraredSourceScale = 0.75f;
const float InfraredOutputValueMinimum = 0.01f;
const float InfraredOutputValueMaximum = 1.0f;

MultiSourceFrameReader multiSourceFrameReader;
FrameDescription frameDescription;
Body[] bodies;

FaceFrameSource[] faceFrameSources;
FaceFrameReader[] faceFrameReaders;
FaceFrameResult[] faceFrameResults;

WriteableBitmap infraredBitmap;
ushort[] infraredPixels;
DrawingGroup drawingGroup;
DrawingImage drawingImageSource;
List<Brush> faceBrushes;
```

In Listing 5-21, we declare arrays in which to contain the frame sources, readers, and results for all the faces that can be simultaneously tracked at an instant. Note that we also added an array for tracked bodies. While tracking body data is not a prerequisite to tracking facial data, in practice you will probably always do so anyway. Among other reasons, this is so we can update the FaceFrameSources to ensure that they are tracking faces from bodies visible in the scene.

Listing 5-22. Constructor for App Using Face Data

```
public MainWindow()
{
    kinect = KinectSensor.GetDefault();

    InfraredFrameSource infraredFrameSource = kinect.InfraredFrameSource;

    multiSourceFrameReader = kinect.OpenMultiSourceFrameReader(FrameSourceTypes.Infrared |
    FrameSourceTypes.Body);
    multiSourceFrameReader.MultiSourceFrameArrived += MultiSource_FrameArrived;

    frameDescription = infraredFrameSource.FrameDescription;

    bodies = new Body[kinect.BodyFrameSource.BodyCount];

    FaceFrameFeatures faceFrameFeatures =
        FaceFrameFeatures.BoundingBoxInInfraredSpace
        | FaceFrameFeatures.PointsInInfraredSpace
        | FaceFrameFeatures.RotationOrientation
        | FaceFrameFeatures.FaceEngagement
        | FaceFrameFeatures.Glasses
```

```
        | FaceFrameFeatures.Happy
        | FaceFrameFeatures.LeftEyeClosed
        | FaceFrameFeatures.RightEyeClosed
        | FaceFrameFeatures.LookingAway
        | FaceFrameFeatures.MouthMoved
        | FaceFrameFeatures.MouthOpen;

    faceFrameSources = new FaceFrameSource[6];
    faceFrameReaders = new FaceFrameReader[6];
    faceFrameResults = new FaceFrameResult[6];

    faceBrushes = new List<Brush>()
    {
        Brushes.Pink,
        Brushes.Orange,
        Brushes.BlueViolet,
        Brushes.Aqua,
        Brushes.LawnGreen,
        Brushes.Chocolate
    };

    for (int i = 0; i < 6; i++)
    {
        faceFrameSources[i] = new FaceFrameSource(kinect, 0, faceFrameFeatures);
        faceFrameReaders[i] = faceFrameSources[i].OpenReader();
        faceFrameReaders[i].FrameArrived += Face_FrameArrived;
    }

    infraredBitmap = new WriteableBitmap(frameDescription.Width,
        frameDescription.Height,
        96.0, 96.0,
        PixelFormats.Gray32Float,
        null);
    infraredPixels = new ushort[frameDescription.LengthInPixels];
    drawingGroup = new DrawingGroup();
    drawingImageSource = new DrawingImage(drawingGroup);

    kinect.Open();

    DataContext = this;

    InitializeComponent();
}
```

With the Face API, we need to declare in advance which features we plan to extract from the FaceFrame. This is why we assign the desired features in the FaceFrameFeatures enum in Listing 5-22. In addition to the features listed, there are also FaceFrameFeatures_BoundingBoxInColorSpace and FaceFrameFeatures_PointsInColorSpace. We are using the infrared space equivalents, which is why we did not include them.

■ **Tip** The Kinect's face-detection algorithms are able to recognize faces using the infrared camera. They use infrared because it is lighting independent, and thus faces are bound to be accurately recognized in any lighting condition. Feature coordinates are internally translated by the API to color space using the `CoordinateMapper` class.

We create the arrays for the face frame sources, readers, and results for the number of faces that we are interested in tracking—in our case, all six. We then loop through the `FaceFrameSource` and `FaceFrameReader` arrays and create the `FaceFrameSources` and set up their readers. The constructor method for `FaceFrameSource`, new `FaceFrameSource(kinect, 0, faceFrameFeatures)`, takes the Kinect sensor, initial tracking ID, and desired face frame features as inputs. The tracking ID is a property of the Body class, and we can use it to set the `FaceFrameSource` to track the face of a specific body.

For our sample app making use of face data, we will display the features on top of an infrared video stream. For this, we have to construct our `ImageSource` slightly differently. Instead of displaying the infrared stream's `WriteableBitmap` directly, we create a `DrawingGroup`, which can be converted to an `ImageSource` by creating a `DrawingImage` out of it. As this is not a book on .NET or WPF, the peculiarities are not important. Just know that we can add multiple visual elements to a `DrawingGroup` and that `DrawingImage` is a type of `ImageSource`. Instead of returning the `WriteableBitmap` in the getter for our public `ImageSource`, we return the `DrawingImage` object, as follows:

```
public ImageSource ImageSource
{
    get
    {
        return drawingImageSource;
    }
}
```

Listing 5-23. FaceFrameReader Event Handler

```
private void Face_FrameArrived(object sender, FaceFrameArrivedEventArgs e)
{
    using (FaceFrame faceFrame = e.FrameReference.AcquireFrame())
    {
        if (faceFrame != null)
        {
            int index = GetFaceSourceIndex(faceFrame.FaceFrameSource);

            faceFrameResults[index] = faceFrame.FaceFrameResult;
        }
    }
}
```

In Listing 5-23, we save the `FaceFrameResult`, which is a property of `FaceFrame`, to our `faceFrameResults` array. We want to keep track of its index so that we can work with its respective frame source, frame reader, and body. Since we have a reference to its `FaceFrameSource` in the `FaceFrame`, we take the initiative of finding its index using `GetFaceSoureIndex(FaceFrameSource faceFrameSource)`, as shown in Listing 5-24. The `Face_FrameArrived` event handler is applied to each reader.

Listing 5-24. Obtaining the Index of a FaceFrameSource

```
private int GetFaceSourceIndex(FaceFrameSource faceFrameSource)
{
    int index = -1;

    for (int i = 0; i < 6; i++)
    {
        if (faceFrameSources[i] == faceFrameSource)
        {
            index = i;
            break;
        }
    }

    return index;
}
```

Finding the index is a straightforward matter. We loop through our faceFrameSources array and check whether any of them match the FaceFrameSource of the obtained FaceFrame. If they do, we know what index to put our FaceFrameResult in.

Listing 5-25. MultiSourceFrameReader Event Handler for Face-Detector App

```
private void MultiSource_FrameArrived(object sender, MultiSourceFrameArrivedEventArgs e)
{
    MultiSourceFrame multiSourceFrame = e.FrameReference.AcquireFrame();

    using (InfraredFrame infraredFrame = multiSourceFrame.InfraredFrameReference.
AcquireFrame())
    {
        if (infraredFrame != null)
        {
            using (KinectBuffer infraredBuffer = infraredFrame.LockImageBuffer())
            {
                ProcessInfraredFrameData(infraredBuffer.UnderlyingBuffer, infraredBuffer.Size);
            }
        }
    }

    using (BodyFrame bodyFrame = multiSourceFrame.BodyFrameReference.AcquireFrame())
    {
        if (bodyFrame != null)
        {
            bodyFrame.GetAndRefreshBodyData(bodies);

            using (DrawingContext dc = drawingGroup.Open())
            {
                dc.DrawImage(infraredBitmap, new Rect(0, 0,
                    infraredBitmap.Width,
                    infraredBitmap.Height));
```

141

```
                    for (int i = 0; i < 6; i++)
                    {
                        if (faceFrameSources[i].IsTrackingIdValid)
                        {
                            if (faceFrameResults[i] != null)
                            {
                                DrawFace(i, faceFrameResults[i], dc);
                            }
                        }
                        else
                        {
                            if (bodies[i].IsTracked)
                            {
                                faceFrameSources[i].TrackingId = bodies[i].TrackingId;
                            }
                        }
                    }
                }
            }
        }
}

private unsafe void ProcessInfraredFrameData(IntPtr infraredFrameData, uint
infraredFrameDataSize)
{
    //infrared frame data is a 16-bit value
    ushort* frameData = (ushort*)infraredFrameData;

    //lock the target bitmap
    this.infraredBitmap.Lock();

    //get the pointer to the bitmap's back buffer
    float* backBuffer = (float*)this.infraredBitmap.BackBuffer;

    //process the infrared data
    for (int i = 0; i < (int)(infraredFrameDataSize / this.frameDescription.BytesPerPixel); ++i)
    {
        //since we are displaying the image as a normalized grayscale image, we need to convert from
        //the ushort data (as provided by the InfraredFrame) to a value from
        //[InfraredOutputValueMinimum, InfraredOutputValueMaximum]
        backBuffer[i] = Math.Min(InfraredOutputValueMaximum,
        (((float)frameData[i] / InfraredSourceValueMaximum * InfraredSourceScale) *
        (1.0f - InfraredOutputValueMinimum)) + InfraredOutputValueMinimum);
    }

    //mark the entire bitmap as needing to be drawn
    this.infraredBitmap.AddDirtyRect(new Int32Rect(0, 0, this.infraredBitmap.PixelWidth,
    this.infraredBitmap.PixelHeight));

    //unlock the bitmap
    this.infraredBitmap.Unlock();
}
```

In Listing 5-25, the event handler for the `MultiSourceFrameReader` collects infrared and body data. Seeing as this represents a single event that is fired repeatedly, as opposed to six different events as with the `FaceFrameReader`, it is better to include the drawing code in here. This way, we avoid having multiple unordered methods trying to access the `DrawingGroup` in a similar timeframe, all trying to overwrite it.

We obtain the infrared data in the same we did in Listing 3-14 in Chapter 3. The XAML front end is also the same as in Chapter 3. Inside of the `using (BodyFrame bodyFrame = multiSourceFrame.BodyFrameReference.AcquireFrame()) { [...] }` block we access the `DrawingContext` of `drawingGroup`. `DrawingContext` is a .NET construct that allows us to draw various types of media in a `DrawingGroup` (among other drawing-related objects). We start off by drawing our infrared image data as the background, and then we loop through the `faceFrameResults` and draw any facial data with the `DrawFace(int index, FaceFrameResult faceFrameResult, DrawingContext drawingContext)` method (which will be defined later). Keep in mind that we only do this if the `faceFrameSources[i].IsTrackingIdValid` property returns true. If not, we consider whether a body is being checked for index i and, if so, assign its `TrackingId` to its respective `FaceFrameSource` (`faceFrameSources[i].TrackindId = bodies[i].TrackingId`).

At this point, if you were to run the code, minus the `DrawFace(...)` method, it would obtain facial data and display an infrared feed to boot, but it would not actually indicate that it has obtained facial data, let alone do anything with it. We thus venture to the first Face API feature, detection, to inform the user that the face has been found by the Kinect.

Detection

In the Face API, *detection* refers to whether any faces belonging to bodies are visible in the Kinect's field of view. In practice, the Face API achieves this by framing a face inside of a bounding box, as seen in Figure 5-13. This bounding box is a property of `FaceFrameResult` called either `FaceBoundingBoxInInfraredSpace` or `FaceBoundingBoxInColorSpace` depending on the image feed you are interested in. It is a `RectI` struct—a rectangle with integer coordinates. The `RectI`'s *Top* and *Bottom* fields each represent a pixel coordinate on the vertical axis, and the *Right* and *Left* fields represent the same on a horizontal axis. Thus, a `RectI` with the following fields would represent a rectangle whose top-left coordinate is (0, 10), and its size would be 13 pixels wide and 10 pixels in height.

```
RectI box = new RectI(); //has no constructor
Box.Top = 10;
Box.Bottom = 20;
Box.Left = 0;
Box.Right = 13;
```

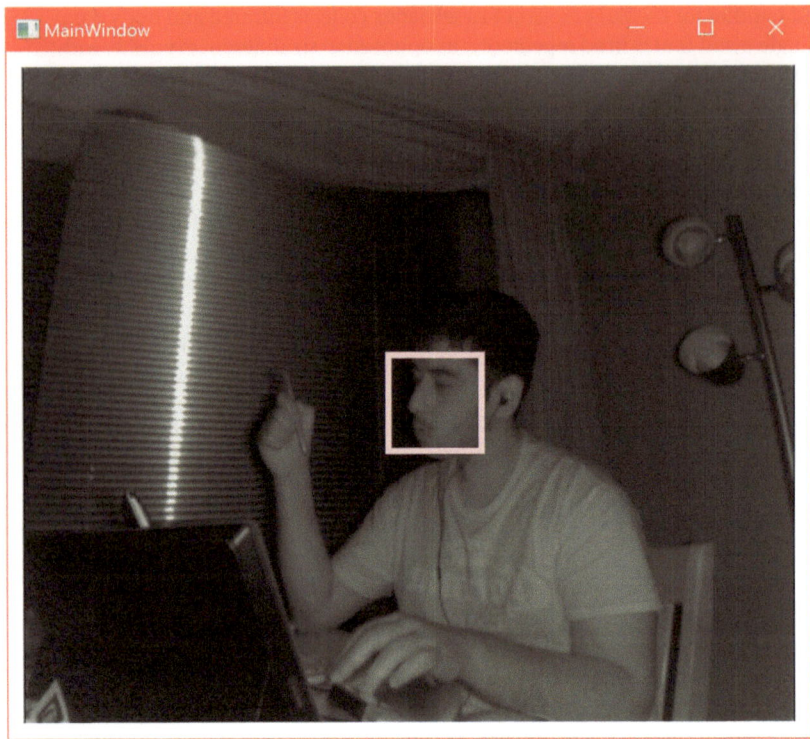

Figure 5-13. *Bounding box representation (FaceBoundingBoxInInfraredSpace) around a detected face. Note that the box drawn on screen is thicker than the actual bounding box, given that the pen used to draw it has a thickness.*

Before we can draw the bounding box on our infrared image, it would be prudent for us to ensure that the box fits within the image. Hence, when new `FaceFrameResults` are collected in `Face_FrameArrived(...)`, we should reject results whose bounding boxes do not fit. This involves a small edit to `Face_FrameArrived(...)`, whose initial version we saw in Listing 5-23.

Listing 5-26. Face Bounding Box Validation Method Call

```
if (faceFrame != null)
{
    int index = GetFaceSourceIndex(faceFrame.FaceFrameSource);

    if (ValidateFaceBoundingBox(faceFrame.FaceFrameResult))
    {
        faceFrameResults[index] = faceFrame.FaceFrameResult;
    }
    else
    {
        faceFrameResults[index] = null;
    }
}
```

In Listing 5-26, we check to see if the bounding box is within the confines of the image with our custom ValidateFaceBoundingBox (FaceFrameResult faceFrameResult) method and reject the FaceFrameResult if it is not.

Listing 5-27. Face Bounding Box Validation

```
private bool ValidateFaceBoundingBox(FaceFrameResult faceFrameResult)
{
    bool isFaceValid = faceFrameResult != null;

    if (isFaceValid)
    {
        RectI boundingBox = faceFrameResult.FaceBoundingBoxInInfraredSpace;
        if (boundingBox!= null)
        {
            isFaceValid = (boundingBox.Right - boundingBox.Left) > 0 &&
                          (boundingBox.Bottom - boundingBox.Top) > 0 &&
                          boundingBox.Right <= frameDescription.Width &&
                          boundingBox.Bottom <= frameDescription.Height;
        }
    }

    return isFaceValid;
}
```

In Listing 5-27, we immediately reject empty FaceFrameResults. We then check to see if the box has meaningful coordinates, e.g., the right side of the bounding box is to the right of the box's left side ((boundingBox.Right - boundingBox.Left) > 0). If it does, we approve of the FaceFrameResult, and it can be used in the DrawFace(...) method to draw the bounding box.

Listing 5-28. Drawing a Bounding Box

```
private void DrawFace(int index, FaceFrameResult faceFrameResult, DrawingContext
drawingContext)
{
    Brush drawingBrush = faceBrushes[0];
    if (index < 6)
    {
        drawingBrush = faceBrushes[index];
    }

    Pen drawingPen = new Pen(drawingBrush, 4);

    RectI faceBoxSource = faceFrameResult.FaceBoundingBoxInInfraredSpace;
    Rect faceBox = new Rect(faceBoxSource.Left, faceBoxSource.Top,
    faceBoxSource.Right - faceBoxSource.Left, faceBoxSource.Bottom - faceBoxSource.Top);
    drawingContext.DrawRectangle(null, drawingPen, faceBox);

}
```

Listing 5-28 features the DrawFace(...) method we called earlier in the MultiFrameSourceReader event handler in Listing 5-25. We start by picking a Brush, which contains our desired color for the boxes drawn around faces. The brushes are from a list initialized in the app constructor as such:

```
faceBrushes = List<Brush>()
{
    Brushes.Pink,
    Brushes.Orange,
    [...] //Add six brushes in total with differing colors
};
```

The brush is then used to create a Pen that includes the desired thickness of the box's outline as well. Finally, we create a rectangle based on the bounding box in FaceFrameResult and draw the rectangle on the DrawingContext, which will then display it to the front end. The compiled result will resemble the previously shown Figure 5-13.

Alignment

Alignment refers to Face API's ability to easily detect five facial landmarks on a face. These landmarks, or points, consist of both eyes, the tip of the nose, and both corners of the mouth. They can be obtained from the FacePointsInInfraredSpace or FacePointsInColorSpace properties of FaceFrameResult (again, the difference between both is which image feed the landmarks map to). FacePointsInXSpace is an IReadOnlyDictionary<FacePointType, PointF>, where FacePointType refers to the title of the point (e.g., MouthCornerLeft) and PointF is a 2D coordinate with float members.

Generally, the points will be found within the bounding box, but it is better to ensure that they fit within the image confines anyway. The ValidateFaceBoundingBox(...) method from Listing 5-27 can be altered to accommodate this check.

Listing 5-29. Face Bounding Box and Points Validation

```
private bool ValidateFaceBoxAndPoints(FaceFrameResult faceFrameResult)
{
    bool isFaceValid = faceFrameResult != null;

    if (isFaceValid)
    {
        RectI boundingBox = faceFrameResult.FaceBoundingBoxInInfraredSpace;
        if (boundingBox != null)
        {
            [...] //Check if bounding box is valid

            if (isFaceValid)
            {
                var facePoints = faceFrameResult.FacePointsInInfraredSpace;
                if (facePoints != null)
                {
```

```
            foreach (PointF pointF in facePoints.Values)
            {
                bool isFacePointValid = pointF.X > 0.0f &&
                pointF.Y > 0.0f &&
                pointF.X < frameDescription.Width &&
                pointF.Y < frameDescription.Height;

                if (!isFacePointValid)
                {
                    isFaceValid = false;
                    break;
                }
            }
        }
    }
}
    return isFaceValid;
}
```

In Listing 5-29, we modified the ValidateFaceBoundingBox(...) method to conduct this check. The check consists of looping through the dictionary of face points and confirming that their x and y values fit in the image. On a side note, the method signature was also altered to reflect its new duties. Make sure to update the method call in Face_FrameFrameArrived(...).

Listing 5-30. Indicating Facial Landmarks on an Image Feed

```
...
drawingContext.DrawRectangle(null, drawingPen, faceBox);
//rest of DrawFace(...) body

if (faceFrameResult.FaceBoundingBoxInInfraredSpace != null)
{
    foreach (PointF pointF in faceFrameResult.FacePointsInInfraredSpace.Values)
    {
        drawingContext.DrawEllipse(null, drawingPen, new Point(pointF.X, pointF.Y), 0.4, 0.4);
    }
}
```

In Listing 5-30, we loop through every face point in `faceFrameResult` and draw an ellipse on the location of the landmark on the image. The final result looks like Figure 5-14.

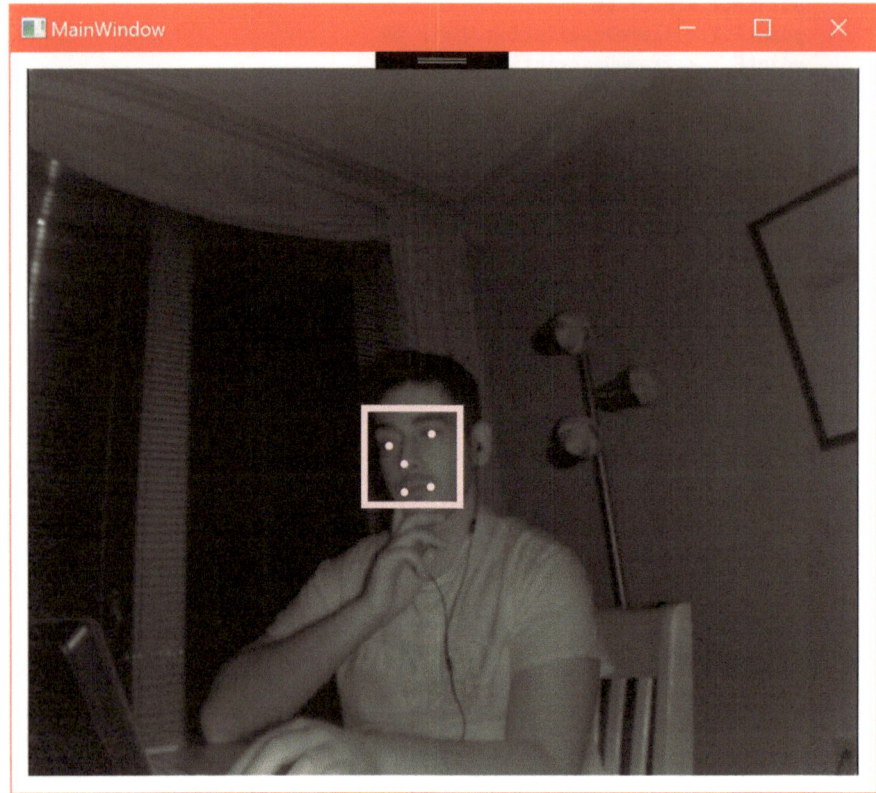

Figure 5-14. *Facial landmark locations drawn on an infrared image*

Orientation

The Face API gives us access to the face's orientation, facing the Kinect, in 3D space. It can be accessed through `FaceFrameResult`'s `FaceRotationQuaternion` property. It is defined with a Vector4. The face orientation is given as a quaternion, similar to joint orientations. This was again done to avoid gimbal lock. Face orientation can be useful in many circumstances, but gaze tracking deserves special mention. It is not possible to determine where a user is looking with the Kinect API, but knowing in which direction their face is pointing is good enough for most use cases and is the solution that most unspecialized commercial systems employ.

The Kinect SDK Browser's *FaceBasics-WPF* sample includes a convenient method by which to obtain a Euler angle in degrees from the face-rotation quaternion. If you are using a game engine or DirectX, you might choose to simply feed the quaternion into them directly, but if you are making a relatively simple app that does something like checking whether a head is tilting, using Euler angles with a heuristic system might be less work for you.

Listing 5-31. Method to Obtain Euler Angles from Face-Rotation Quaternion

```
/**usage:
int pitch, yaw, roll; ExtractFaceRotationInDegrees(faceFrameResult.FaceRotationQuaternion,
out pitch, out yaw, out roll);
**/
private static void ExtractFaceRotationInDegrees(Vector4 rotQuaternion, out int pitch, out
int yaw, out int roll)
{
    double x = rotQuaternion.X;
    double y = rotQuaternion.Y;
    double z = rotQuaternion.Z;
    double w = rotQuaternion.W;

    double yawD, pitchD, rollD;
    pitchD = Math.Atan2(2 * ((y * z) + (w * x)), (w * w) - (x * x) - (y * y) + (z * z)) /
    Math.PI * 180.0;
    yawD = Math.Asin(2 * ((w * y) - (x * z))) / Math.PI * 180.0;
    rollD = Math.Atan2(2 * ((x * y) + (w * z)), (w * w) + (x * x) - (y * y) - (z * z)) /
    Math.PI * 180.0;

    double increment = 5.0;
    pitch = (int)(Math.Floor((pitchD + ((increment / 2.0) * (pitchD > 0 ? 1.0 : -1.0))) /
    increment) * increment);
    yaw = (int)(Math.Floor((yawD + ((increment / 2.0) * (yawD > 0 ? 1.0 : -1.0))) /
    increment) * increment);
    roll = (int)(Math.Floor((rollD + ((increment / 2.0) * (rollD > 0 ? 1.0 : -1.0))) /
    increment) * increment);
}
```

The formulas in Listing 5-31 might look complex to the uninitiated, but you should not have to worry about their specifics. The first three formulas that calculate pitchD, yawD, and rollD are responsible for converting our quaternion into a Euler angle. The next three, used to calculate pitch, yaw, and roll, are clamping functions. They basically round the Euler angle values to the nearest multiple of the increment value. For example, if the pitch is actually 53.01666666 degrees and we set the increment to 5, it would appear as 55 degrees.

■ **Tip** Unfamiliar by what is conveyed here by *pitch*, *yaw*, and *roll*? Chances are you have seen these terms used to describe the direction of a plane, but might have forgotten which one refers to which axis. Pitch describes rotation around the x-axis, such as when a person is nodding in agreement; yaw describes rotation around the y-axis, such as when a person is shaking their head from side to side in disagreement; and roll describes rotation around the z-axis, such as when a person is tilting their head. When the pitch, yaw, and roll are all 0 degrees, the user is facing the Kinect camera directly.

Expressions

Expressions constitute the final portion of the Face API's capabilities. You might have noticed that the Body class has a handful of members, namely Activities, Appearance, Engaged, and Expressions, that are unusable. These were included for cross-compatibility with the Xbox Kinect API. However, we are able to access the same capabilities though the Face API. The Face API has a series of binary classifiers that can be used to determine whether a person is happy, wearing glasses, has their left or right eye closed, is looking away, is engaged with the application, if their mouth has moved, and if their mouth is open. In the case of looking away and engagement, the classifiers assume that the user is facing toward the Kinect. Thus, if you use these classifiers in your app, make sure that the Kinect is above or beneath the screen or location that the user is supposed to be looking at.

The expressions can be accessed from the FaceProperties property of FaceFrameResult. FaceProperties is an IReadOnlyDictionary<FaceProperty, DetectionResult>. FaceProperty is an enum that contains the title of the expression (e.g., Happy or Engaged), and DetectionResult is an enum that be Yes, No, Maybe, or Unknown to indicate whether the expression has been detected.

To demonstrate, let us display a smiley face on the screen when the user is happy.

Listing 5-32. Displaying a Smiley Face When a Face Indicates Happiness

```
if (faceFrameResult.FaceProperties != null)
{
    if (faceFrameResult.FaceProperties[FaceProperty.Happy] == DetectionResult.Yes)
    {
        Point nosePoint = new Point(faceFrameResult.FacePointsInInfraredSpace[FacePointType.Nose].X,
                    faceFrameResult.FacePointsInInfraredSpace[FacePointType.Nose].Y);
        drawingContext.DrawText(new FormattedText(
            "☺",
            System.Globalization.CultureInfo.GetCultureInfo("en-us"),
            FlowDirection.RightToLeft,
            new Typeface("Segoe UI"),
            68,
            drawingBrush),
            nosePoint);
    }
}
```

Listing 5-32 is a snippet of code that can be include in the DrawFace(...) method to draw a ☺ character on the screen when the DetectionResult of the Happy classifier holds a Yes value. We use the coordinates of the Nose face point to draw the character. The character's top-right corner, which consists of empty space, is set at the location of the nose. DrawingContext's DrawText(...) method requires a FormattedText and Point to output text onto the image. In practice, a TextBlock XAML control can offer more customization options than can those available through FormattedText, but it requires a bit more wiring. The completed work should look like Figure 5-15.

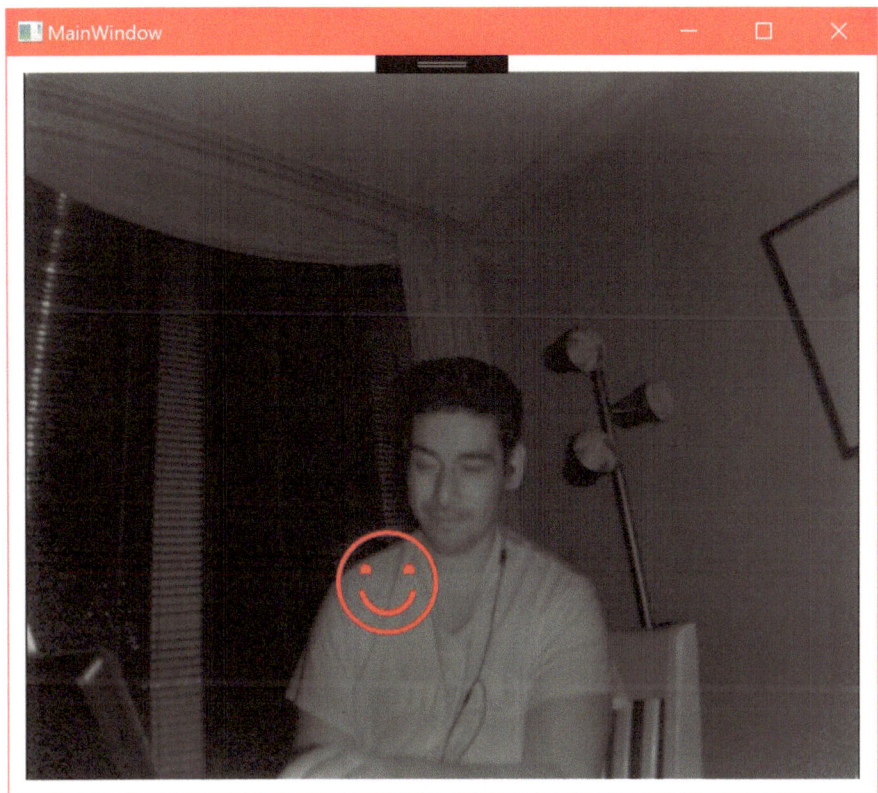

Figure 5-15. *Smiley face character displayed when face appears to be happy. The bounding box and facial landmark visualizations were omitted in this picture.*

■ **Tip** Avoid showing your teeth and looking directly toward the camera in an infrared picture so as to avoid looking like a heinous creature.

HD Face

For casual usage, the Face API will typically suffice. For precision work like modeling or tracking very specific parts of the face in realtime, the HD Face API will better serve you. Whereas the regular Face API performs tracking and classification work in 2D, HD Face is entirely in 3D and provides its coordinates as `CameraSpacePoints`. Its range is more limited than that of the Face API, at 0.5m to 2m (1.64 ft. to 6.56 ft.), and, like the Face API, facial tracking of only one user at a time per reader is supported.

The face tracked by HD Face is represented through 1,347 vertices, as portrayed in Figure 5-16. Anyone who has used face tracking in Kinect for Windows SDK v1 will remember that there used to be a mere hundred or so vertices and that we used to build entire commercial-grade applications out of them!

Figure 5-16. *HD Face vertices recreating a detected face*

Some of the more common vertices can be accessed through the `HighDetailFacePoints` enum and are listed in Table 5-1. This provides a manner in which to refer to vertices of interest (e.g., tip of nose) without resorting to remembering their index number in the vertices list.

Table 5-1. *Vertices Referable Through the HighDetailFacePoints Enumeration*

Key (preface with HighDetailFacePoints_)	Vertex Index
EyeLeft	0
LefteyeInnercorner	210
LefteyeOutercorner	469
LefteyeMidtop	241
LefteyeMidbottom	1104
RighteyeInnercorner	843
RighteyeOutercorner	1117
RighteyeMidtop	731
RighteyeMidbottom	1090
LefteyebrowInner	346
LefteyebrowOuter	140
LefteyebrowCenter	222
RighteyebrowInner	803
RighteyebrowOuter	758
RighteyebrowCenter	849
MouthLeftcorner	91
MouthRightcorner	687
MouthUpperlipMidtop	19
MouthUpperlipMidbottom	1072

(*continued*)

Table 5-1. (*continued*)

Key (preface with HighDetailFacePoints_)	Vertex Index
MouthLowerlipMidtop	10
MouthLowerlipMidbottom	8
NoseTip	18
NoseBottom	14
NoseBottomleft	156
NoseBottomright	783
NoseTop	24
NoseTopleft	151
NoseTopright	772
ForeheadCenter	28
LeftcheekCenter	412
RightcheekCenter	933
Leftcheekbone	458
Rightcheekbone	674
ChinCenter	4
LowerjawLeftend	1307
LowerjawRightend	1327

▪ **Note** The HighDetailFacePoint keys must be prefaced with HighDetailFacePoints_ when being used. For example, EyeLeft must be referred to as HighDetailFacePoints_EyeLeft in your code.

The face can also be recreated through 94 *shape units*. This entails taking a "standard" face model and applying shape unit deformations as well as a scaling factor. A standard face model is essentially a face model averaged from the faces of many, many individuals. This averaging has already been done by the Kinect team, so we can use the standard face model without further calibration. Shape units are numeric weights, typically around –2f to +2f, that indicate to the HD Face API how to deform the face in specific areas.

Other than tracking models of the face, HD Face has a few other nifty tricks up its sleeves. Like the Face API, it can provide the orientation of the face as a rotational quaternion. Additionally, it can provide the pivot point of the head. This is analogous to the head joint in the regular Body skeleton, but has a different vertical point to better describe the point at which the head pivots around. The hair color and skin color of a user are also provided by the Kinect as a property of the face model.

Finally, we have access to *animation units*. These are comparable to the expression classifiers for the Face API, but they describe a range instead of simply providing a binary classification. Typically, they provide values between 0f to 1f that depend on the progression of the animation. For example, for the JawOpen animation unit, a value of 0.2f would indicate that the jaw is barely open, whereas 0.93f would mean that that the jaw is practically open to the greatest possible extent. Three of the animation units—JawSlideRight, RightEyebrowLowerer, and LeftEyebrowLowerer—have numeric weights between –1f and +1f. This is because they describe motions that can be reversed in another direction. For example, when JawSlideRight has a value of –0.76f, the jaw is slid far along toward the left. The available animation units are: JawOpen, LipPucker,

JawSlideRight, LipStretcherRight, LipStretcherLeft, LipCornerPullerLeft, LipCornerPullerRight, LipCornerDepressorLeft, LipCornerDepressorRight, LeftcheekPuff, RightcheekPuff, LefteyeClosed, RighteyeClosed, RighteyebrowLowerer, LefteyebrowLowerer, LowerlipDepressorLeft, and LowerlipDepressorRight. Animation units can be applied on the face model to alter its appearance. Performing gestures on the face that are not covered by animation units will not be reflected in the face model. For example, since there is no animation unit for crinkling your nose, a 3D visualization of the face like the one in Figure 5-16 will not be shown to be crinkling its nose.

Obtaining HD Face data

Data procurement for HD Face is akin to that for any other Data Source. As you could have expected, there are a reader and a Data Source for HD face data, HighDefinitionFaceFrameReader and HighDefinitionFaceFrameSource, respectively, from which we collect frames containing data. The resulting HighDefinitionFaceFrame is responsible for the two central pieces of HD Face data: FaceModel and FaceAlignment.

FaceModel contains the structural details of the face. Namely, this includes the FaceShapeDeformations property, which is an IReadOnlyDictionary<FaceShapeDeformations, float> describing the shape units applied to the face. It also includes a TriangleIndices property, an IReadOnlyList<uint> that lists the 7,890 indices of the triangles used to create the face mesh for modeling; SkinColor and HairColor, which are 32-bit unsigned integers that describe a 4-byte ARGB color describing the skin color or hair color; Scale, a float representing the scaling factor of the face; and CalculateVerticesForAlignment(FaceAlignment faceAlignment), a method that furnishes a list of CameraSpacePoints representing facial vertices calculated from the FaceAlignment.

■ **Note** Hair and skin color can only be obtained after the face model has been "captured" using FaceModelBuilder. The process to do so is described in the "Face Capture" section that follows the current section.

On a side note, converting the uint values of SkinColor and HairColor to colors involves shifting their bits to expose the ARGB data. The following code will accomplish this:

```
uint color = faceModel.SkinColor;
Byte A = (byte) (color >> 24);
Byte B = (byte) (color >> 16);
Byte G = (byte) (color >> 8);
Byte R = (byte) (color >> 0);
Color skinColor = Color.FromArgb(A, R, G, B);
```

FaceAlignment contains orientation and animation unit data. Its members include AnimationUnits, which is an IReadOnlyDictionary<FaceShapeAnimations, float>, FaceShapeAnimations being an enum describing the animation unit FaceOrientation, a Vector4 representing a quaternion, and HeadPivotPoint, which is a CameraSpacePoint. Since the locations of vertices depend on how the face model is rotated, in practice it is more critical to our application than is FaceModel, for which we can just use the default average face model provided by the Kinect.

Listing 5-33. Private Variable Declaration to Obtain HD Face Data

```
private KinectSensor kinect;

private BodyFrameReader bodyFrameReader;

private HighDefinitionFaceFrameSource HDFaceFrameSource;
private HighDefinitionFaceFrameReader HDFaceFrameReader;
private FaceModel faceModel;
private FaceAlignment faceAlignment;
```

Listing 5-33 features the essential private variables for HD face tracking. We see that body data is to be tracked in addition to face data. While this was imperative for regular face tracking, it is downright necessary for HD face tracking. Since only one face can be tracked in HD, the TrackingId of the face must be manually picked from one of the available bodies.

Listing 5-34. Initializing HD Face Data Source and Reader

```
HDFaceFrameSource = new HighDefinitionFaceFrameSource(Kinect);
HDFaceFrameReader = HDFaceFrameSource.OpenReader();
HDFaceFrameReader.FrameArrived += HDFace_FrameArrived;

faceAlignment = new FaceAlignment();
```

In Listing 5-34, HD Face's Data Source is shown to be perhaps the simplest no-frills Data Source to get running. Additionally, we create an instance of the FaceAlignment object to hold its data throughout the lifetime of the application. Like with an array of Bodies, FaceAlignment will be updated instead of being re-initialized on each HD Face frame.

Listing 5-35. BodyFrameReader Event Handler for HD Face Tracking

```
private void Body_FrameArrived(object sender, BodyFrameArrivedEventArgs e)
{
    using (BodyFrame bodyFrame = e.FrameReference.AcquireFrame())
    {
        if (bodyFrame != null)
        {
            Body[] bodies = new Body[bodyFrame.BodyCount];
            bodyFrame.GetAndRefreshBodyData(bodies);

            Body body = bodies.Where(b => b.IsTracked).FirstOrDefault();

            if (!HDFaceFrameSource.IsTrackingIdValid)
            {
                if (body != null)
                {
                    HDFaceFrameSource.TrackingId = body.TrackingId;
                }
            }
        }
    }
}
```

155

In Listing 5-35, our goal is to determine which body's face to track. We otherwise do not need to touch the skeleton data. Using a bit of LINQ, we pick the first available body, and if the HighDefinitionFaceFrameSource is not currently tracking a body, we assign it a body to track.

Listing 5-36. HighDefinitionFaceFrameReader Event Handler

```
private void HDFAce_FrameArrived(object sender, HighDefinitionFaceFrameArrivedEventArgs e)
{
    using (HighDefinitionFaceFrame HDFrame = e.FrameReference.AcquireFrame())
    {
        if (HDFrame!= null && HDFrame.IsFaceTracked)
        {
            HDFrame.GetAndRefreshFaceAlignmentResult(faceAlignment);
            if (faceModel == null) {
                faceModel = HDFrame.FaceModel;

            }
            ProcessHDFaceData();
        }
    }
}
```

In Listing 5-36, we see that FaceAlignment is collected through the GetAndRefereshFaceAlignmentRe sult(FaceAlignment faceAlignment) method of HighDefinitionFaceFrame. If we so choose, we can also grab the FaceModel, but this is not necessary. The standard face model provided by the Kinect could be initialized in the constructor and coupled with the FaceAlignment to provide a decent representation of a user's face. After this point, we are free to process our data for whatever purpose.

It should be noted that HighDefinitionFaceFrame has an additional property, FaceAlignmentQuality, which is an enum with *High* and *Low* values to describe how accurate the obtained FaceAlignment is.

Listing 5-37. Processing HD Face Data

```
private void ProcessHDFaceData()
{
    if (faceModel != null)
    {
        var vertices = faceModel.CalculateVerticesForAlignment(faceAlignment);
        //Perform work with vertices, such as displaying on screen
    }
}
```

Other than the vertices, the rest of HD Face's data can be gleaned from FaceModel and FaceAlignment directly. As mentioned earlier, the position of the vertices in a 3D space depends on how their constituent model is rotated in 3D space. Hence, in Listing 5-37 we must calculate them with the CalculateVertic esForAlignment(FaceAlignment faceAlignment) method. This is not a static method; it does depend on the data of a FaceModel instance. However, it can be used on the data in the default initialization of FaceModel without gathering anything from HighDefinitionFaceFrames. Despite this possibility, HD Face unsurprisingly shines brightest when it gets a chance to reconstruct a FaceModel from an analysis of a user's face, which we will cover next.

Face Capture & Modeling

With the vertices provided from FaceModel and FaceAlignment, we can create a 3D mesh of the user's face by connecting them into triangles. To get a more accurate representation, however, we have to use the FaceModelBuilder class to produce a FaceModel with shape unit deformations applied. FaceModelBuilder needs to capture 16 frames from all sides of the face. Thus, with the help of the API, you will have to inform users of your app when to move their faces to be properly captured. Your users should take off any accessories on or near their faces, such as glasses or hats.

To learn how to use FaceModelBuilder and create a mesh with or without it, we will turn to the *HDFaceBasics-WPF* sample in the SDK Browser. This app (shown in Figure 5-17) creates a 3D visualization of a user, and once face capture has been completed, this visualization is deformed to look more like the user's face.

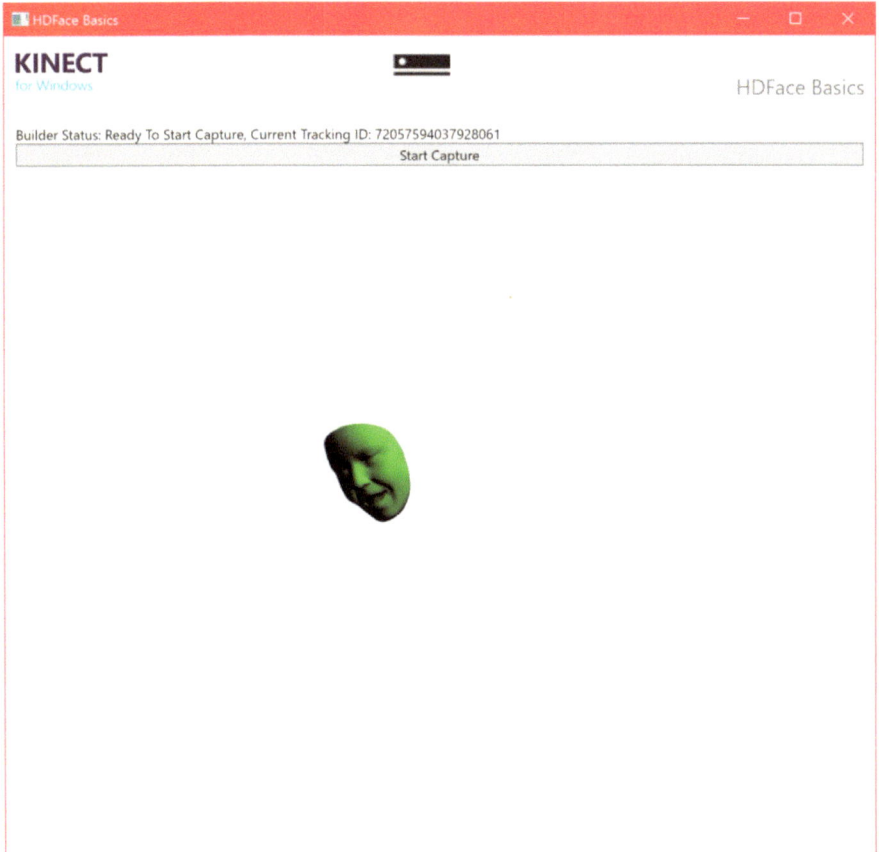

Figure 5-17. *3D visualization created by HD Face. User face has not yet been captured*

In Figure 5-17, we see that HD Face is *Ready To Start Capture*. Pressing *Start Capture* will fire off FaceModelBuilder, which will instruct the user in which direction to tilt their face in the *Builder Status* Text Block.

157

In the `InitializeHDFace()` method of `MainWindow.xaml.cs`, the process of creating a 3D mesh is started off with the `InitializeMesh()` and `UpdateMesh()` methods. These methods take the vertices, create triangles from them, and then apply them to a `MeshGeometry3D` object that is responsible for the 3D visualization we see on the front end. `MeshGeometry3D` is a part of the `System.Windows.Media.Media3D` namespace that allows us to build simple 3D applications in WPF (the functionalities of this library are sometimes colloquially referred to as *WPF 3D*).

Listing 5-38. Initializing a 3D Mesh from HD Face Data

```
private void InitializeMesh()
{
    var vertices = this.currentFaceModel.CalculateVerticesForAlignment(this.currentFaceAlignment);

    var triangleIndices = this.currentFaceModel.TriangleIndices;

    var indices = new Int32Collection(triangleIndices.Count);

    for (int i = 0; i < triangleIndices.Count; i += 3)
    {
        uint index01 = triangleIndices[i];
        uint index02 = triangleIndices[i + 1];
        uint index03 = triangleIndices[i + 2];

        indices.Add((int)index03);
        indices.Add((int)index02);
        indices.Add((int)index01);
    }

    this.theGeometry.TriangleIndices = indices;
    this.theGeometry.Normals = null;
    this.theGeometry.Positions = new Point3DCollection();
    this.theGeometry.TextureCoordinates = new PointCollection();

    foreach (var vert in vertices)
    {
        this.theGeometry.Positions.Add(new Point3D(vert.X, vert.Y, -vert.Z));
        this.theGeometry.TextureCoordinates.Add(new Point());
    }
}
```

In Listing 5-38, we have all the code necessary to turn vertices into a 3D mesh. After grabbing the vertices and triangle indices of the standard model face, we add the triangle indices to an `Int32Collection` for use by a `MeshGeometry3D` instance. This will indicate which vertices create which triangles.

When adding the triangle indices to the `Int32Collection`, we add them in reverse order (i.e., index03, index02, index01). This is because WPF 3D creates front-facing mesh triangles in counterclockwise order. Setting them in clockwise would make the mesh triangles face backward.

After setting the indices to the `MeshGeometry3D` (known here as `theGeometry` and initialized in the XAML code), we add the vertices. The z coordinate of the vertices is made negative so that the face appears behind the screen, as the positive z-axis sticks out of the screen. `TextureCoordinates` are set as null values, as we do not intend to texture our face.

Listing 5-39. Updating a 3D Mesh from HD Face Data

```
private void UpdateMesh()
{
    var vertices = this.currentFaceModel.CalculateVerticesForAlignment(this.currentFaceAlignment);

    for (int i = 0; i < vertices.Count; i++)
    {
        var vert = vertices[i];
        this.theGeometry.Positions[i] = new Point3D(vert.X, vert.Y, -vert.Z);
    }
}
```

Since the triangle indices are already set into the MeshGeometry3D, we only need to update the position of the vertices over time. To accomplish this, UpdateMesh() (Listing 5-39) is called every time HighDefinitionFaceFrameReader provides a new frame.

Listing 5-40. Initializing FaceModelBuilder

```
private void StartCapture()
{
    this.StopFaceCapture();

    this.faceModelBuilder = null;

    this.faceModelBuilder = this.highDefinitionFaceFrameSource.OpenModelBuilder(FaceModelBui
    lderAttributes.None);

    this.faceModelBuilder.BeginFaceDataCollection();

    this.faceModelBuilder.CollectionCompleted += this.HdFaceBuilder_CollectionCompleted;
}
```

In Listing 5-40, the StartCapture() method encapsulates the initializing process of FaceModelBuilder. StopFaceCapture() disposes of any existing FaceModelBuilder so that we can start fresh. We assign the FaceModelBuilder by opening it from the HighDefinitionFaceFrameSource's OpenModelBuilder(FaceMode lBuilderAttribute enabledAttributes) method. The other parameters we can include are HairColor and SkinColor. If we wanted both, we could do OpenModelBuilder(FaceModelBuilderAttribute.HairColor | FaceModelBuilderAttribute.SkinColor).

The FaceModelBulder is started with its BeginFaceDataCollection() method, and we attach an event handler to its CollectionCompleted event so that we can tell it to deform the face after the new FaceModel has been created from it.

Listing 5-41. Creating a New FaceModel from FaceModelBuilder's Analysis

```
private void HdFaceBuilder_CollectionCompleted(object sender,
FaceModelBuilderCollectionCompletedEventArgs e)
{
    var modelData = e.ModelData;

    this.currentFaceModel = modelData.ProduceFaceModel();

    this.faceModelBuilder.Dispose();
    this.faceModelBuilder = null;

    this.CurrentBuilderStatus = "Capture Complete";
}
```

159

Once the `FaceModelBuilder` has finished capturing all its requisite frames (Listing 5-41), it provides us with a `FaceModelData` object whose `ProduceFaceModel()` method creates the more tailored `FaceModel` that we desire. It will be used by `UpdateMesh()` automatically to alter the mesh.

Listing 5-42. Determining FaceModelBuilder's Status

```
private void CheckOnBuilderStatus()
{
    if (this.faceModelBuilder == null)
    {
        return;
    }

    string newStatus = string.Empty;

    var captureStatus = this.faceModelBuilder.CaptureStatus;
    newStatus += captureStatus.ToString();

    var collectionStatus = this.faceModelBuilder.CollectionStatus;

    newStatus += ", " + GetCollectionStatusText(collectionStatus);

    this.CurrentBuilderStatus = newStatus;
}
```

Two final interesting bits of `FaceModelBuilder` functionality are its `CaptureStatus` and `CollectionStatus` properties (Listing 5-42). They provide capture-quality warnings and details as to what frames are left to capture, respectively. The `FaceModelBuilderCaptureStatus` enum contains the `OtherViewsNeeded`, `LostFaceTrack`, `FaceTooFar`, `FaceTooNear`, `GoodFrameCapture`, and `SystemError` values. The `FaceModelBuilderCollectionStatus` enum contains the `MoreFramesNeeded`, `FrontViewFramesNeeded`, `LeftViewsNeeded`, `TiltedUpViewsNeeded`, and `Complete` values. These can be used to provide tips to users on how they should position their faces to complete face capture.

Conclusion

In this chapter, we conducted a comprehensive overview of the various body- and face-tracking features of the Kinect. We learned how these tracking features were segmented into the `BodyFrameSource`, `BodyIndexFrameSource`, `FaceFrameSource`, and `HighDefinitionFaceFrameSource` Data Sources. While we have not yet touched upon gestures, by learning about the properties and members of the various Data Sources, we can see how they might come to be used heuristically. Fortunately, the Kinect includes some tools that will help us avoid such tedious and manual labor.

This particular conclusion is more than just the parting words of Chapter 5. It marks a symbolic culmination of what can be considered Part I of this book. We spent the last several chapters understanding and extricating data from the various sensors of the Kinect. These are the fundamental building blocks of Kinect applications, and if you were to stop reading at this point (please don't), you would be able to build your own Kinect solutions from scratch with knowledge of practically the full API.

Building Kinect applications consists of more than just what is in the Kinect, however. How its data gets processed by other technologies, such as computer vision, machine learning, and game engines, will be more responsible for whether your solution is a simple Kinect data viewer or a full-fledged consumer- or commercial-grade application. In the next chapter, we will begin looking at computer-vision and image-processing techniques using the Kinect so that we can derive meaning from otherwise flat images.

CHAPTER 6

▦ ▦ ▦

Computer Vision & Image Processing

Hopefully, by this point some of the Kinect's magic has worn off and you can see the machine for what it really is: two cameras with varying degrees of sophistication and a laser pointer. Barring the exorbitant cost of fielding a time of flight (ToF) camera, recreating a device that is conceptually similar to the Kinect in your own garage is not impossible. Getting the color, depth, and infrared streams from it could be technically challenging, but is essentially a solved problem. What sets the Kinect apart from such a hobbyist device, however, other than its precision manufacturing and marginally superior components, is its capability to look at the depth feed and extract bodies and faces from it.

"So what?" you say. "Is that not why we ordered a Kinect from the Microsoft Store instead of digging out our Arduinos and requesting chip and sensor samples from TI?" What if I told you that you too, could track bodies, faces, dogs, cats, stop signs, paintings, buildings, golden arches, swooshes, other Kinects, and pretty much anything else you can imagine without having to rely on some XDoohickeyFrameSource? That the 25 detected body joints are merely software abstractions that you could have devised yourself? There is no sacred sensor or chip that can track things like human bodies on its own. These are all capabilities that you can code yourself, which you should in fact learn how to do if you intend to get the most out of your Kinect.

What the Kinect is, aside from its various sensors, is essentially a bag of *computer vision* and *image processing* algorithmic tricks. Not to disparage its hardware (it is nothing to scoff at), but these algorithms are the reason we write home about the Kinect.

When we say computer vision, we mean algorithms that help computer programs form decisions and derive understanding from images (and by extension, videos). Are there soccer balls in this picture, and if so, how many? Is the Indian elephant in the video clip crying or merely sweaty? Do the seminiferous tubule cross-sections in the microscope slides have a circularity > 0.85, and if so, can they be circled?

▦ **Note** Elephants do not sweat, because they do not have sweat glands. This is why they are often seen fanning their ears and rolling in the mud to cool down instead.

Image processing, on the other hand, is the application of mathematical operations on images to transform them in a desired manner. Such algorithms can be used to apply an Instagram filter, perform red-eye removal, or subtract one video frame from another to find out what areas of a scene have changed over time. Image processing techniques are themselves often used in computer vision procedures.

© Mansib Rahman 2017

M. Rahman, *Beginning Microsoft Kinect for Windows SDK 2.0*, DOI 10.1007/978-1-4842-2316-1_6

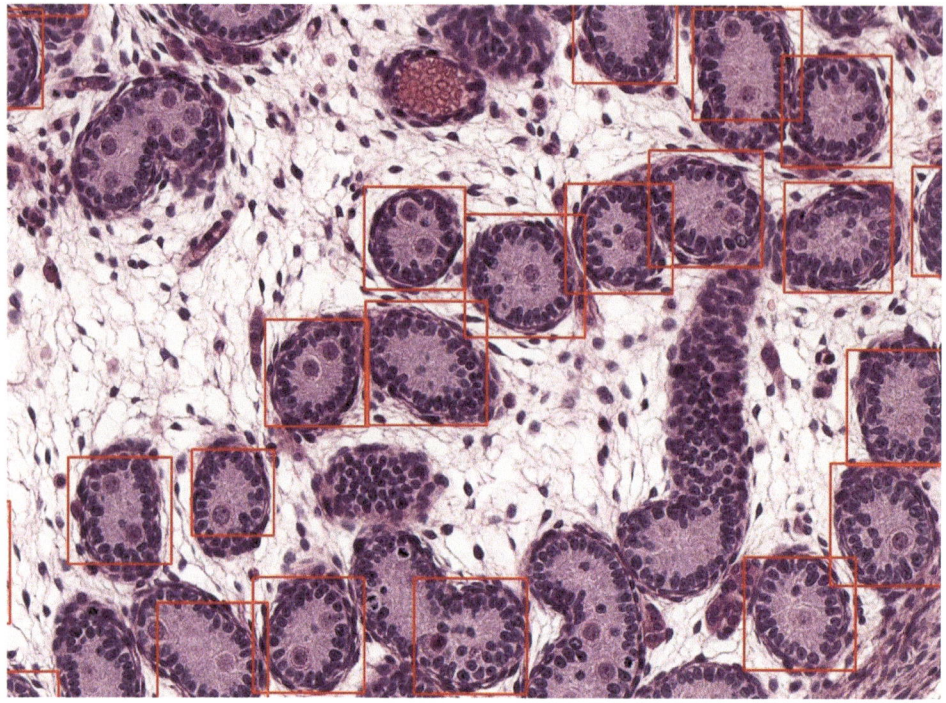

Figure 6-1. *Seminiferous tubule cross-sections in male rat reproductive organs with circularities > 0.85 highlighted using computer vision techniques (base image provided courtesy of the lab of Dr. Bernard Robaire at McGill University's Faculty of Medicine)*

Algorithms and mathematical operations may cause consternation for some, but there is no need to worry; you will not immediately need to know how to convolute a kernel around an image matrix (though it helps). Researchers and developers have pooled together to create numerous open source libraries that implement a wide variety of computer vision and image processing techniques for us to freely use. The most popular one is likely *OpenCV*, Intel's BSD-licensed library, which is what we will be using in this chapter. There are many other excellent libraries that can be used, however, such as *AForge.NET* and *SimpleCV*. Although the exact implementations will be different among the libraries, the general idea can be replicated with any of them.

OpenCV's primary interface is in C++, but there are bindings for Python and Java. For .NET usage, we must rely on a wrapper. We will be using the *Emgu CV* wrapper, which in addition to C# supports VB, VC++, IronPython, and other .NET-compatible languages and is cross-platform, supporting not only the typical desktop OS, but also iOS, Android, Windows Phone, Windows Store, and frameworks such as Unity and Xamarin. Whenever we refer to Emgu CV code and structures in this chapter, assume there are equivalents in OpenCV.

■ **Tip**   When we say Emgu CV is a .NET wrapper for OpenCV, *wrapper* refers to a set of classes that can be used to call unmanaged code from a managed context. This allows us to call C++ OpenCV-like method signatures from within our C# code.

Computer vision is a massive topic, one that can span entire bookshelves and university courses. This chapter can only cover the equivalent of a tuft of hair on a western lowland gorilla's armpit's worth. While you might not be able to create a system that captures a couple of dozen body joints from a cat as fluidly as the Kinect does with a human body, at least not off the bat, hopefully this will whet your appetite for computer vision and image processing and inspire you to develop applications with basic tracking-analysis capabilities that you might not have been able to accomplish with just heuristic methods.

■ **Note**   Seeing as it is impossible to cover the entire OpenCV/Emgu CV library in one chapter, I recommend that you consult the official Emgu CV API reference at `http://www.emgu.com/wiki/files/3.1.0/document/index.html` to augment your learning. Additional material on OpenCV can be found at `http://docs.opencv.org/3.1.0`.

Setting Up Emgu CV

The process of setting up Emgu CV as a dependency is perhaps its biggest pain point. Like bullet casings and unexploded ordinance littering former battlefields, you will find scores of unanswered questions on Stack Overflow, the Emgu CV forums, and MSDN, often on the issue of "The `type initializer for 'Emgu.CV.CvInvoke' threw an exception`," which is code for "You bungled your Emgu CV dependencies." *C'est la vie d'un développeur*. Fortunately, we have *decent* technical books to guide us through the process.

System Specs

Neither Emgu CV nor OpenCV has much in the way of publicly available system specifications. You would think that algorithms that are iterating dozens of times on multi-million-pixel image frames would be very CPU intensive, and you would not be wrong. But OpenCV has been around since the Windows 98 era and can run just fine on an iPhone 4's 800MHz ARM CPU. Generally, you will want to use something above the system specs for the Kinect for Windows SDK v2 so that there are CPU cycles to spare for your image processing work after expending the necessary system resources for the Kinect. If you plan to use OpenCV's CUDA module, you will need to have a CUDA-enabled GPU, which means NVIDIA.

■ **Tip**   *CUDA*, which formerly meant Compute Unified Device Architecture but is now no longer an acronym, is a platform and API that allows supported GPUs to be used for many tasks typically handled by the CPU. This is very nifty for image processing tasks because of the high number of cores on GPUs. CUDA allows the GPU to be exploited through a C/C++ or Fortran API. OpenCV (and by extension Emgu CV) has a module that allows us to parallelize many tasks with CUDA-enabled GPUs.

Installation & Verification

Emgu CV can be downloaded from SourceForge at `https://sourceforge.net/projects/emgucv/`.

At the time of writing, the current latest version of Emgu CV is **3.1.0**, and that is the version the book's code and instructions will be tested against. There *should* be no breaking changes in future versions, but your mileage may vary. The particular file downloaded from SourceForge was `libemgucv-windesktop-3.1.0.2282.exe`, and the installer was run with all its default configuration options.

Once the installer finishes, verify that it put everything in the right place. Visit the installation folder (the default location is C:\Emgu). Then, navigate to emgucv-windesktop 3.1.0.xxxx ➤ Solution ➤ VS2013-2015 and open the Emgu.CV.Example Visual Studio solution.

By default, the Hello World sample program should be loaded (if not, you should skip down to Figure 6-3 to see how you can load it). After compiling and running, you should see a blue screen emblazoned with the words "Hello, world," as in Figure 6-2.

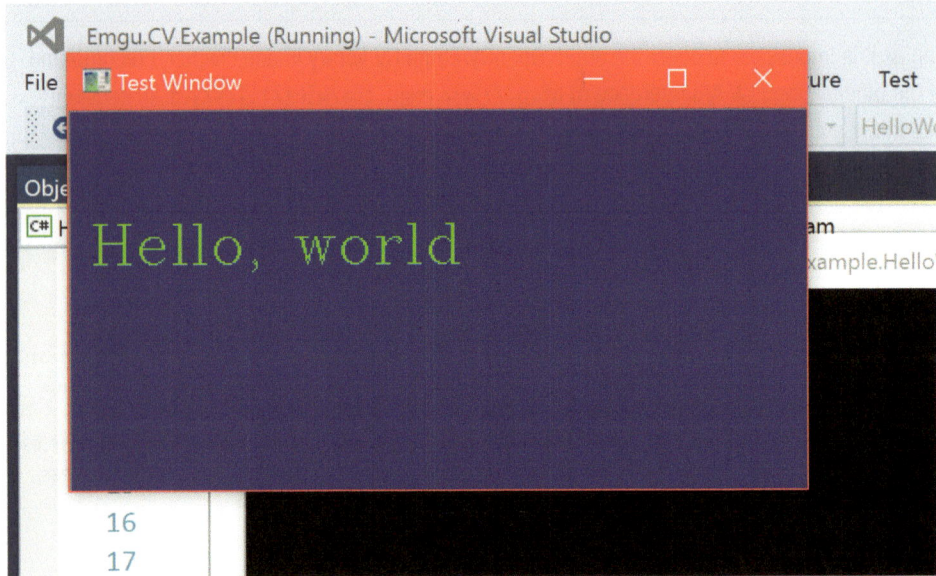

Figure 6-2. *Hello World sample in Emgu CV*

The sample might look like something that could have been achieved trivially with WPF or WinForms, but all of it is in fact made from calls to the Emgu CV API. The blue background is actually an Emgu CV image, which we will discuss further in the next section.

In the example solution, we have other samples that can be experimented with. Let us run the **CameraCapture** project to see some basic image processing techniques at work. To change the active project in Visual Studio, visit the *Startup Projects* picker on the toolbar and select your desired project, as in Figure 6-3.

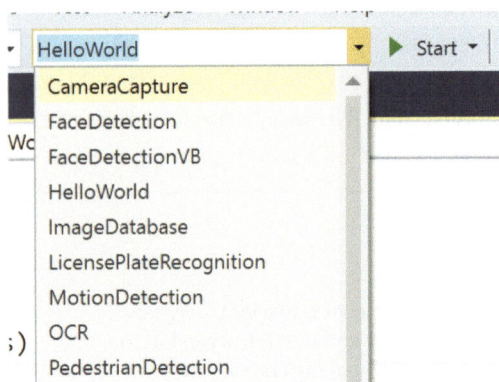

Figure 6-3. *Choosing the CameraCapture project from among others in the Startup Projects picker*

Your computer will have to have a webcam configured in order for the project to run. For devices such as the Surface Book or Surface Pro, the sample works out of the box. Unless you hacked your Kinect to be used as a webcam, Emgu CV does not get screens from it directly, because we have not connected it to the sample project. The compiled result of the CameraCapture sample should look something like Figure 6-4.

Note   As of Windows 10, it is actually possible to use the Kinect as a regular webcam. As a bonus, with the Kinect's IR capabilities, we can use it for Windows Hello (though, if you ask me, it is easier to simply use an IR-enabled laptop camera like those found in the Surface line of products).

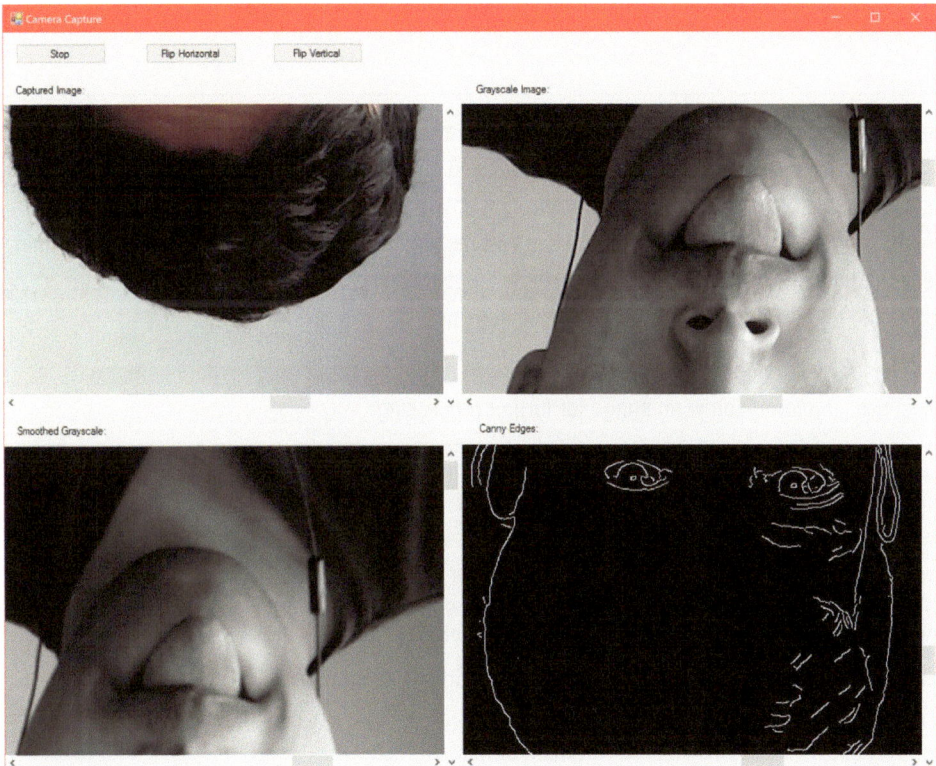

Figure 6-4. *Playing with CameraCapture in the Emgu CV samples. Because of the large resolution of the image, there are scroll bars to pan it.*

In Figure 6-4, we see a snapshot of the application livestreaming footage from a webcam. The captured video frames are processed with different techniques and presented for our viewing pleasure. On the top right, we have the regular webcam image in grayscale. Converting a color image to grayscale is one of the most basic image processing techniques. If you recall in Chapter 3, manipulating images involved iterating through pixels of the image and applying functions to alter their color components. With Emgu CV, one can just call `CvInvoke.CvtColor(initialImage, finalImage, ColorConversion.Bgr2Gray)` to accomplish the same thing.

> ■ **Tip** Typically, when we grayscale an image we have to go from a three-color channel image (RGB) to one with only one (Gray). One method of doing this is by summing the RGB components of a pixel and dividing them by three. Human eyes do not perceive the three colors equally, however, so it is better to weigh the three components differently. A common approach is $gray\ pixel\ intensity = \dfrac{0.3 \times R + 0.59 \times G + 0.11 \times B}{3}$, where R, G, and B represent the intensities of the Red, Green, and Blue pixel color components, respectively.

The smoothed gray image on the bottom left is the regular grayscale image blurred, *downsampled*, blurred, and then *upsampled*. The bottom right image has the *Canny edge detector* algorithm applied to the regular image with the detected edges highlighted in the output image. We will explore similar algorithms in more depth later in the chapter. If the effects on your images look similar to the ones found in this book, then your installation should be fine.

> ■ **Tip** Downsampling is essentially a way to reduce the information in an image so that we can shrink the image itself. Upsampling is increasing the size of the image based on an interpolation of what colors the new pixels should be. It can never quite recover details lost through downsampling. The simplest way to explain the difference between *resampling* an image and *resizing* it is that resampling changes the number of pixels within an image, whereas resizing changes the size of the image without changing the number of pixels. In other words, upsizing an image magnifies it, whereas as resizing does not.

The Canny edge detector is an algorithm used to detect edges in an image, often used in more complex algorithms to select salient information in an image for further processing.

Including Emgu CV as a Dependency

Things have gotten easier with the advent of the 3.1.x releases of Emgu CV, as many of the DLLs have been consolidated, but it still has the potential to trip you up. For the current version, the following instructions should set everything up correctly, but these may be subject to change in future versions.

1. Create a new blank WPF project in Visual Studio and give it a name.

2. Set the project's platform to x64 (Emgu CV works with x86 too, however). See the **Note** for Figure 5-12 in Chapter 5 if you need help recalling how.

3. Compile the solution.

4. Grab `Emgu.CV.UI.dll` and `Emgu.CV.World.dll` from the *emgucv-windesktop 3.1.0.2282\bin* folder and copy them to the *<ProjectName>\bin\x64\Debug* folder.

5. Copy the contents of the *emgucv-windesktop 3.1.0.2282\bin\x64* folder and paste it into the *<ProjectName>\bin\x64\Debug* folder.

6. Open the Reference Manager window from Solution Explorer, then browse and add `Emgu.CV.UI.dll` and `Emgu.CV.World.dll` as references.

7. In `MainWindow.xaml.cs`, ensure that you include the `Emgu.CV` namespace.

■ **Note** If you change from a Debug to a Release configuration, you will need to ensure that all the DLLs are in the *Release* folder as well.

To verify that everything has installed properly, try running the code in Listing 6-1. To run it successfully, you need to include any random image named img.jpg in the project's *x64\Debug* folder.

Listing 6-1. Load an Image into WPF with Emgu CV

```
using System.Windows;
using Emgu.CV;

namespace KinectEmguCVDependency
{
    public partial class MainWindow : Window
    {
        public MainWindow()
        {

            Mat img = CvInvoke.Imread("img.jpg", Emgu.CV.CvEnum.LoadImageType.AnyColor);
            /*
            in later versions, the correct code might be: Mat img = CvInvoke.Imread
            ("img.jpg", Emgu.CV.CvEnum.ImreadModes.AnyColor);
            */
        }
    }
}
```

The code in Listing 6-1 does not do anything except load an image to be used for whatever purpose. Instead of relying on the default WPF classes, however, it uses an Emgu CV call: Cv.Invoke.Imread(...). If you get no errors and a blank screen, this means everything is in place.

Manipulating Images in Emgu CV

As we saw in Chapter 3, it is entirely possible—and, in fact, not too difficult—to fiddle with the individual data bytes of an image. For more advanced usage with computer-version libraries such as OpenCV, however, we must rely on somewhat more abstracted versions of byte arrays that allow algorithms to take advantage of the image's metadata.

Understanding Mat

OpenCV has been around since the turn of the millennia; back then, it was still all in C. Images were primarily represented through the IplImage struct, otherwise known as Image<TColor, TDepth> in Emgu CV. With the advent of C++, OpenCV saw IplImage gradually phased out in favor of the object-oriented Mat class (also known as Mat in Emgu CV). A lot of old code and textbooks rely on the IplImage, but as of OpenCV 3.0, Mat is officially preferred.

Byte arrays representing images are essentially matrices, hence the shorthand *Mat* for OpenCV's image byte array representation. `Mat` contains two parts, a matrix header that contains metadata (e.g., how many color channels, rows, cols, etc.) and a pointer to the location of the actual matrix. This distinction is made because computer vision work naturally entails performing very memory-intensive tasks, and it would be very inefficient to copy the actual matrices all the time. Hence, if you create a `Mat` from another `Mat`'s data, they share a reference to the same `Mat`. A `Mat` must be explicitly duplicated with its `Clone()` method to establish two separate references.

Using Mat with Kinect Image Data

The old `Image<TDepth, TColor>` class required a handful of extension methods to get working with the Kinect. Using `Mat` is considerably easier and only requires a few extra lines as opposed to an entire class's worth of methods like before.

Listing 6-2. Converting Kinect Image Data to Mat Format

```
using System;
using System.Windows;
using System.Windows.Media;
using System.Windows.Media.Imaging;
using System.Runtime.InteropServices;
using Microsoft.Kinect;
using Emgu.CV;

namespace KinectImageProcessingBasics
{
    public partial class MainWindow : Window
    {
        [DllImport("kernel32.dll", EntryPoint = "CopyMemory", SetLastError = false)]
        public static extern void CopyMemory(IntPtr dest, IntPtr src, uint count);

        private KinectSensor kinect;
        private FrameDescription colorFrameDesc;
        private WriteableBitmap colorBitmap;

        public MainWindow()
        {
            kinect = KinectSensor.GetDefault();
            ColorFrameSource colorFrameSource = kinect.ColorFrameSource;
            colorFrameDesc = colorFrameSource.FrameDescription;
            ColorFrameReader colorFrameReader = colorFrameSource.OpenReader();
            colorFrameReader.FrameArrived += Color_FrameArrived;
            colorBitmap = new WriteableBitmap(colorFrameDesc.Width,
                colorFrameDesc.Height,
                96.0,
                96.0,
                PixelFormats.Bgra32,
                null);

            DataContext = this;
```

```csharp
        kinect.Open();

        InitializeComponent();
    }

    public ImageSource ImageSource
    {
        get
        {
            return colorBitmap;
        }
    }

    private void Color_FrameArrived(object sender, ColorFrameArrivedEventArgs e)
    {
        using (ColorFrame colorFrame = e.FrameReference.AcquireFrame())
        {
            if (colorFrame != null)
            {
                if ((colorFrameDesc.Width == colorBitmap.PixelWidth) &&
                (colorFrameDesc.Height == colorBitmap.PixelHeight))
                {
                    using (KinectBuffer colorBuffer = colorFrame.LockRawImageBuffer())
                    {
                        colorBitmap.Lock();

                        Mat img = new Mat(colorFrameDesc.Height, colorFrameDesc.Width,
                        Emgu.CV.CvEnum.DepthType.Cv8U, 4);
                        colorFrame.CopyConvertedFrameDataToIntPtr(
                        img.DataPointer,
                        (uint)(colorFrameDesc.Width * colorFrameDesc.Height * 4),
                        ColorImageFormat.Bgra);
                        //Process data in Mat at this point
                        CopyMemory(colorBitmap.BackBuffer, img.DataPointer, (uint)
                        (colorFrameDesc.Width * colorFrameDesc.Height * 4));

                        colorBitmap.AddDirtyRect(new Int32Rect(0, 0, this.colorBitmap.
                        PixelWidth, this.colorBitmap.PixelHeight));

                        colorBitmap.Unlock();
                        img.Dispose();
                    }
                }
            }
        }
    }
}
```

MainPage.xaml remains the same, however:

```
<Grid>
        <Image Source="{Binding ImageSource}" Stretch="UniformToFill" />
</Grid>
```

In Listing 6-2, we have an entire application that grabs a color image frame from the Kinect, transposes its data to a Mat, and then pushes the Mat's data into a WriteableBitmap for display. Naturally, we must include the Emgu.CV namespace, but we also include the System.Runtime.InteropServices namespace because we will be relying on a C++ function to deal with Mat data.

The function in question is CopyMemory(IntPtr dest, IntPtr src, uint count), which will copy image data from our Mat to the WriteableBitmap's BackBuffer. To use it, we have to call unmanaged code from the WIN32 API. We thus rely on the DllImport attribute to retrieve the CopyMemory(...) function from kernel32.dll in [DllImport("kernel32.dll", EntryPoint = "CopyMemory", SetLastError = false)].

A new Mat object is created every time the Color_FrameArrived(...) event handler is called, as opposed to reusing the same one, because we want to prevent access to its data by two concurrent calls of the event handler. Mat has nine different constructors, which allows it to be initialized from an image saved to disk, or simply be empty with some metadata. In our case, its parameters in order are rows, columns, Emgu.CV.CvEnum.DepthType, and channels. Rows and columns are essentially equivalent to the height and width of the image in pixels. DepthType refers to how much data each channel of a pixel can hold. Cv8U refers to eight unsigned bits, also known as bytes. The channels value is set to 4 for R, G, B, and A. With the CopyConvertedFrameDataToIntPtr(...) method, we copy the Kinect's color image data directly from its buffer to the Mat's buffer.

In this example, we directly copy data from the Mat to the WriteableBitmap with CopyMemory(...). In a production scenario, we would have performed some processing or algorithmic work first. Ideally, this work would have been performed on a background worker thread for performance reasons, including writing to a WriteableBitmap, but for simplicity this was avoided here. On an i5 Surface Book, the code results in performance comparable to the official *ColorBasics-WPF* sample, with an additional ~50 MB of RAM usage. After we are done with the Mat object, we make sure to dispose of it to prevent a memory leak.

Although the code described in Listing 6-1 showed the basic usage of the Mat class for a Kinect application, we did not even have to use it to the extent that we did. We can also display the Mat directly with Emgu CV's GUI API. In fact, it is not necessary to render the algorithmic results from the Mat to any image at all. Once we obtain our data or decision, we can throw out the Mat and render the results or decisions on a Canvas or to the Kinect image data directly with a DrawingContext. Combined with a polling and background worker-thread approach, this would allow us to display a color image feed constantly without having to slow it down with algorithmic work. We do not have to apply the computer vision algorithms as frequently as we receive frames. If we are trying to determine a user's gender, for example, we could run the algorithm on every tenth frame, as the gender is unlikely to change during operation of the app. It is always best to be judicious in our use of computer vision and image processing techniques to avoid overtaxing the host computer.

Basic Image Processing Techniques

I do not think anyone would disagree with me when I say that the cardinal rule of working with data is that scrubbing and cleaning the data is the most important step in processing it. In the case of images, the best computer vision algorithms in the world on the fastest machines are completely useless if the image data is incomprehensible. As you can imagine, we cannot discern contours in a blurry image, and we have a difficult time tracking something like a red ball in an image full of other objects with varying tones of red. With the help of a few key image processing techniques, we can drastically simplify the image for decision-making algorithms. This is by no means a comprehensive overview of even all the basic image processing techniques available for use, but rather a sampling to induct the reader into a mindset to go uncover more.

Color Conversion

In casual parlance, we describe the world's colors with a system of words and, for the most part, understand each other without too much ambiguity. The sky is blue, the dirt is brown, and the leaves are green. Computers, on the other hand, have several options when it comes to describing colors, all of which have stringent definitions for each of their colors. These color description systems, known as *color spaces*, are all suitable for different applications. RGB, for example, facilitates representation of colors on digital systems such as televisions and monitors, and CMYK (**C**yan, **M**agenta, **Y**ellow, and Blac**k**) is ideal for color printing. In computer vision, we may often need to go in between color spaces to take advantage of certain algorithms or to extract information. There are in fact over 150 color conversions we can make in OpenCV, though you will initially only need to know a select few.

In Emgu CV, we change an image's color space from one to another in one method call:

```
CvInvoke.CvtColor(sourceImage, destinationImage, ColorConversion.Bgra2Gray);
```

The sourceImage and destinationImage inputs are Mats, and ColorConversion is an Emgu CV enum that indicates from and to which color space we are converting. The source and destination images need to be the same size. You should keep in mind that image information will be lost in many of these conversions. If you convert a color image to grayscale, for example, the color channels will be stripped and that data will be irrecoverable. When the grayscale image is converted back to color, it will have four channels, but the color channels (and not the opacity channel, which will be set to max) will all hold the same values as the prior gray channel.

More on Grayscale

Grayscale is a color space that deserves special mention because of its pervasive use in computer vision. It is typically used when something needs to be measured on a single scale, but still needs to be visualized. The most obvious example is when you want to measure the intensity of light in an image. Darker gray regions in the image would naturally represent darker colors, as in Figure 6-5. We are not limited to only grayscaling RGB images, however. We can grayscale an individual color channel to see the intensity of that color in the image. Depth images are grayscaled to measure distance. A force-sensing capacitive touch screen, such as the one found in the iPhone 6s or the scuttled Nokia McLaren project, could provide a grayscale image of which regions of the screen receive more pressure. Grayscale in Emgu CV is measured with 8 bits, where 255 indicates pure white and 0 indicates pure black, with varying degrees of gray in between.

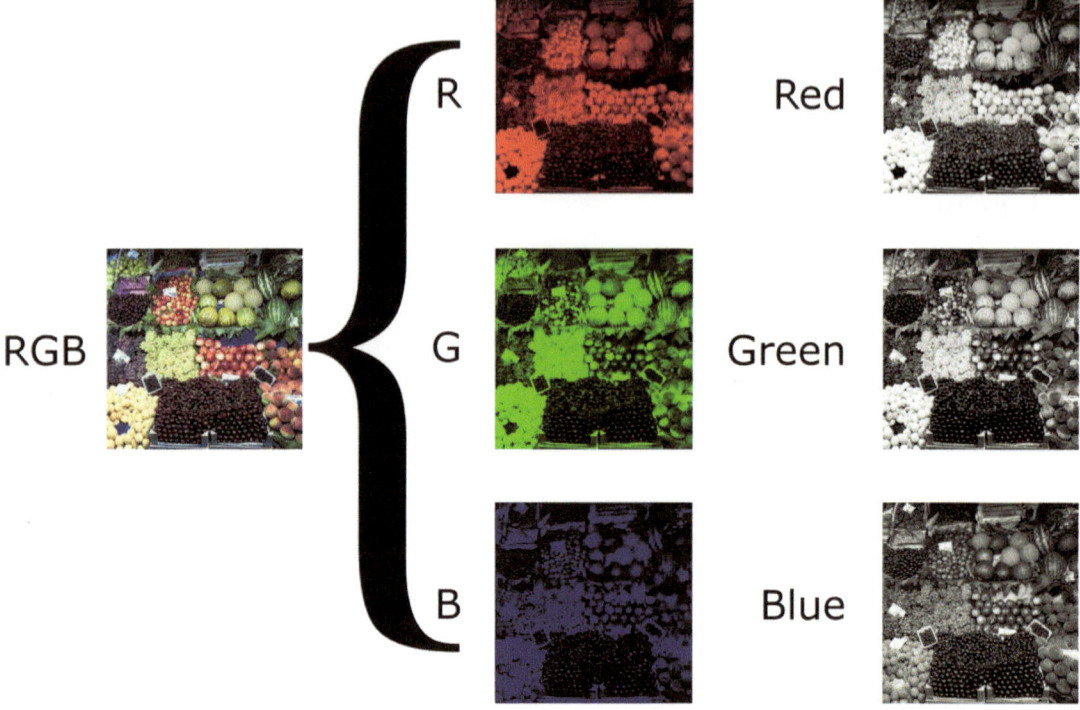

Figure 6-5. *An RGB image split into color channels that are then grayscaled to better help visualize their intensities in the original image. White areas represent where the color channel's intensity is the highest (© Nevit Dilmen)*

Harkening back to our Kinect sample with Mat, converting to grayscale requires a couple of alterations. Since there is less data, we need to adjust the size of the buffers to accommodate this.

Listing 6-3. Converting Kinect Color Image to Grayscale

```
...
//inside body of Color_FrameArrived(...)
colorFrame.CopyConvertedFrameDataToIntPtr(img.DataPointer, (uint)(colorFrameDesc.Width *
colorFrameDesc.Height * 4), ColorImageFormat.Bgra);
CvInvoke.CvtColor(img, img, Emgu.CV.CvEnum.ColorConversion.Bgra2Gray);
CopyMemory(colorBitmap.BackBuffer, img.DataPointer, (uint)(colorFrameDesc.Width *
colorFrameDesc.Height));
...
```

In Listing 6-3, we make a call to CvtColor(...) and rewrite img as a grayscale image. Additionally, we alter the buffer-size argument in CopyMemory(...). Grayscale images are a single byte; thus, we no longer need to multiply by four as we did with RGBA images.

Listing 6-4. Setting WriteableBitmap Format to Gray8

```
colorBitmap = new WriteableBitmap(colorFrameDesc.Width,
                colorFrameDesc.Height,
                96.0,
                96.0,
                PixelFormats.Gray8,
                null);
```

The changes to grayscale need to be reflected to the WriteableBitmap as well. In Listing 6-4, we set the PixelFormats of our WriteableBitmap to Gray8. The compiled result should look similar to Figure 6-6.

Figure 6-6. *Kinect color stream in grayscale*

Thresholding

Thresholding is an image processing operation that sets the pixels of an image to one of two colors, typically white and black, depending on whether the pixel's intensity meets a certain threshold. The final result is a binary image where pixels' values are essentially described as either true (white) or false (black). This helps us segment an image into regions and create sharply defined borders. From a computer vision perspective, this can help us extract desired features from an image or provide a basis from which to apply further algorithms.

OpenCV has five basic thresholding functions that can be used: *binary thresholding, inverted binary thresholding, truncating, thresholding to zero,* and *inversely thresholding to zero.* These are described in detail in Table 6-1 and shown in Figure 6-7. While our goal can usually be achieved by more than one of the thresholding functions, using the right one can save extra steps and processing. Some of these functions take a MaxValue argument, which will be the intensity value that is assigned to a true or false pixel. Thresholding should typically be applied to grayscale images, but can be applied to images with other color formats as well.

Table 6-1. *Basic Thresholding Operations in OpenCV*

Threshold Type	Description
Binary Threshold	If the intensity of a pixel is greater than the threshold, set its value to MaxValue. Else, set it to 0. ThresholdType enum for use in Emgu CV is Binary.
Inverted Binary Threshold	The opposite of binary thresholding. If the intensity of a pixel is greater than the threshold, set its value to 0. Else, set it to MaxValue. ThresholdType enum for use in Emgu CV is BinaryInv.
Truncate	If the intensity of a pixel is greater than the threshold, set it to the threshold's value. Else, the intensity of the pixel stays the same. ThresholdType enum for use in Emgu CV is Trunc.
Threshold to Zero	If the intensity of a pixel is less than the threshold, set its value to 0. Else, the intensity of the pixel stays the same. ThresholdType enum for use in Emgu CV is ToZero.
Inverted Threshold to Zero	The opposite of thresholding to zero. If the intensity of a pixel is greater than the threshold, set its value to 0. Else, the intensity of the pixel stays the same. ThresholdType enum for use in Emgu CV is ToZeroInv.

Figure 6-7. *Tokyo Tower at night with different threshold techniques applied. Thresholding applied on top row from left to right: none, binary, inverted binary. Thresholding applied on bottom row from left to right: truncation, threshold to zero, inverted threshold to zero. A value of 100 was used for the threshold and 255 for the maximum value.*

Listing 6-5. Thresholding a Kinect Color Feed Image

```
colorFrame.CopyConvertedFrameDataToIntPtr(img.DataPointer, (uint)(colorFrameDesc.Width *
colorFrameDesc.Height * 4), ColorImageFormat.Bgra);
CvInvoke.CvtColor(img, img, Emgu.CV.CvEnum.ColorConversion.Bgra2Gray);
CvInvoke.Threshold(img, img, 220, 255, Emgu.CV.CvEnum.ThresholdType.Binary);
CopyMemory(colorBitmap.BackBuffer, img.DataPointer, (uint)(colorFrameDesc.Width *
colorFrameDesc.Height));
```

175

In Listing 6-5, we make a call to Threshold(...), and, as with CvtColor(...), the first two arguments are the input and output images, respectively. This is followed by the threshold value (double), the max value (double), and the ThresholdType (enum). ThresholdType includes the five basic thresholding operations, as well as the more advanced Otsu and Mask. By setting the threshold to 220 and the max value to 255 and applying a binary threshold, we are telling our application to blacken all but the brightest pixels, which should be made white. The result looks something akin to Figure 6-8, where only the natural light from the outside, as well as its reflection on the computer monitor and table, is considered bright.

Figure 6-8. *Kinect color feed with binary thresholding applied*

Smoothing

Smoothing (or blurring) is generally used to remove undesirable details within an image. This might include noise, edges, blemishes, sensitive data, or other fine phenomena. It also makes the transition between colors more fluid, which is a consequence of the image's edges being less distinct. In a computer vision context, this can be important in reducing false positives from object-detection algorithms.

■ **Note** Blurring sensitive data is an inadequate form of protection against prying eyes. This is especially true with alphanumeric characters. Data can be interpolated from the blurred result; it is better to completely erase sensitive data on an image altogether by overwriting its pixels with junk (0-intensity pixels).

There are four basic smoothing filters that can be used in OpenCV. These are *averaging, Gaussian filtering, median filtering,* and *bilateral filtering.* As with thresholding operations, there are filters in each scenario that are most appropriate to use. Some of these scenarios are described in Table 6-2.

Table 6-2. *Basic Smoothing Filters in OpenCV*

Smoothing Filter Type	Description
Averaging	Basic smoothing filter that determines pixel values by averaging their neighboring pixels. Can be called in Emgu CV with CvInvoke.Blur(Mat src, Mat dst, System.Drawing.Size ksize). src and dst are the input and output images, respectively. ksize is an odd-number-sized box matrix (e.g., (3, 3) or (5, 5) or (7, 7), etc.) otherwise known as a kernel. A higher ksize will result in a more strongly blurred image. The average filter is sometimes known as a mean filter or a normalized box filter.
Gaussian Filtering	A filter that is effective at removing *Gaussian noise* in an image. Gaussian noise is noise whose values follow a Gaussian distribution. In practice, this means sensor noise in images caused by bad lighting, high temperature, and electric circuitry. Images put through Gaussian filtering look like they are behind a translucent screen. The filter can be called with CvInvoke.GaussianBlur (Mat src, Mat dst, Size ksize, double sigmaX, double sigmaY = 0, BorderType borderType = BorderType.Reflect101). For amateur applications, the sigma values can be left at 0, and OpenCV will calculate the proper values from the kernel size. BorderType is an optional value and can be ignored as well.
Median Filtering	A filter that is effective at removing *salt-and-pepper noise* in an image. Salt-and-pepper noise typically consists of white and black pixels randomly occurring throughout an image, like the static noise on older TVs that have no channel currently playing. The noise does not necessarily have to be white and black, however; it can be other colors. For lesser levels of Gaussian noise, it can be substituted in place of the Gaussian filter to better preserve edges in images. Can be called with CvInvoke.MedianBlur(Mat src, Mat dst, int ksize). See Figure 6-9.
Bilateral Filtering	A filter that is effective at removing general noise from an image while preserving edges. Slower than other filters and can lead to *gradient reversal,* which is the introduction of false edges in the image. Can be called with CvInvoke.BilateralFilter(Mat src, Mat dst, int diameter, double sigmaColor, double sigmaSpace, BorderType borderType = BorderType. Reflect101). Values of diameters larger than five tend to be very slow, thus five can be used as a default parameter initially. Both sigma values can share the same value for convenience. Large sigma values, such as 150 or more, will lead to images that appear cartoonish.

Figure 6-9. *Salt-and-pepper image before and after median filtering is applied. Top-left: original image, top-right: image with salt-and-pepper noise applied, bottom: image after median filtering has been applied. A 5 × 5 kernel was used.*

Sharpening

There is no built-in sharpening filter in OpenCV. Instead, we can use the Gaussian filter in a procedure that is perhaps ironically known as *unsharp masking*. Unsharp masking is the most common type of sharpening, and you have probably encountered if you have used Photoshop or GIMP. The filter works by subtracting a slightly blurry version of the photo in question from said photo. The idea is that the area that gets blurred is where the edges are and that removing a similar quantity of blur from the original will increase the contrast between the edges. In practice, this effect can be replicated in two lines in Emgu CV or OpenCV.

Listing 6-6. Implementing an unsharp mask in Emgu CV

```
CvInvoke.GaussianBlur(image, mask, new System.Drawing.Size(3, 3), 0)
CvInvoke.AddWeighted(image, 1.3, mask, -0.4, 0, result);
```

image, mask, and result in Listing 6-6 are all Mats; the second input of AddWeighted(...) is the weight accorded to the elements of the image Mat, whereas the fourth input is weight accorded to those of the mask Mat (the second image input in the function). The fifth input is a scalar value that can be added to the combined intensity values of image and mask. After we get the Gaussian blur of image in the mask Mat, we add it with negative weight (essentially subtraction) to image with the AddWeighted(...) method. The resulting sharpened image is in the result Mat. The parameters can be tinkered with to alter the degree of sharpening applied to the image. Increasing the kernel size of the Gaussian blur and the weights in favor of the mask to be subtracted will result in a more exaggerated sharpening of the image.

Morphological Transformations

Without going into too much detail, morphological transformations are techniques used to process the image based on its geometry. You will generally come to use them extensively in computer vision work to deal with noise and to bring focus to certain features within an image. While there are numerous morphological transformations that we can make use of, there are four basic ones that should cover your bases initially: *erosion, dilation, opening,* and *closing*. These are described in Table 6-3 and demonstrated in Figure 6-10.

Note & Tip   Morphological transformations in OpenCV are applied in respect to the high-intensity pixels in the image. In a binary image, this means all the white pixels. So, when we say we are dilating an image, we are almost literally dilating the brighter regions in the image. For this reason, we strive to keep whatever we are interested in (the foreground) in white for image processing operations. That being said, all morphological operations have an equivalent reverse operation; thus, we can also keep the objects of interest in black and the background in white if convenient and apply the reverse operation.

Table 6-3. *Basic Morphological Transformations in OpenCV*

Morphological Transformation	Description
Erosion	The erosion operator thins the bright regions of an image (while growing the darker regions). It is one of the two basic morphological operators and is the opposite of the dilation operator. Some of its uses include separating foreground objects, eradicating noise, and reducing highlights in an image. `CvInvoke.Erosion(Mat src, Mat dst, Mat element, Point anchor, int iterations, borderType borderType, MCvScalar borderValue);`. `IntPtr.Zero` can be used to use the default 3 x 3 structuring element. `Point` can be set to the default value of `(-1, -1)`. `borderType` and `borderValue` can be set to default and 0 respectively.
Dilation	The dilation operator thickens the bright regions of an image (while shrinking the darker regions). It is one of the two basic morphological operators and is the opposite of the erosion operator. Some of its uses include combining foreground objects and bringing focus to certain areas of the image. Dilation can be applied with `CvInvoke.Dilate (Mat src, Mat dst, Mat element, Point anchor, int iterations, borderType borderType, MCvScalar borderValue);`.
Opening	Opening is a morphological operation that essentially consists of eroding an image and then dilating it. It is the sister of the closing morphological operation. The result is an image with small foreground noise reduced. It is less destructive than removing noise through erosion. Opening is applied with the `CvInvoke.MorphologyEx(Mat src, Mat dst, MorphOp operation, Mat kernel, Point anchor, int iterations, BorderType borderType, MCvScalar borderValue);`. `MorphOp` is an enum representing the name of the morphological operator—Open, in our case.
Closing	Closing is a morphological operation that essentially consists of dilating an image and then eroding it. It is the sister of the opening morphological operation. The result is an image with small dark holes filled in (or small background noise reduced). It is more precise than trying to fill holes through dilation alone. Closing can be applied with `CvInvoke. MorphologyEx(Mat src, Mat dst, MorphOp operation, Mat kernel, Point anchor, int iterations, BorderType borderType, MCvScalar borderValue);`. The `MorphOp` value for Closing is `Close`.

▪ **Note** `Erode` and `Dilate` can also be called with the `CvInvoke.MorphologyEx(...)` method. The relevant `MorphOp` values are `Erode` and `Dilate`.

Figure 6-10. *An evening view of Moriya, Ibaraki Prefecture, Japan, with different morphological operators applied. From top to bottom: original, erosion, dilation, opening, closing. Each operator was applied on the original image for 20 iterations using a 5 × 5 kernel.*

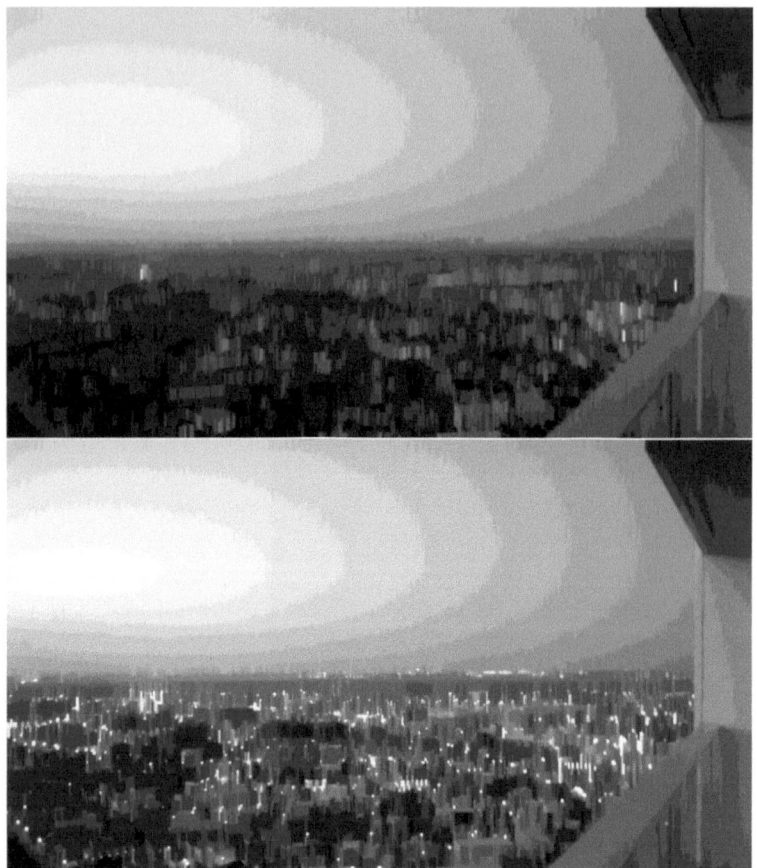

Figure 6-10. (*continued*).

Highlighting Edges with Morphological Operators

At times, you may want to visually highlight the contours of an object in an image. This might be to bring attention to an object your algorithm detected, or perhaps to add some type of visual effect (e.g., make someone look like a superhero in a Kinect camera feed by giving them a colored glow around their body). One way to achieve this effect is through morphological operators.

The process consists of dilating the area of interest and then subtracting the dilation from the original image.

Listing 6-7. Highlighting an Edge Using Morphological Operators

```
Mat img = new Mat("orange.jpg", LoadImageType.Grayscale);
Mat thresholdedImg = new Mat();
CvInvoke.Threshold(img, thresholdedImg, 240, 255, ThresholdType.BinaryInv);
Mat dest = thresholdedImg.Clone();
CvInvoke.Dilate(dest, dest, new Mat(), new Point(-1, -1), 5, BorderType.Default, new MCvScalar(0));
CvInvoke.BitwiseXor(thresholdedImg, dest, dest);
CvInvoke.BitwiseXor(img, dest, img);
```

In Listing 6-7, we choose to highlight the contours of an orange. We start off by taking in an image of an orange (Figure 6-11a) inside of a new Mat. We choose to work with a grayscale version (Figure 6-11b) for simplicity's sake. We binary threshold it (Figure 6-11c) using `CvInvoke.Threshold(img, thresholdedImg, 240, 255, ThresholdType.BinaryInv);`, turning all the white background pixels dark, as they are of no interest, and the darker orange pixels white, since that is the region of interest. On a copy of the thresholded image, we dilate the bright regions (Figure 6-11d) with OpenCV's dilate method: `CvInvoke.Dilate(dest, dest, new Mat(), new Point(-1, -1), 5, BorderType.Default, new MCvScalar(0));`. All the default arguments are used, except for the iterations value, which is set to 5. *Iterations* refers to how many times you want to apply the morphological operator. In our case, the more times `dilate` is called, the larger the edge highlighting will ultimately be. We then XOR the thresholded image with its dilated result (Figure 6-11e). Since the XOR operation only results in a true, or 255, brightness value for each pixel that is different between both corresponding `Mat`s, only the regions in an image that have changed between both images will be highlighted in the resulting image. This region is the part of the image that has been dilated beyond the original thresholded image. Finally, we XOR this resulting image with the original image of the grayscale orange to obtain the highlighted contour of the orange (Figure 6-11f).

Figure 6-11a. *Standard orange image asset (United States Department of Health and Human Services)*

Figure 6-11b. *Orange image, grayscaled*

Figure 6-11c. *Orange grayscale image with thresholding applied*

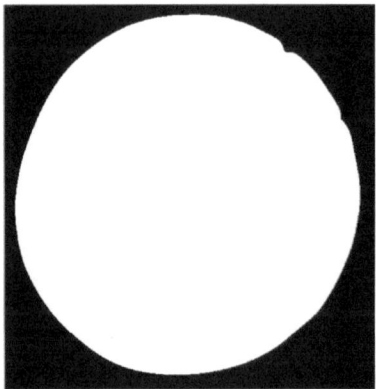

Figure 6-11d. *Image after having dilation applied. While the image looks similar to Figure 6-11c, under closer inspection, you will notice that the white area takes up a slightly greater portion of the image. A couple of black pixels near the top of the white area have also been filled in.*

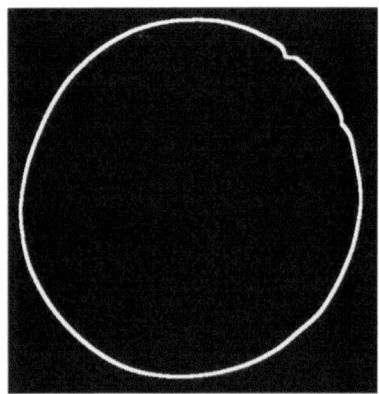

Figure 6-11e. *The XOR'd result of Figure 6-11c and Figure 6-11d. This is the contouring that will be applied to the final image.*

Figure 6-11f. *The XOR'd result of Figure 6-11b and Figure 6-11e. The contour obtained from dilation drawn on the initial image.*

We applied the contour highlighting to a grayscale image, but had we wanted, we could have used the same technique to apply, say, a red or blue contour around the original image of the orange. This would entail taking pixel values of Figure 6-11e and applying them to the relevant color channels of Figure 6-11a in a weighted manner.

Bitwise & Arithmetical Operations

In the previous section, we briefly dwelled on the bitwise XOR operator. It was a quick way for us to determine the difference between two pictures. *Bitwise* and *arithmetical* operations such as the XOR operator are commonly used in image processing. Since images are ultimately arrays filled with numbers, on a numerical level, such operations work as you would expect. On a visual level, however, it might not immediately be obvious which operation to use for which result. Bitwise and arithmetical operations fall under the broader category of *array operations*. There are dozens of such operations in OpenCV. We covered a sparse few already, such as the inRange(...) method. While in time you will come to learn all of them, for now we will focus on the elementary ones.

The arithmetical and bitwise operators are add, subtract, divide, multiply, AND, NOT, OR, and XOR. Again, their functionality is self-explanatory. Adding, for example, adds each pixel between the corresponding array indices of two images. We will rely on visual representations of these operations to gain a better understanding of them. Figures 6-12a and 6-12b are two base images we will use to demonstrate the use of these operations.

Figure 6-12a. *The first source image*

Figure 6-12b. *The second source image (the red borders are not a part of the image)*

Addition

Adding, as explained earlier, adds each corresponding pixel in both images together. When adding two pixels results in a value larger than 255 in any color channel, the excess is cut off. This results in the big white circle being the most prominent artefact in Figure 6-13. Although it's mathematically clear, it can still be weird to wrap your head around the fact that adding gray to black results in gray in the world of image processing, as opposed to black in, say, a paint app or the real world.

Figure 6-13. *Addition of Figure 6-12b to Figure 6-12a*

Subtraction

As you would imagine, subtracting returns the opposite result of adding (Figure 6-14).

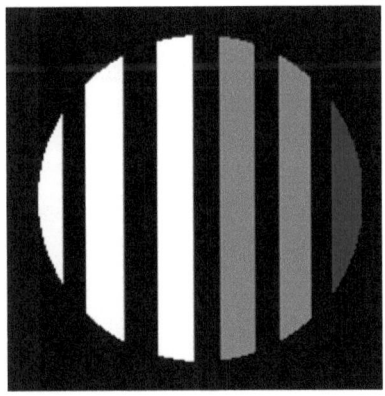

Figure 6-14. *Subtraction of Figure 6-12b from Figure 6-12a*

Multiplication

Multiplying two images together is not often too useful, given that multiplying any two pixel values above 16 will result in a value above 255. One way to make use of it would be to intensify certain areas of an image using a mask (Figure 6-15).

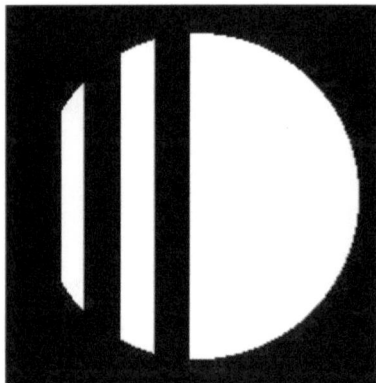

Figure 6-15. *Multiplication of Figure 6-12b by Figure 6-12a*

We are not obliged to multiply two images together, however. We can multiply an image by a scalar. This uniformly brightens an image. Figure 6-16 features Figure 6-12b brightened by 50 percent. This is achieved by multiplying the source image by a scalar value of 1.5. With a positive value smaller than 1, we can achieve the opposite effect.

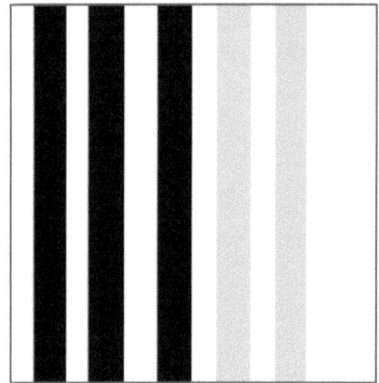

Figure 6-16. *Multiplication of Figure 6-12b by a scalar value of 1.5*

Note how the 25 percent gray stripe all the way on the right has turned white. Gray at 25 percent has a 75 percent intensity value, or 191.25. 191.25 × 1.5 = 286.75, which is larger than 255; hence, it becomes white.

Division

Division possesses characteristics similar to those of multiplication (Figure 6-17).

Figure 6-17. *Division of Figure 6-12a by Figure 6-12b*

Likewise, for division, operating with small scalar values tends to be more useful (Figure 6-18).

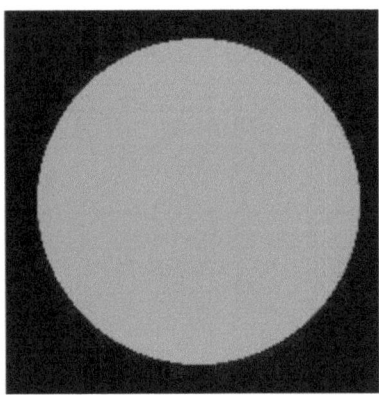

Figure 6-18. *Division of Figure 6-12a by a scalar of 2*

There is a way to benefit from dividing two images, however. Dividing any number (other than 0) by itself results a value of 1, a value close enough to 0 from an image processing context. We can use this property to determine areas that have changed between two images. Areas that remain the same will appear black, whereas areas that are not the same will report different values. This is particularly useful in removing glare from images when they were taken from multiple angles, though this is a more advanced technique that will not be discussed here.

Bitwise AND

The bitwise AND operator works as you might have expected adding two images would have. Lighter colors such as the color white act as a background to darker colors. Adding two light colors together results in a darker color (Figure 6-19).

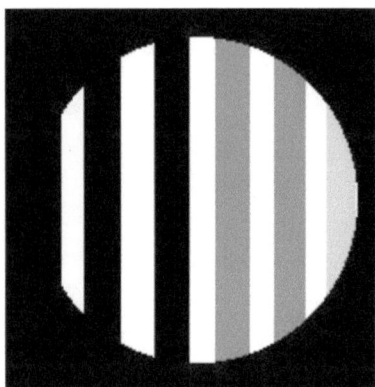

Figure 6-19. *Bitwise conjunction of Figure 6-12a and Figure 6-12b*

■ **Note** I just referred to white as a color, and that is sure to ruffle some feathers (everything does these days!). Many of us learned early in school, whether it be from friends, parents, or teachers, that white and/or black are not colors. The answer will depend on whether you look at the additive or subtractive theory of colors, or whether you look at color as light or as pigmentation. Personally, I prefer Wikipedia's definition best: White, gray, and black are *achromatic* colors—colors without a hue.

Bitwise NOT

Bitwise NOT is a bit interesting in that it is a unary operation. Thus, it operates only on a single source image. It is my favorite operator because it is the one with the most predictable results: it inverses the colors in an image (Figures 6-20a and 6-20b).

Figure 6-20a. *Logical negation of Figure 6-12a*

Figure 6-20b. *Logical negation of Figure 6-12b*

Bitwise OR

Looking at Figure 6-21, you would think that the bitwise operator is equivalent to the additive arithmetical operation. While they work similarly in certain cases, they are in fact different. See Figure 6-21.

Figure 6-21. *Bitwise disjunction of Figure 6-12a and Figure 6-12b*

To see how they differ, let's rotate Figure 6-12b by 90 degrees and apply itself in its original orientation with the bitwise OR and addition operations (Figure 6-22).

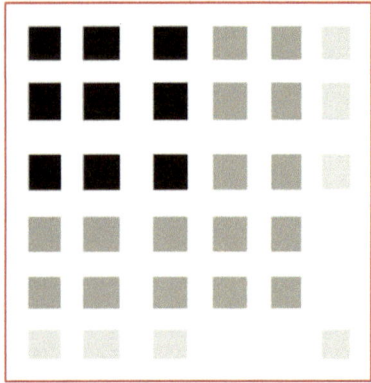

Figure 6-21a. *Bitwise disjunction of Figure 6-12b and Figure 6-12b rotated by 90°*

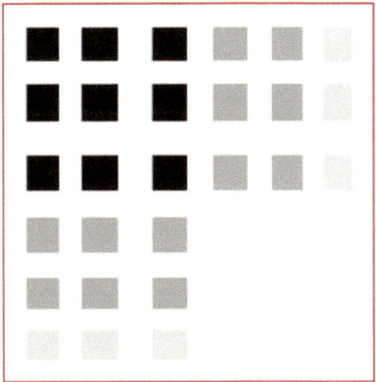

Figure 6-21b. *Addition of Figure 6-12b and Figure 6-12b rotated by 90°*

Interestingly, the OR operation results in regions staying the same color if both of the source regions had the same color previously. This is unlike the addition operation, which just maxes the regions to white. Mathematically, this makes sense. 1000 0000 | 1000 0000 (128, the intensity of Gray 50 percent) results in 1000 0000. All in all, the OR operation, being so permissive, will usually result in an image that is much brighter, if not mostly white. Most bits in the resulting image will be switched to ones, unless both of their source bits were zeroes. A great use case for this property is to see which parts two images have in common.

◾ **Tip** It is important to investigate array operations on different corner cases; they often produce unexpected results. Better yet, thresholding the images to binary will yield more reliable results.

Bitwise XOR

We already saw the bitwise XOR operator in action. We used it to get the difference between two binary images. It is regularly used for such. We should avoid relying on XOR unless the image is binary, however. You would expect two images with highly differing intensities to result in an image with a high intensity. Thus, it will surprise some to learn that when we XOR an intensity 128 with an intensity of 100, as opposed to a dull gray, the resulting image would be nearly white! This is because 128 in binary is 1000 0000, which is a digit larger than the decimal number 100, 110 0100 in binary. For this reason, it is recommended to use subtraction when possible. See Figure 6-22.

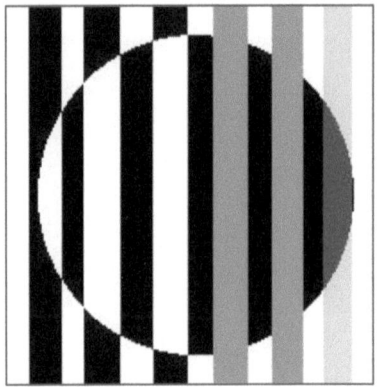

Figure 6-22. *Exclusive bitwise disjunction of Figure 6-12a and Figure 6-12b*

Using Arithmetic & Bitwise Operators

To use any of the arithmetic for the operators in the app, simply call them through CvInvoke.Add(src1, src2, dst); with Add being replaced by the relevant arithmetic operation: Subtract, Divide, or Multiply. src2 can also be a ScalarArray(double value) if you wish to operate with a scalar value.

Likewise, with bitwise operators, call CvInvoke.BitwiseAnd(src1, src2, dst);, replacing the And portion of the signature with the relevant bitwise operator: Not, Or, or Xor. BitwiseNot(...) will of course only take one src argument.

Visualizing Movement Through the Use of Arithmetic Operators

Arithmetic and bitwise operators are often used to compare and contrast two images. One practical purpose of this capability is to detect movement between frames. This can be employed to develop a simplistic video-surveillance or -monitoring system.

Our system will consist of an image-viewing app that displays a binary image. White pixels will indicate movement, whereas black pixels will indicate the lack of movement. The final app will resemble Figure 6-23a. See Listing 6-8.

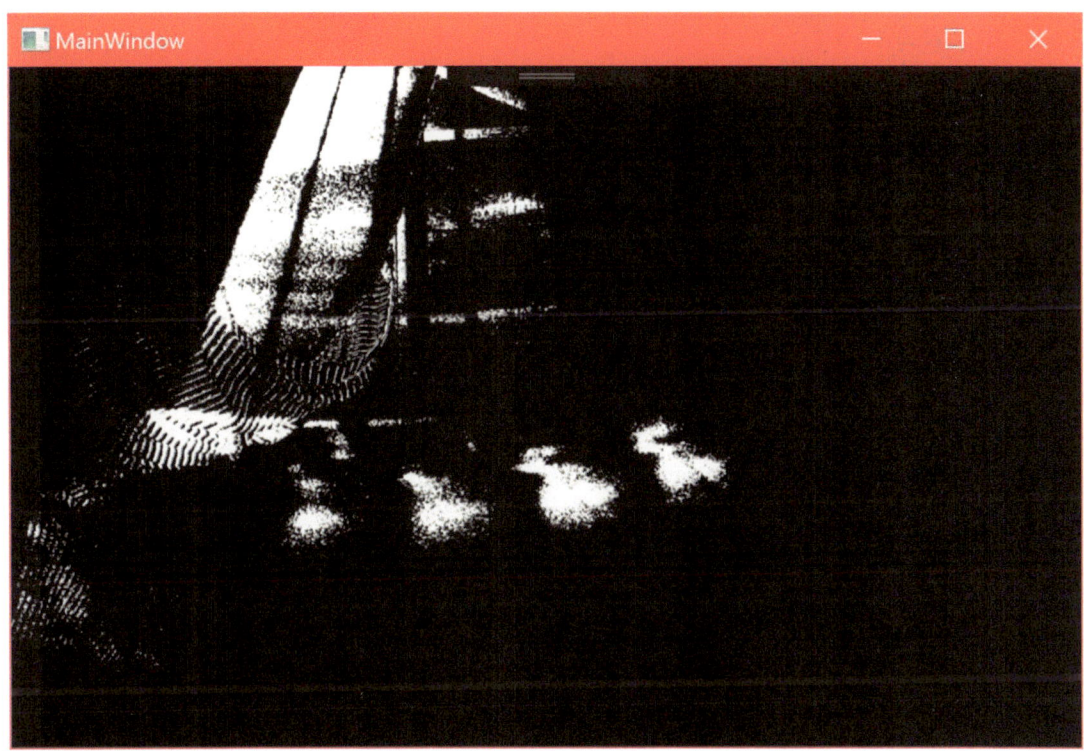

Figure 6-23a. *The Kinect Motion Detector app visualizing the motion of a swinging pendulum made from a 5¥ coin and a length of floss*

Figure 6-23b. *The precursor scene to Figure 6-24a in color*

Listing 6-8. Detecting Motion with Arithmetic Operators

```
[...] //Declare Kinect and Bitmap variables

Mat priorFrame;
Queue<Mat> subtractedMats = new Queue<Mat>();
Mat cumulativeFrame;

[...] //Initialize Kinect and WriteableBitmap in MainWindow constructor

private void Color_FrameArrived(object sender, ColorFrameArrivedEventArgs e)
{
    using (ColorFrame colorFrame = e.FrameReference.AcquireFrame())
    {
        if (colorFrame != null)
        {
            if ((colorFrameDesc.Width == colorBitmap.PixelWidth) &&
            (colorFrameDesc.Height == colorBitmap.PixelHeight))
            {
                using (KinectBuffer colorBuffer = colorFrame.LockRawImageBuffer())
                {

                    Mat img = new Mat(colorFrameDesc.Height, colorFrameDesc.Width,
                    Emgu.CV.CvEnum.DepthType.Cv8U, 4);
                    colorFrame.CopyConvertedFrameDataToIntPtr(
                    img.DataPointer, (uint)(colorFrameDesc.Width * colorFrameDesc.Height *
                    4), ColorImageFormat.Bgra);
                    CvInvoke.CvtColor(img, img, Emgu.CV.CvEnum.ColorConversion.Bgra2Gray);

                    if (priorFrame != null)
                    {
                        CvInvoke.Subtract(priorFrame, img, priorFrame);
                        CvInvoke.Threshold(priorFrame, priorFrame, 20, 255,
                        Emgu.CV.CvEnum.ThresholdType.Binary);
                        CvInvoke.GaussianBlur(priorFrame, priorFrame, new System.Drawing.
                        Size(3, 3), 5);
                        subtractedMats.Enqueue(priorFrame);
                    }
                    if (subtractedMats.Count > 4)
                    {
                        subtractedMats.Dequeue().Dispose();

                        Mat[] subtractedMatsArray = subtractedMats.ToArray();
                        cumulativeFrame = subtractedMatsArray[0];

                        for (int i = 1; i < 4; i++)
                        {
                            CvInvoke.Add(cumulativeFrame, subtractedMatsArray[i],
                            cumulativeFrame);
                        }
                        colorBitmap.Lock();
```

```
                    CopyMemory(colorBitmap.BackBuffer, cumulativeFrame.DataPointer,
                    (uint)(colorFrameDesc.Width * colorFrameDesc.Height));
                    colorBitmap.AddDirtyRect(new Int32Rect(0, 0, colorBitmap.PixelWidth,
                    colorBitmap.PixelHeight));

                    colorBitmap.Unlock();
                }
                priorFrame = img.Clone();
                img.Dispose();
            }
        }
    }
}
}
```

In Listing 6-8, we repurposed the **Kinect Image Processing Basics** project to develop our movement-detector app. The concept is pretty straightforward. Subtracting one frame from another shows the regions that have changed over the span of $1/30^{th}$ of a second. It could suffice to simply display these frames as they are, but for many applications, it would be preferable to sum these frames to show a protracted motion. This would enable a very slight movement—say, the twitch of a finger or the rising and falling of the chest—to be much more perceptible and distinguishable from mere background noise. In our program, we sum four of these frames together to get a cumulative frame representing a motion lasting $2/15^{th}$ of a second ($4 \times 1/30^{th}$ of a second differences). The nice thing is that we are always using the last five frames taken by the Kinect to view these changes; thus, the resulting video feed still plays at 30 frames per second.

Listing 6-9. Kinect Motion Detector – Subtracting Frames

```
CvInvoke.CvtColor(img, img, Emgu.CV.CvEnum.ColorConversion.Bgra2Gray);

if (priorFrame != null)
{
    CvInvoke.Subtract(priorFrame, img, priorFrame);
    CvInvoke.Threshold(priorFrame, priorFrame, 20, 255, Emgu.CV.CvEnum.ThresholdType.
    Binary);
    CvInvoke.GaussianBlur(priorFrame, priorFrame, new System.Drawing.Size(3, 3), 5);
    subtractedMats.Enqueue(priorFrame);
}
```

In Listing 6-9, we have the portion of the `Color_FrameArrived(...)` method that deals with the subtraction of one frame from another. Although the app could conceivably work in color, we grayscale all images so that the final result is easier to work with for any object-detection tools and to make it easier to understand for any user.

A `priorFrame` `Mat` holds a reference to the image taken last time `Color_FrameArrived(...)` was fired. We need two images to apply subtraction, so we wait until we have a second image before starting. After subtraction, we threshold so that most minor changes between frames appear white. This tends to cause noise as a result of the limits of the Kinect hardware, but we can apply a Gaussian blur to rectify this. The parameters for the thresholding and blurring can be tweaked to your liking. Finally, we save the image containing the difference of the current and prior frames into the `subtractedMats` queue. This queue contains the last few differences, which will be summed together in the next step (Listing 6-10).

Listing 6-10. Kinect Motion Detector – Summing Frame Differences

```
if (subtractedMats.Count > 4)
{
    subtractedMats.Dequeue().Dispose();

    Mat[] subtractedMatsArray = subtractedMats.ToArray();
    cumulativeFrame = subtractedMatsArray[0];

    for (int i = 1; i < 4; i++)
    {
        CvInvoke.Add(cumulativeFrame, subtractedMatsArray[i], cumulativeFrame);
    }
    colorBitmap.Lock();

    CopyMemory(colorBitmap.BackBuffer, cumulativeFrame.DataPointer, (uint)(colorFrameDesc.
    Width * colorFrameDesc.Height));
    colorBitmap.AddDirtyRect(new Int32Rect(0, 0, colorBitmap.PixelWidth, colorBitmap.
    PixelHeight));

    colorBitmap.Unlock();
}

priorFrame = img.Clone();
img.Dispose();
```

We only display an image if our queue has five images (four after junking the oldest one). These are summed together in a for loop and then displayed. The current image is copied into the priorFrame variable for reuse in the next frame's event handler call.

A final note: we make sure to dispose of the Mat that is dequeued and the one that was just obtained. Not doing so would cause the application to eventually run out of memory.

Although this is a very basic motion detector, it can serve as the foundation for a more complex project. For example, with the use of blob-detection techniques, it can be used to track the velocity of cars moving down a certain stretch of road. It can be placed near a hospital bed and be used to determine the breathing rate of a patient or to see if they are twitching or having seizures (this can be enhanced with the use of the Kinect's skeletal-tracking abilities). The possibilities are limitless, yet to start you only require an understanding of arithmetic and 2D grids.

Object Detection

Object detection is probably the most touted capability in computer vision introductions. Detecting objects, after all, is how self-driving cars can attain humanlike prescience. Companies like Google, Microsoft, and Tesla spend millions of dollars and human hours developing and collating object-recognition algorithms for various robotics and artificial-intelligence endeavors (the Kinect, of course, being a notable example). Fortunately, we do not have to spend a dime to start using some of these algorithms. OpenCV has some object-recognition tools that can be used out of the box, and we can build around them further to achieve most of our goals.

It is worth giving some attention to the concept of *features* and feature detection. Features are essentially geometric points of interest in an image's foreground. These might include corners, blobs, and edges, among other phenomena. *Feature detectors*, algorithms that detect certain features, can be strung together with image processing techniques to detect objects in an image. The topic can be expansive; thus, we will focus on a couple of out-of-the-box object-recognition techniques in this chapter.

> **Note**   A blob refers to a concentration of uniformly textured pixels. For example, the thresholded image of an orange in Figure 6-11c consists of one big white blob. In a sense, the black background region could also be described as a blob. What we consider a blob depends on algorithmic parameters.

Simple Blob Detector

As previously mentioned, objects are typically detected with a series of feature detectors and image processing techniques. OpenCV bundles one such series together in a class called the *Simple Blob Detector*. Simple Blob Detector returns all the blobs detected in an image filtered by the desired area, darkness, circularity, convexity, and ratio of their inertias.

> **Note**   Do not fret if you do not remember your college physics! Inertia in this context refers to how likely the blob is to rotate around its principal axis. For all practical intents, this translates to the degree of elongation of the blob, with higher inertia values referring to lesser degrees of elongation (more inertia means the blob will be less susceptible to rotation). Convexity, on the other hand, refers to how dented a blob is. Higher convexity equates to less denting, whereas a blob with low convexity (in other words, more concavity) has one or more larger dents in its shape.

Figure 6-24. *When compared to a similarly sized banana, the apple has a higher inertia than the banana. Looking from a 2D perspective, the banana's contour is less convex than the apple's (neglecting the apple's stem).*

To demonstrate the use of Simple Blob Detector, we will make a program that detects oranges among a group of similarly shaped fruits and vegetables, like those in Figure 6-25.

Figure 6-25. *An assortment of fruits on a bed cover*

■ **Note**   The following project is a console project. I would copy an existing EmguCV sample such as Hello World to avoid having to set up the dependencies and platform settings from scratch.

Listing 6-11. Detecting Oranges with Simple Blob Detector

```
using System;
using System.Drawing;
using Emgu.CV;
using Emgu.CV.Util;
using Emgu.CV.CvEnum;
using Emgu.CV.Structure;
using Emgu.CV.Features2D;

namespace OrangeDetector
{
   class Program
   {
      static void Main(string[] args)
      {
            String win1 = "Orange Detector"; //The name of the window
            CvInvoke.NamedWindow(win1); //Create the window using the specific name
```

```
        MCvScalar orangeMin = new MCvScalar(10, 211, 140);
        MCvScalar orangeMax = new MCvScalar(18, 255, 255);

        Mat img = new Mat("fruits.jpg", ImreadModes.AnyColor);
        Mat hsvImg = new Mat();
        CvInvoke.CvtColor(img, hsvImg, ColorConversion.Bgr2Hsv);

        CvInvoke.InRange(hsvImg, new ScalarArray(orangeMin), new ScalarArray(orangeMax),
        hsvImg);

        CvInvoke.MorphologyEx(hsvImg, hsvImg, MorphOp.Close, new Mat(), new Point
        (-1, -1), 5, BorderType.Default, new MCvScalar());

        SimpleBlobDetectorParams param = new SimpleBlobDetectorParams();
        param.FilterByCircularity = false;
        param.FilterByConvexity = false;
        param.FilterByInertia = false;
        param.FilterByColor = false;
        param.MinArea = 1000;
        param.MaxArea = 50000;

        SimpleBlobDetector detector = new SimpleBlobDetector(param);
        MKeyPoint[] keypoints = detector.Detect(hsvImg);
        Features2DToolbox.DrawKeypoints(img, new VectorOfKeyPoint(keypoints), img, new
        Bgr(255, 0, 0), Features2DToolbox.KeypointDrawType.DrawRichKeypoints);

        CvInvoke.Imshow(win1, img); //Show image
        CvInvoke.WaitKey(0); //Wait for key press before executing next line
        CvInvoke.DestroyWindow(win1);
    }
  }
}
```

Listing 6-11 features all the code necessary to detect oranges in an image. The general process is that we first filter our image by the HSV values of our oranges so that only orange-colored regions in our image show up in a binary image. We then use Simple Blob Detector to highlight these regions.

The first step is performed outside of Visual Studio. Using a tool such as GIMP or Photoshop (or even MS Paint), we obtain the HSV range for the oranges. For those not familiar with HSV, it is a color profile like RGB. It refers to hue, saturation, and value. Hue refers to the color (e.g., orange, red, blue, etc.), saturation refers to the intensity of the color, and value refers to the brightness of the color.

The HSV model is not standardized between applications and technologies, so the HSV values in GIMP, as depicted in Figure 6-26, will not match those in OpenCV. We will have to translate the values mathematically.

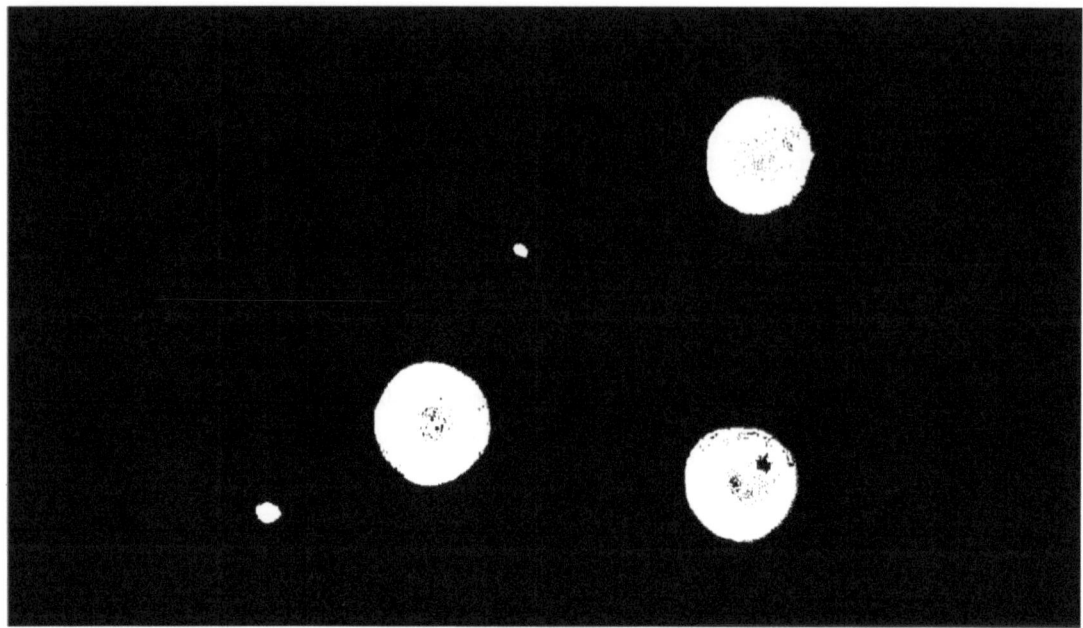

Figure 6-26. *HSV values taken by GIMP's color picker*

The hue values for our oranges range from 20 to 36 in GIMP, the saturation from 83 to 100, and the value from 55 to 100. In GIMP, the entire range of the hue is 0 to 360, starting and ending at the color red; the saturation ranges from 0 to 100; and the value ranges from 0 to 100. The ranges in OpenCV depend on the color spaces. Since we will be performing a BGR-to-HSV transformation, our hue will range from 0 to 179, starting and ending at the color red. The saturation and value will range from 0 to 255. To determine the hue range, we simply divide our GIMP values by 2. The saturation and value can be obtained by getting their percentages and multiplying those by 255. The end result is H: 10–18, S: 211–255, V: 140–255.

■ **Note** Had we done RGB to HSV, our hue would have started and ended at the color blue. Had we done RGB to HSV Full, our hue would have gone from 0 to 255. As you can imagine, improper color spaces can be cause for much consternation in image processing work.

■ **Note** We did not take lighting into consideration in our algorithm. Generally, you will have to use techniques such as *histogram equalization* to minimize the effects of lighting on your detection tasks. While the technique itself is not very complicated, its proper use is somewhat beyond the scope of this book. Visit `http://docs.opencv.org/3.1.0/d5/daf/tutorial_py_histogram_equalization.html` to learn more.

Now that we have our HSV range, we create two MCvScalars to describe their lower and upper limits:

```
MCvScalar orangeMin = new MCvScalar(10, 211, 140);
MCvScalar orangeMax = new MCvScalar(18, 255, 255);
```

MCvScalar is simply a construct for holding single, double, triple, or quadruple tuples.

We then load our fruit image and convert from BGR to HSV. We could have gone from RGB as well; the native image color space is sRGB. The only real difference between BGR and RGB, however, is a matter of interpretation, and interpreting BGR in HSV was easier in our case (see earlier note).

```
Mat img = new Mat("fruits.jpg", LoadImageType.AnyColor);
Mat hsvImg = new Mat();
CvInvoke.CvtColor(img, hsvImg, ColorConversion.Bgr2Hsv);
```

We then select all HSV values within our desired range (between orangeMin and orangeMax):

```
CvInvoke.InRange(hsvImg, new ScalarArray(orangeMin), new ScalarArray(orangeMax), hsvImg);
```

Our MCvScalars have to be included as the single elements in two ScalarArrays, whose function is eponymous. In essence, the InRange(...) method is another way of applying thresholds to our image. In our case, all values within the HSV range we set will appear as white on the image, and all other values will appear as black. The resulting image will look like Figure 6-27.

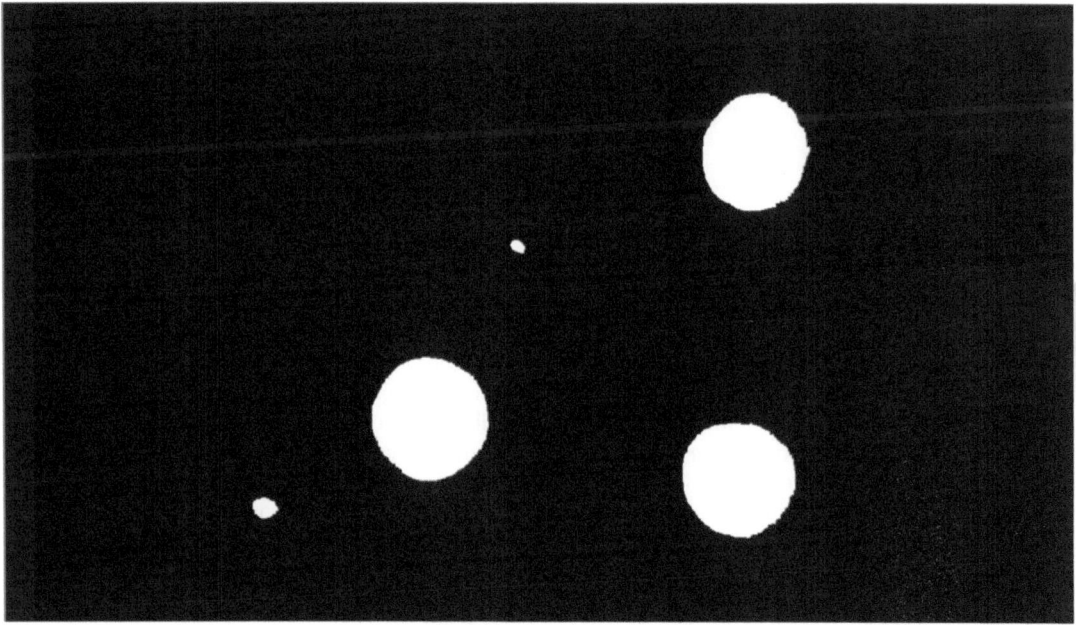

Figure 6-27. *Filtering the image by orange HSV values*

In Figure 6-27, there are a lot of tiny black specks in the white blobs representing the oranges. There is also a larger single black spot in the bottom right orange, which is where its green stem is situated. We will use the Closing morphological operation to eliminate these artefacts:

```
CvInvoke.MorphologyEx(hsvImg, hsvImg, MorphOp.Close, new Mat(), new Point(-1, -1), 5,
BorderType.Default, new MCvScalar());
```

We used five iterations of the closing operation with default parameters. The processed result is depicted in Figure 6-28.

Figure 6-28. *Our HSV-thresholded image with the closing morphological operation applied*

In addition to the three larger white blobs, which represent the location of our oranges (which are really tangerines, by the way), we have two smaller white splotches in the image. If you guessed that they represented portions of the tomatoes tinged orange due to the lighting, you would be correct. We will simply filter them out in Simple Blob Detector by ignoring all blobs under 1,000 square pixels in area. In a production project, we would have a more advanced set of criteria to eliminate false positives (the Kinect in particular makes this easier because of its ability to measure depth and in turn measure heights and widths). For our simple project, however, we will stick to what comes out of the box for the purpose of demonstrating Simple Blob Detector's parameters. We configure these parameters with the use of the SimpleBlobDetectorParams class:

```
SimpleBlobDetectorParams param = new SimpleBlobDetectorParams();
param.FilterByCircularity = false;
param.FilterByConvexity = false;
param.FilterByInertia = false;
param.FilterByColor = false;
param.MinArea = 1000;
param.MaxArea = 50000;
```

We set filtering to false for all parameters other than area, as we do not need them. The parameters have default values when set to true, and instead of setting their values to something more appropriate for detecting oranges, it is easier to simply turn them off. Had we tried to separate the oranges from, say, carrots, some of them would have been useful enough to have kept on.

The final bit of code to get our orange detector working consists of the initialization of `SimpleBlobDetector` itself and its circling of the blobs:

```
SimpleBlobDetector detector = new SimpleBlobDetector(param);
MKeyPoint[] keypoints = detector.Detect(hsvImg);
Features2DToolbox.DrawKeypoints(img, new VectorOfKeyPoint(keypoints), img, new Bgr
(255, 0, 0), Features2DToolbox.KeypointDrawType.DrawRichKeypoints);
```

The constructor for `SimpleBlobDetector` takes our parameter object. We then call the instantiated object's `Detect(Mat img)` method, which returns a series of *keypoints* (`MKeyPoints` in Emgu; the M is for "managed"). This is inputted along with our original color image in the `DrawKeypoints(...)` method, which draws circles around our blobs. The inputs for this method are the input image, the keypoints (which must be in a `VectorOfKeyPoint` object), the output image, the color of the circles to be drawn (blue in this case, for contrast), and the `KeypointDrawType`. The `KeypointDrawType` dictates how elaborately the keypoints should be drawn on the image. `DrawRichKeypoints` indicates that we should circle the entire keypoint as opposed to just indicating the center with a dot (which is the default). It should be noted that the `MKeyPoints` object contains properties such as the size and property of the detected blob. The final result is shown in Figure 6-29.

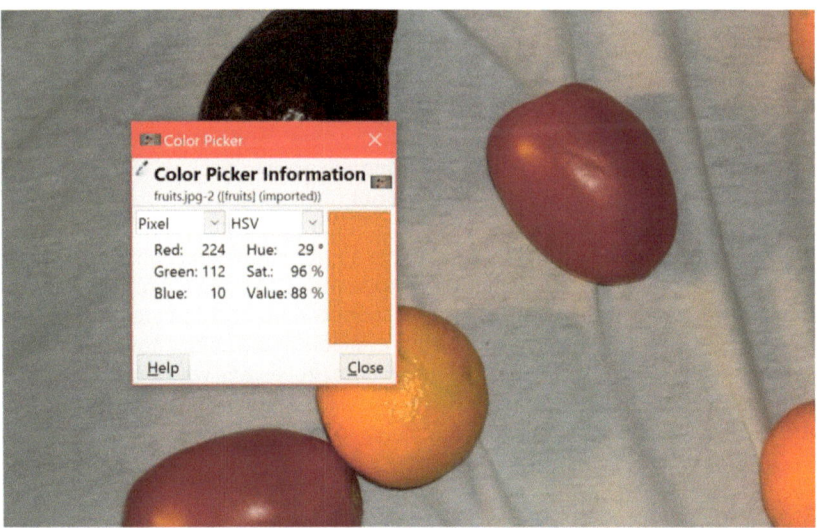

Figure 6-29. *Detected oranges in the assortment of fruits and vegetables*

The blue circles fail to completely encircle a couple of the oranges. This is because the HSV range did not include the darkest shades of the orange, which are near the edge of the oranges where the light does not shine. I left it like this intentionally. It is actually trivial to fix; you can try altering the HSV thresholds and `SimpleBlobDetector` area parameters to see this. The goal of this is to demonstrate that computer vision algorithms are rarely perfect, but often are merely "good enough." In our case, we could have achieved something close to perfection, but this is not a complicated project. There are tradeoffs we always have to consider. Allowing for a greater HSV range, for example, would make the tomatoes more susceptible to being detected, and we would need more aggressive filtering in that regard.

The example we developed is somewhat contrived. Had we taken the picture from a different angle, a different camera, with different lighting and different fruits and/or vegetables, or had we adjusted a million other factors differently, our code would have probably failed to detect every orange or would have detected false oranges. Additionally, this code would need to be further altered to support the detection of other fruits or vegetables. This is a common predicament you will come across in computer vision work. Recognizing objects in one specific scenario is often not that difficult, but enabling your code to work with a high success rate over many different environments is where the challenge lies.

■ **Tip** Whenever possible, try to reduce environmental inconsistencies before the Kinect even starts filming. Standardize as much of your lighting and scene as possible. Refrain from having too much motion and noise in the background. Remove any unnecessary artefacts in the foreground. This will render your computer vision and image processing tasks less arduous.

Conclusion

There are some optimizations applied to the Kinect by Microsoft that you will never be able to replicate without designing your own Kinect, but for most projects an elementary knowledge of computer vision and image processing techniques will take you far. For those who are the types to dream big, the Kinect need not merely be a way to track skeletal joints. The Kinect is but a starter kit, with sensors and certain computer vision and machine-learning abstractions baked in. If there is something you want to track or analyze that the Kinect cannot do on its own already, let your software be the limit, not your hardware.

The techniques discussed in this chapter will take you only so far, but hopefully this gave you a taste of what can be achieved with some matrices and your intuition. All the large, commercial computer vision projects out there are still built with many of these simple blocks, whether it be Amazon Go or Google Car.

Before moving forward, I would like to include a word of caution about computer vision algorithms, particularly about those in the OpenCV library. Not all the algorithms available in the library are free to use commercially. So, it is recommended that you perform due diligence before proceeding with a complex computer vision project. Additionally, certain frameworks, such as Emgu CV, require your code to be released as open source or that you purchase a commercial license.

CHAPTER 7

Game Development with Unity

The original use case for the Kinect, and perhaps still the most popular, is game development. It is difficult to have a discussion on game development these days without bringing up Unity. Most readers will already be familiar with Unity. For those who are not acquainted, know that Unity is a cross-platform game engine that targets various APIs. These include the ever-fashionable Direct3D as well as the pretender to the throne, OpenGL. It is not limited to PCs, however. It also supports mobile, Windows Store, VR/AR, websites, and consoles. Unity apps are primarily developed in C# (though there is also JavaScript support), thus Unity does not require too much of a context switch from typical Kinect programming. It is free, is easy to get started with, and, importantly, has third-party support for the Kinect. In this chapter, we will cover the basics of integrating the Kinect with Unity.

Getting Started

The first step is to download Unity from `https://store.unity.com`. The free Personal version will suffice. You should be using Unity 5 and onward (this chapter was written using Unity 5.4), but the Kinect for Windows v2 will also work with Unity 4 Pro. Unity should be installed by following the installer prompts. Make sure that **Windows Build Support** is enabled in the Download Assistant, as in Figure 7-1.

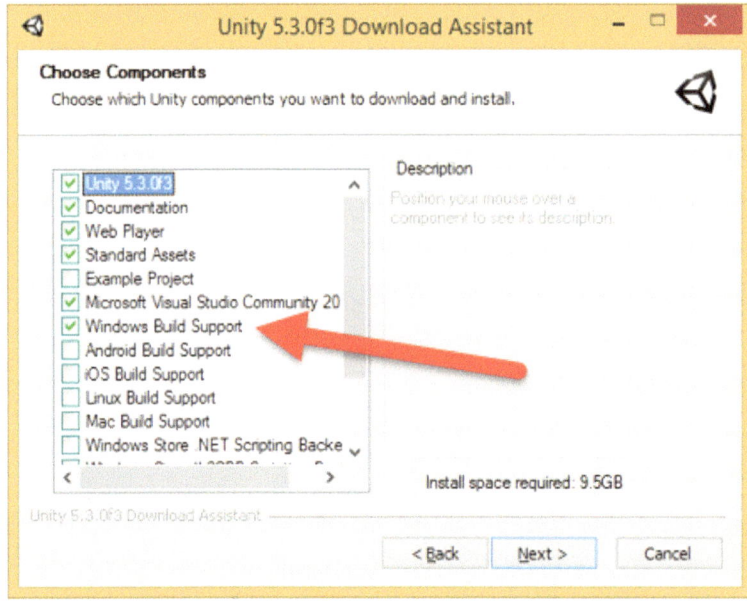

Figure 7-1. *Windows Build Support in the Download Assistant*

© Mansib Rahman 2017
M. Rahman, *Beginning Microsoft Kinect for Windows SDK 2.0*, DOI 10.1007/978-1-4842-2316-1_7

■ **Note**   If you require further assistance with installing Unity, visit https://docs.unity3d.com/Manual/
InstallingUnity.html

You will also need to download the Kinect plugins for Unity from Microsoft. You can get them in a ZIP
file from https://go.microsoft.com/fwlink/p/?LinkId=513177.

After unzipping the plugin files, you will have three Unity package files and two sample Unity project
folders named *KinectView* and *GreenScreen.* The Unity package files must be imported into Unity to enable
Kinect functionality. The primary plugin with all the data streams is Kinect.2.0.1410.19000. There is also
Kinect.Face.2.0.1410.19000 for face and Kinect.VisualGestureBuilder.2.0.1410.1900 for use with
Visual Gesture Builder gestures (which will be covered in chapter 8).

■ **Note**   The numbers at the end of the Unity package files may be different for you, but are unlikely to
change at this point.

To import these plugins, we need a Unity project. We will create a new one from scratch. Start Unity
and proceed to create one by clicking on *New project.* You will be presented with options to title your
project, set its location, and a couple of others, as in Figure 7-2. Give your project a name, choose a folder,
and then click on *Create project.*

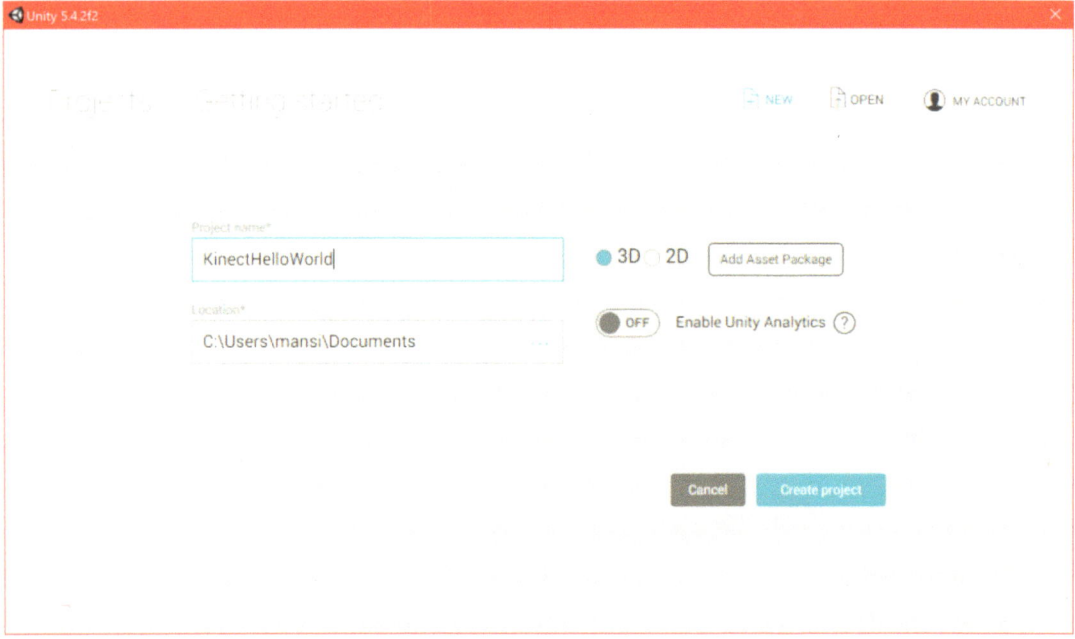

Figure 7-2. *Creating a new project in Unity*

The Unity editor will now load, as in Figure 7-3. Head over to the *Assets* tab on the menu bar and click
on *Import Package,* followed by *Custom Package,* as in Figure 7-4.

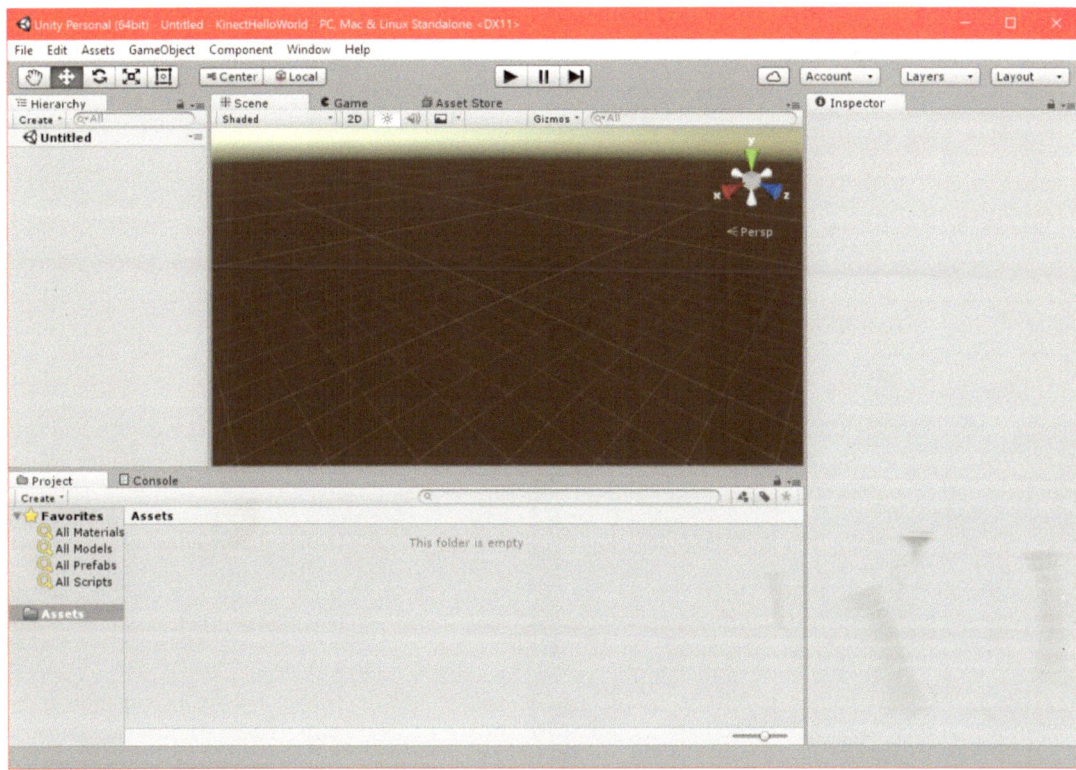

Figure 7-3. *The Unity editor*

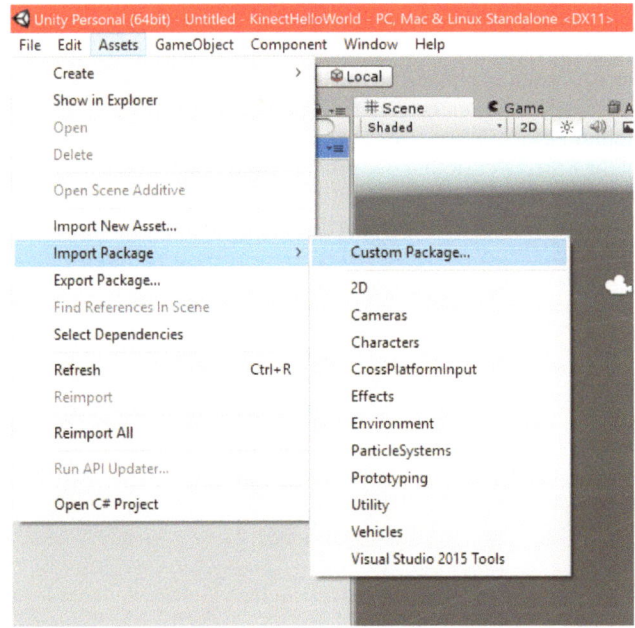

Figure 7-4. *Importing a custom package in Unity*

Choose the `Kinect.2.0.1410.19000` file that was uncompressed earlier. In the Import Unity Package window that shows up (Figure 7-5), click on *All* if all the boxes are not ticked already, and then finally click on *Import*.

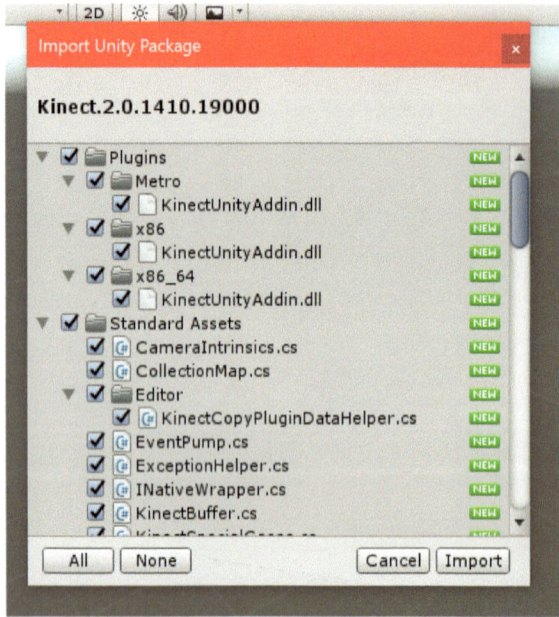

Figure 7-5. *The Import Unity Package window*

The selected folders and files should have all been imported and should now be visible in the Assets window on the bottom of the Unity Editor window. All the game objects used in a Unity project (e.g., models, cameras, shaders, etc.) and their associated scripts will be found here.

Let us test out a sample project to see if everything fell into place. In the Project window, toward the bottom of your screen, right-click on the *Assets* folder and select *Show in Explorer*. This will open the project's folder in Windows File Explorer. In another File Explorer window, navigate to the folder where you unzipped the Kinect Unity plugin and copy the *KinectView* folder back to the *Assets* folder of the *KinectHelloWorld* project. Once you have done this, return to Unity.

In Unity, a popup (Figure 7-6) should appear saying that an API update is required in order for the sample to work in Unity. Click on *I Made a Backup. Go Ahead!*

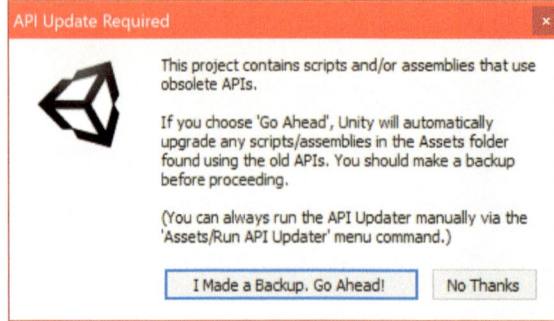

Figure 7-6. *Select "I Made a Backup. Go Ahead!" in the API Update Required Popup*

As files are added and removed from the folder structure, Unity watches and dynamically compiles and updates the Unity project artifacts accordingly. Unity has to be the active window in Windows for this to occur.

■ **Note**   Clicking on the **MainScene** Unity scene file in the *KinectView* folder from the File Explorer will open Unity, but will not open the scene or treat the file as if it were a part of the overall project. The scene must be copied into the appropriate folder structure, and then the API and artifacts will be updated through the popup.

After the API updates are fulfilled, there should be a *KinectView* folder in the Assets section of Unity. Click on it and then double-click on *MainScene*. The sample should load in the Unity editor's *scene* viewer (the large center window). To test it out, ensure that the Kinect device is plugged in and the `KinectMonitor.exe` service is running. Next, press **Ctrl** + **P** or press the play button at the center top. Unity should start the "game" at this point, and the *game* viewer will display the Kinect feeds, as in Figure 7-7.

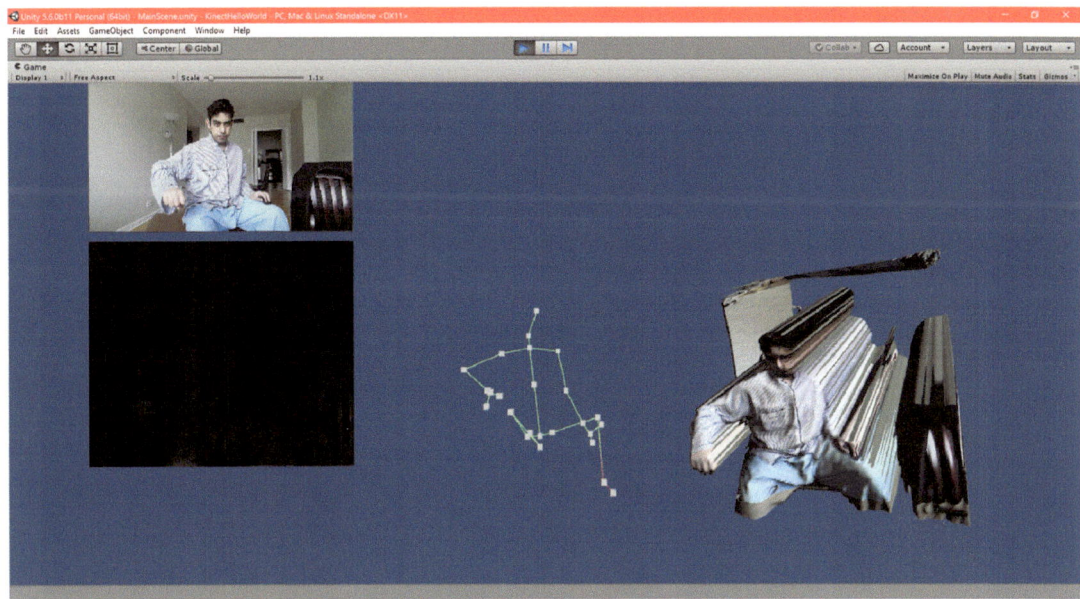

Figure 7-7. *The Unity game viewer displaying feeds from the Kinect*

The `KinectView` scene features the color, infrared, and skeletal feeds from the Kinect. The depth feed is rendered as a point cloud. Obviously, the space between an object in the foreground and the wall behind cannot be visualized, as the Kinect's line of sight there is obstructed. Despite this, the accuracy of the Kinect's rendering of the folds on the clothing is still pretty impressive.

This really is not obvious, but you can rotate the depth point cloud using the keyboard arrow keys or the ASDW keys (I would not have found out had I not looked at the source).

Another noteworthy feature of the scene is that you can choose whether to synchronize the frames received through the Kinect for the depth visualization by switching between using one `MultiSourceFrameReader` or individual `SourceFrameReaders`. This can be done by clicking the screen or pressing the Left Ctrl key (or **Spacebar**, depending on your Unity configuration).

Understanding Kinect for Windows v2 for Unity Through KinectView

The Kinect for Windows v2 API for Unity is in many ways very similar to the traditional Kinect APIs for Windows and Windows Store. To understand how they vary, we have to understand the fundamental differences between programming a typical .NET WPF app and a Unity game. One such distinguishing factor is Unity's reliance on the *game loop* programming pattern. As you might recall, in Chapter 2 we briefly discussed an events-based approach to Kinect development as opposed to a polling-based approach. While we have used events exclusively throughout the book, I did mention that polling was better suited for certain cases. Well, that case has now reared its face in all its glory. Game loops make extensive use of polling, and Unity is no exception. To accommodate this, the Kinect API in Unity is geared toward polling-based data collection. Additionally, WPF constructs such as WriteableBitmaps are not relevant in Unity, so we must find analogues to display video frames inside of a game.

▪ **Note** Similar polling methods are actually available in WPF Kinect apps, but we elected to focus on events, as they were better suited to the WPF paradigm.

To better understand the structure of the Kinect API for Unity, we will explore the source code of the KinectView app.

A Primer on Unity

When opening the KinectView scene, we are greeted by some rectangular prisms suspended in 3D space. If you cannot see all of them within the view-box as they are shown in Figure 7-8, try moving around with the keyboard arrow keys and panning with the right-click button. These prisms (among other things) are known in Unity as *game objects*. Game objects are base containers that encapsulate entities and decide how they are portrayed and how they behave within a scene. The four prisms in the scene that render the camera, infrared, skeletal, and depth data are all game objects of our scene.

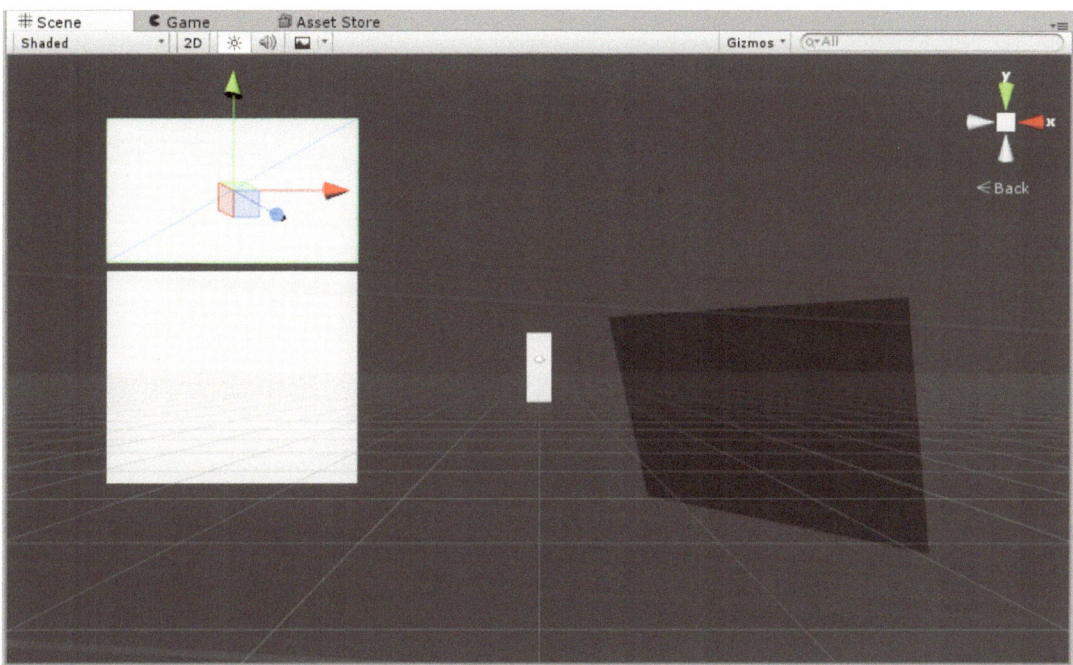

Figure 7-8. *Unity's scene editor displaying some of the KinectView game objects*

Each game object is made up of various properties that can be viewed in the *Inspector* panel (Figure 7-9), which is typically found on the right side of the scene editor. These properties are known in Unity as *components*. Components can include meshes, which decide the shape of the object (if it even has one); colliders, which decide the object's physical interactions with other objects; scripts, which dictate the object's behavior; and many other things. By default, a game object only has one component: the *transform* component. This dictates where a game object is placed in the 3D world, along with its orientation and size. Other components can be added depending on what you want to achieve with the game object.

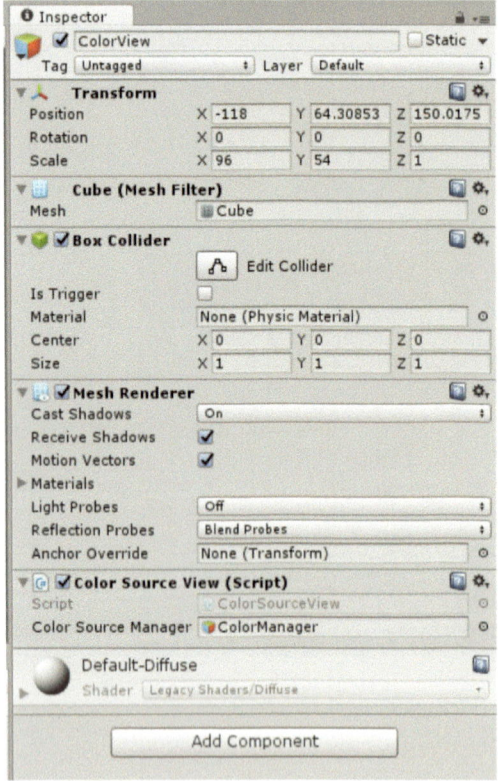

Figure 7-9. *Inspector panel for the color feed window in the KinectView app*

Not every game object is clearly visualized in the scene editors. Game objects such as lights or cameras might have small icons that are barely visible among the other objects in the editor, and scripts are invisible altogether. Additionally, certain objects are sub-objects of other objects, and the hierarchy will not always be clear from what is seen in the editor. To facilitate the tracking of objects that fall in the aforementioned categories, and all game objects in general, we can refer to the *Hierarchy* panel (Figure 7-10).

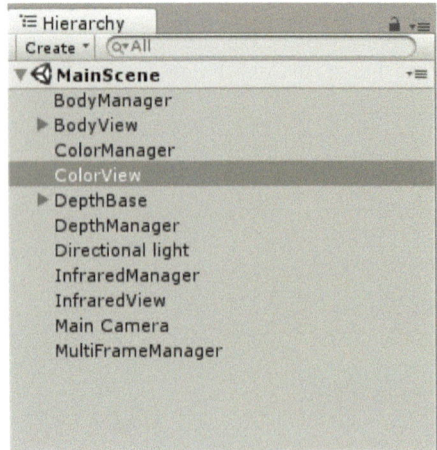

Figure 7-10. *The Hierarchy panel listing all the game objects in the KinectView app*

If you were to click on an object on the list, like BodyManager or MultiFrameManager, its position in the 3D Unity game scene would be highlighted in the scene editor (if you do not see it, double click on the list item to bring the object into your field of view).

Color and Infrared

The color and infrared feeds function quite similarly and are hence grouped together. The infrared data requires a bit of extra processing since we need to implement its visualization from raw data. The two Data Sources begin with the top-level game objects ColorManager and InfraredManager. Neither of them contain anything but their eponymous Data Source management scripts: ColorSourceManager.cs and InfraredSourceManager.cs. Both of these scripts are initialized and begin running as soon as the game starts. Let us take a look at ColorSourceManager.cs.

Listing 7-1. ColorSourceManager.cs

```
using UnityEngine;
using System.Collections;
using Windows.Kinect;

public class ColorSourceManager : MonoBehaviour
{
    public int ColorWidth { get; private set; }
    public int ColorHeight { get; private set; }

    private KinectSensor _Sensor;
    private ColorFrameReader _Reader;
    private Texture2D _Texture;
    private byte[] _Data;

    public Texture2D GetColorTexture()
    {
        return _Texture;
    }

    void Start()
    {
        _Sensor = KinectSensor.GetDefault();

        if (_Sensor != null)
        {
            _Reader = _Sensor.ColorFrameSource.OpenReader();

            var frameDesc = _Sensor.ColorFrameSource.CreateFrameDescription(ColorImageFormat.Rgba);
            ColorWidth = frameDesc.Width;
            ColorHeight = frameDesc.Height;

            _Texture = new Texture2D(frameDesc.Width, frameDesc.Height, TextureFormat.RGBA32, false);
            _Data = new byte[frameDesc.BytesPerPixel * frameDesc.LengthInPixels];
```

```
            if (!_Sensor.IsOpen)
            {
                _Sensor.Open();
            }
        }
    }

    void Update()
    {
        if (_Reader != null)
        {
            var frame = _Reader.AcquireLatestFrame();

            if (frame != null)
            {
                frame.CopyConvertedFrameDataToArray(_Data, ColorImageFormat.Rgba);
                _Texture.LoadRawTextureData(_Data);
                _Texture.Apply();

                frame.Dispose();
                frame = null;
            }
        }
    }

    void OnApplicationQuit()
    {
        if (_Reader != null)
        {
            _Reader.Dispose();
            _Reader = null;
        }

        if (_Sensor != null)
        {
            if (_Sensor.IsOpen)
            {
                _Sensor.Close();
            }

            _Sensor = null;
        }
    }
}
```

Listing 7-1 should look awfully familiar, unless you have skipped the rest of the book until this point. Unsurprisingly, we start off by referencing the UnityEngine namespace. This contains all Unity-related code and must be included in any script that wants to interact with the Unity game world. We see that the Kinect's namespace is different; we must reference Windows.Kinect instead of the usual Microsoft.Kinect. In our class declaration, we then proceed to inherit from the MonoBehaviour class. MonoBehaviour is the base class that every Unity script derives from. It contains, among other things, the Start(), Update(), and OnApplicationQuit() function callbacks. These are responsible for script (de)initialization and hooking onto the Unity game loop. Additionally, we have methods such as GetComponent<Type>(), which allows the script to interact with game objects.

The script's variable declarations are very similar to those of a WPF Kinect project:

```
private KinectSensor _Sensor;
private ColorFrameReader _Reader;
private Texture2D _Texture;
private byte[] _Data;

public Texture2D GetColorTexture()
{
    return _Texture;
}
```

KinectSensor and ColorFrameReader are both present. Instead of Bitmaps, however, we have a Texture2D. This is the standard class for handling texture assets. Just like we had public getters for all the WriteableBitmaps we used previously, we now make one for the Texture2D.

Start() is a function invoked by the Unity Engine and is called as soon as a script is enabled; at the start of the game, in our case. It is the equivalent of a class initialization in our WPF examples. The code is again remarkably similar to our other examples in this book. Perhaps only the Texture2D initialization requires special attention:

```
_Texture = new Texture2D(frameDesc.Width, frameDesc.Height, TextureFormat.RGBA32, false);
```

The first three arguments are width, height, and color format. The fourth argument is whether the texture gets *mipmapped*. This is a technique used to attain a more seamless 2D representation of depth in a graphical application. It is not really pertinent to our application, so we can turn it off.

■ **Tip**   Mipmapping (Figure 7-11) refers to an image-processing technique that scales a high-resolution image into several progressively lower-resolution images. Each level of scale features an image that is a power of two smaller than the last. This increases performance by allowing the rendering engine to render lower-resolution copies of the image in certain scenarios, such as when an image is being rendered far away in a 3D world. It also reduces image noise in motion scenes (when a character is moving in relation to the image) and increases scene realism. The drawback of mipmapping is that it requires more space and memory, as there are now multiple prerendered copies of the image.

Figure 7-11. *An image of Lena with mipmapping applied. Lena is a historical and controversial test image widely used in the computer-vision and image-processing field. The original picture was taken by Dwight Hooker for the November 1972 issue of Playboy magazine.*

The Update() call is fired on every game frame. While in practice it is possible to use it as a measure of time, we should use Time.deltaTime to get the time passed from the last Update call instead, because lengthy code can slow down the Update invocations, which would yield incorrect time measurements. For our purposes, we can use it to substitute the functionality of our FrameArrived event handlers, as follows:

```
void Update()
{
    if (_Reader != null)
    {
        var frame = _Reader.AcquireLatestFrame();

        if (frame != null)
        {
            frame.CopyConvertedFrameDataToArray(_Data, ColorImageFormat.Rgba);
            _Texture.LoadRawTextureData(_Data);
            _Texture.Apply();

            frame.Dispose();
            frame = null;
        }
    }
}
```

After getting the latest frame from the _Reader.AcquireLatestFrame() call and copying its data with frame.CopyCovertedFrameDataToArray(byte[] arr, ColorImageFormat fmt), all we have to do is load it in our texture with _Texture.LoadRawTexture(byte[] data) and save the new data onto the texture for display with _Texture.Apply(). The using keyword was not used like it was in our other samples, so the frames are disposed of manually.

To tidy things up, we dispose of the sensor and reader in the OnApplicationQuit() method as we would in an WPF app. As the name states, the method is called when the Unity application itself quits.

While the aforementioned code collects and saves color feed data, we have to apply a script on one of the prisms to actually display it. For this reason, the `ColorView` game object has the `ColorSourceView.cs` script attached to it (Listing 7-2).

Listing 7-2. ColorSourceView.cs

```
using UnityEngine;
using System.Collections;
using Windows.Kinect;

public class ColorSourceView : MonoBehaviour
{
    public GameObject ColorSourceManager;
    private ColorSourceManager _ColorManager;

    void Start()
    {

gameObject.GetComponent<Renderer>().material.SetTextureScale("_MainTex", new Vector2(-1, 1));
    }

    void Update()
    {
        if (ColorSourceManager == null)
        {
            return;
        }

        _ColorManager = ColorSourceManager.GetComponent<ColorSourceManager>();
        if (_ColorManager == null)
        {
            return;
        }

        gameObject.GetComponent<Renderer>().material.mainTexture = _ColorManager.GetColorTexture();
    }
}
```

There is a public and a private variable for the `ColorSourceManager` instance. We can set it programmatically from another class, but there is also a nifty Unity feature that allows us to set it via the GUI as well. Taking a look at the `ColorView` object's Inspector panel, there is the option to set a Color Source Manager (Figure 7-12).

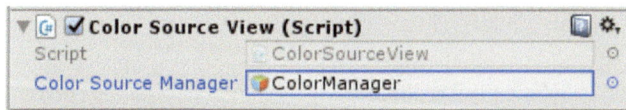

Figure 7-12. *The Color Source Manager option in the ColorView game object*

Clicking on the circle icon beside the option, we can choose any game object in the scene to be our Color Source Manager. Naturally, the `ColorManager` object is an apt choice.

Returning to our code, we have the `gameObject.GetComponent<Renderer>().material.SetTextureScale("_MainTex", new Vector2(-1, 1))` statement in the `Start` method. This sets the scale of the image that will be displayed on the `ColorView` prism. `GetComponent<Type>()`'s type is set to *Renderer*, thus it gets the Renderer component (which you can verify it has through the Inspector window). This component has a material property that we are manipulating on the code side. `SetTextureScale()`'s two arguments are the name of the texture and its scale. The name refers to the name of the texture that the prism's shader (i.e., `Default-Diffuse`) uses, `_MainTex` in our case. Changing the size and signs of the scale vector will change how it fits on the prism surface as well as its direction. For example, setting the vector to (1, 2) flips the color feed image in the horizontal direction and halves the height in the vertical direction, making it repeat twice.

■ **Note** `GetComponent<Type>()` method could have been called without the explicit `gameObject` declaration. In this context, it is the equivalent of having used the `this` keyword.

The two important statements in the `Update` call are `_ColorManager = ColorSourceManager.GetComponent<ColorSourceManager>()` and `gameObject.GetComponent<Renderer>().material.mainTexture = _ColorManager.GetColorTexture()`. The former sets the private `ColorSourceManager` variable. It is a bit redundant to call it every frame, and we could have opted to call it once in the script initialization. The latter statement is used to set the shader texture using the `Texture2D` getter in our `ColorSourceManager` class.

At this point, the prism can get a new texture whenever one is available in the current frame, and the code is complete. Before we take a look at the non-image data feeds, let us inspect how the infrared data is processed (Listing 7-3).

Listing 7-3. InfraredSourceManager.cs Processing Pixel Data

```
...
frame.CopyFrameDataToArray(_Data);

int index = 0;
foreach(var ir in _Data)
{
    byte intensity = (byte)(ir >> 8);
    _RawData[index++] = intensity;
    _RawData[index++] = intensity;
    _RawData[index++] = intensity;
    _RawData[index++] = 255; // Alpha
}

_Texture.LoadRawTextureData(_RawData);
_Texture.Apply();
...
```

As you might recall from Chapter 3, infrared data for a pixel is obtained as a `ushort` (16 bits). Our RGB data is byte-sized, however, so we shift the second byte in the `ushort` forward and use that as our grayscale intensity value. Since our image is RGB and our data is grayscale, we set the three color intensities in each pixel using the same value to make the RGB image still appear grayscale. Other than this brief bit of processing, the infrared image data is obtained and displayed in the same manner as color.

Skeletal

Like with color data, gathering skeletal data from the Kinect in Unity is practically identical to doing so in the regular Kinect API. What changes is how the skeleton is displayed. In my opinion, it is much easier to work with and much more powerful. This is because the rendered skeleton has a 3D footprint in Unity unlike the comparable example in WPF, and this is achieved without requiring us to go into the nitty gritty of DirectX programming.

Looking at the `Update()` loop for `BodySourceManager.cs`, we can quickly confirm that it is virtually identical to the method for getting skeletal data in WPF (Listing 7-4).

Listing 7-4. BodySourceManager.cs Acquiring Skeletal Data

```
void Update()
{
    if (_Reader != null)
    {
        var frame = _Reader.AcquireLatestFrame();
        if (frame != null)
        {
            if (_Data == null)
            {
                _Data = new Body[_Sensor.BodyFrameSource.BodyCount];
            }

            frame.GetAndRefreshBodyData(_Data);

            frame.Dispose();
            frame = null;
        }
    }
}
```

In Listing 7-4, we see that in practice the only difference is with how the frame is disposed of, and that was more of a matter of preference than anything.

Listing 7-5. BodySourceView.cs Tracking and Management of Skeletons

```
void Update ()
{
    if (BodySourceManager == null)
    {
        return;
    }

    _BodyManager = BodySourceManager.GetComponent<BodySourceManager>();
    if (_BodyManager == null)
    {
        return;
    }
```

```
        Kinect.Body[] data = _BodyManager.GetData();
        if (data == null)
        {
            return;
        }

        List<ulong> trackedIds = new List<ulong>();
        foreach(var body in data)
        {
            if (body == null)
            {
                continue;
            }

            if (body.IsTracked)
            {
                trackedIds.Add(body.TrackingId);
            }
        }

        List<ulong> knownIds = new List<ulong>(_Bodies.Keys);

        // First delete untracked bodies
        foreach(ulong trackingId in knownIds)
        {
            if (!trackedIds.Contains(trackingId))
            {
                Destroy(_Bodies[trackingId]);
                _Bodies.Remove(trackingId);
            }
        }

        foreach(var body in data)
        {
            if (body == null)
            {
                continue;
            }

            if (body.IsTracked)
            {
                if (!_Bodies.ContainsKey(body.TrackingId))
                {
                    _Bodies[body.TrackingId] = CreateBodyObject(body.TrackingId);
                }

                RefreshBodyObject(body, _Bodies[body.TrackingId]);
            }
        }
    }
}
```

The Update method in BodySourceView.cs, as shown in Listing 7-5, is responsible for the tracking and management of skeletons. The tracking code is not particularly noteworthy, except for the portion that is responsible for the deletion of existing bodies and the creation of new ones. Destroy(_Bodies[trackingId]) is a Unity method to destroy game objects or components. The actual object, which in our case is a generated skeleton, is only destroyed after the update loop is finished. If a body is detected but not yet tracked a new body is created using the custom CreateBodyObject(ulong id). Another custom method, RefreshBodyObject(Body body, GameBody gameBody), is used to refresh the body on each frame.

Listing 7-6. BodySourceView.cs Generation of New Skeletons

```
private GameObject CreateBodyObject(ulong id)
{
    GameObject body = new GameObject("Body:" + id);

    for (Kinect.JointType jt = Kinect.JointType.SpineBase; jt <= Kinect.JointType.ThumbRight; jt++)
    {
        GameObject jointObj = GameObject.CreatePrimitive(PrimitiveType.Cube);

        LineRenderer lr = jointObj.AddComponent<LineRenderer>();
        lr.SetVertexCount(2);
        lr.material = BoneMaterial;
        lr.SetWidth(0.05f, 0.05f);

        jointObj.transform.localScale = new Vector3(0.3f, 0.3f, 0.3f);
        jointObj.name = jt.ToString();
        jointObj.transform.parent = body.transform;
    }

    return body;
}
```

In Listing 7-6, we create new skeleton objects in Unity (this can be confirmed by taking a glance at the Hierarchy window while a skeleton is being actively tracked). We go through all the JointType enums and create a cube representing a joint along with a bone to connect with its parent joint. The cube is a game object created with a mesh renderer through the Unity GameObject.CreatePrimitive(PrimitiveType shape) method. The bone is similarly a game object with a line renderer added via AddComponent<LineRenderer>(). A bone material is added to the line. This material is provided as a default asset in the example by Microsoft and can be set in the Unity editor in BodySourceViews' public variables.

■ **Note** A *mesh renderer* is a game engine tool used to render an array of triangles into a geometrical shape such as a cube or sphere. A *line renderer* is similarly a tool used to render an array of points into a line in 3D. These are added to game objects as components to describe their perceived physical structure in a 3D or 2D world.

Note that none of the bones are connected in this method. We simply scale the joints and bones appropriately and set their parent transform to that of their body's.

Listing 7-7. BodySourceView.cs Animation of Skeletons

```
private void RefreshBodyObject(Kinect.Body body, GameObject bodyObject)
{
    for (Kinect.JointType jt = Kinect.JointType.SpineBase; jt <= Kinect.JointType.ThumbRight; jt++)
    {
        Kinect.Joint sourceJoint = body.Joints[jt];
        Kinect.Joint? targetJoint = null;

        if (_BoneMap.ContainsKey(jt))
        {
            targetJoint = body.Joints[_BoneMap[jt]];
        }

        Transform jointObj = bodyObject.transform.FindChild(jt.ToString());
        jointObj.localPosition = GetVector3FromJoint(sourceJoint);

        LineRenderer lr = jointObj.GetComponent<LineRenderer>();
        if (targetJoint.HasValue)
        {
            lr.SetPosition(0, jointObj.localPosition);
            lr.SetPosition(1, GetVector3FromJoint(targetJoint.Value));
            lr.SetColors(GetColorForState (sourceJoint.TrackingState),
            GetColorForState(targetJoint.Value.TrackingState));
        }
        else
        {
            lr.enabled = false;
        }
    }
}
```

In Listing 7-7, the joint positions are obtained from the body and are used to position and orientate the joints and bone lines. We go through all the joints and check if they have another joint we can draw a bone to. The bone pairs are stored in a dictionary titled _BoneMap. We set the joint positions with jointObj. localPosition = GetVector3FromJoint(sourceJoint). GetVector3FromJoint(Joint joint) is simply a helper method that grabs the position values from the joints and creates a three-dimensional positional vector with the magnitudes scaled for better viewing. If the joint has a bone pair with another joint whose position is known (tracked or interpolated), we access the line renderer and set its points to that of each joint so as to reposition the bone in-between. Finally, the joint's tracking state is indicated with the GetColorForState(Kinect.TrackingState state) helper method. It simply returns a color based on the tracking state that is then drawn at the ends of each bone. This completes the entire process of displaying skeletal data.

Aside from the scripting side of things, it is interesting to note that the small white prism representing the skeleton in the scene editor (affectionately known as Cube in the Hierarchy view) is not actually integral to the demo. The skeleton will be successfully displayed even if the prism is removed from the scene. In fact, if you check the cube's attached script, the object is simply disabled at the start of the demo. We can assume that the demo developer included it there to indicate where the skeleton should be displayed within the game world.

Depth

Recreating the point cloud of the depth feed is a bit of an involved process that I will not be exploring in depth (no pun intended). Instead, we will take a more general overview.

There is nothing novel in how depth data is gathered, and `MultiSourceFrame` data is not gathered in a particularly noteworthy manner either. It is worth taking a look at how the depth point cloud switches between both Data Sources, however. They both have script game objects in the scene: `DepthManager` and `MultiFrameManager`. The `DepthSourceView` object references both of them through public variables in its own script. The switch between the Data Sources is initiated in the `Update` loop of `DepthSourceView`, as shown in Listing 7-8.

Listing 7-8. DepthSourceView.cs Input Control for Data Source Management

```
void Update()
{
    if (_Sensor == null)
    {
        return;
    }

    if (Input.GetButtonDown("Fire1"))
    {
        if (ViewMode == DepthViewMode.MultiSourceReader)
        {
            ViewMode = DepthViewMode.SeparateSourceReaders;
        }
        else
        {
            ViewMode = DepthViewMode.MultiSourceReader;
        }
    }
...
```

To detect if the user wants to switch Data Sources, we refer to `UnityEngine`'s static `Input` class. It has a `GetButtonDown(string buttonName)` method that returns a boolean indicating whether the button has been pressed. "Fire1" is a default button mapping in Unity to handle common game actions (i.e., firing a gun or making a confirmation). It should be mapped to **Left Ctrl** or **Mouse0** (left-click) unless the user configures this to something else. Once an input indicating a Data Source switch occurs, we simply flip between the `DepthViewMode` enum values and use this to execute the code for the appropriate Data Source.

We also want to notify the user of the current chosen Data Source. This is done by printing the current value of `DepthViewMode` to the GUI. The `MonoBehaviour` class has a method we can use to handle GUI manipulation: `OnGUI()`. It is fired repeatedly, like the `Update` method. The exact frequency may vary depending on whether any event triggers it, but it is often as frequent as, if not more frequent than, the current frame rate. Usage of this method is demonstrated in Listing 7-9.

Listing 7-9. BodySourceView.cs GUI handler

```
void OnGUI()
{
    GUI.BeginGroup(new Rect(0, 0, Screen.width, Screen.height));
    GUI.TextField(new Rect(Screen.width - 250, 10, 250, 20), "DepthMode: " + ViewMode.ToString());
    GUI.EndGroup();
}
```

GUI is another `UnityEngine` static class, and it is responsible for handling the GUI (big surprise there!). The GUI features nesting through the use of *groups*. All GUI controls must be in a group. We can create a group with the `GUI.BeginGroup(Rect position)` method, which takes a rectangle denoting the area covered by the group as a parameter. This must be terminated with a matching `GUI.EndGroup()` call. Within the group created in Listing 7-9, we call `GUI.TextField(Rect position, string text)` to display the Data Source as text on the GUI.

The depth point cloud is rendered with a programmatically generated mesh. The slanted gray rectangle to the right of the game scene is not actually what the mesh is generated in. If you check the `DepthBase` game object, it contains two child game objects: `DepthView` and `DesignerOnlyView`. `DesignerOnlyView` is that rectangle, and it has a script disabling itself at the start of a game. It is merely a visual helper to indicate the location of the point cloud in the scene editor.

The actual mesh is created in the game's `Start` handler with the user-defined `CreateMesh(int width, int height)` method. The mesh is downsampled by four, as it would otherwise have too many vertices for Unity to handle (65,535 is the max, though the error message will say 65,000). `CreateMesh` creates a series of vertices and triangles that initially resembles a rectangle. It also UV maps to the 2D color image feed.

■ **Tip** *UV mapping* is the process of projecting a 2D texture map onto a 3D model. Think of trying to wrap a 2D map of the world onto a sphere to make it into a globe. That is a problem that UV mapping aims to tackle. The 'U' and 'V' in UV mapping simply stands for the X and Y axes in 2D, as 'X', 'Y', and 'Z' are already reserved for 3D. To learn more about UV mapping, refer to learning resources for a 3D modeling tool such as **Blender**.

In every `Update` called, the custom `RefreshData(ushort[] depthData, int colorWidth, int colorHeight)` method is used to reform the mesh. It uses `CoordinateMapper` to map the depth points to color space. Since the point cloud has been downsampled, the depth point that is displayed on the point cloud is actually an average of it and 15 of its neighboring depth points. Finally, the UV map is updated to reflect the new depth points.

Summary

In this chapter, we covered the basics of Unity development using the Kinect for Windows v2. We saw that the Kinect's data is best obtained using polling, as opposed to event handlers.

Although we covered the bases for integrating Unity with the Kinect, we have barely scratched the surface when it comes to Unity development itself. Just like with computer vision and image processing, entire books could be written about Unity development. The scenarios that can be developed are limitless. A HoloLens-based adventure game? A tool to analyze the movement of tennis players to offer fine-tuned corrections? A virtual drum set? These are all things you can develop with Unity. Fortunately, Unity is one of the most accessible game engines and, given enough time and effort, you will be creating amazing Kinect experiences with it.

In this next chapter, we will cover some additional tools to facilitate the development of interactive experiences with the Kinect.

CHAPTER 8

■ ■ ■

Miscellaneous Tools

Much of what we have covered through Chapter 7 has been about how to interact with the Kinect's data purely through code. While this grants us an endless capacity to develop very specialized Kinect applications, we risk reinventing the wheel for many common tasks.

Just consider the undertaking of detecting gestures. The most rudimentary manner to approach it would involve some form of heuristics; for example, if the hand joint is higher than the head joint, the person has their hand raised. It is not difficult to see how this increases in complexity for even the simplest of gestures. A hand-waving gesture would require some way to track the periodicity of the wave to check whether the person is actually waving, just saying stop, or perhaps just has their hand raised idly. Given that each joint has three position properties, our conditional statements stand to be very long and undesirably nested. The situation worsens dramatically if we start considering velocity and acceleration.

The ideal solution is one that incorporates custom machine-learning algorithms. The process involves gathering the data, cleaning and sanitizing the data, labeling the data, training the algorithms to build a model, and then testing. This seemingly complex process has the unique advantage of overcoming discrepancies, such as between how the gesture is performed by two different people or how it is affected by different environmental conditions. These algorithms can also be repurposed for different gestures, drastically reducing development time if there are many to incorporate. All this being said, developing a machine-learning solution might be beyond the approach of an amateur developer, and it might not be economical for a small team on a tight deadline.

Fortunately, Microsoft, in its infinite wisdom, has decided that a gesture-recognition toolkit would indeed be a welcome addition in the arsenal of the everyday developer (and not just that of large Xbox game studios). So instead of developing machine-learning algorithms from scratch, we can use **Visual Gesture Builder** (**VGB**). This should save you A LOT of time, so it is only natural that we give it attention in this book.

Another nifty tool we will be covering is **Kinect Fusion**. Kinect Fusion is essentially a 3D scanning tool. Think a photocopier, but for 3D objects. Finally, there is also **Kinect Ripple**. Kinect Ripple is a projector-based infotainment system that allows us to create interactive presentations with HTML5 in JavaScript. It is only by taking advantage of tools such as these that we can exploit the Kinect's full potential.

Visual Gesture Builder

Visual Gesture Builder is a tool included with the Kinect for Windows v2 SDK that uses machine-learning algorithms to create gesture databases for Kinect applications. These databases are used to perform gesture detection at runtime using the VGB API. VGB uses the *AdaBoost* algorithm to determine when a user performs a certain gesture and uses the *RFRProgress* algorithm to determine a gesture's progress. These gestures are captured with **Kinect Studio** (refer to Chapter 2 for a refresher) and fed into VGB.

© Mansib Rahman 2017
M. Rahman, *Beginning Microsoft Kinect for Windows SDK 2.0*, DOI 10.1007/978-1-4842-2316-1_8

■ **Note** AdaBoost and RFRProgress are short for *Adaptive Boosting* and *Random Forest Regression Progress*, respectively.

VGB "learns" gestures by taking in clips and having you tag the portions of the clips where the desired gesture is visible. On the flip side, the untagged parts of the clips are used as negative footage to indicate what does **not** constitute a gesture. A small number of clips should provide a usable gesture-detection solution, but to get a production-grade system it is best to input hundreds, if not thousands, of examples of the gesture. These should include bodies with different proportions, different camera angles, and different environmental conditions. An order of magnitude increase in the number of required videos is usually what is necessary to go from 95 percent to 99.99 percent, which can be critical in an industrial or commercial setting where the application might be used hundreds of times per day. This is not exactly a VGB issue, more just a general machine-learning drawback and one that is typically present with a heuristics-based approach as well.

Before you get started with VGB, you should make note of the distinction between *discrete* and *continuous* gesture detection. On one hand, you have discrete detection, which relies on AdaBoost, a binary classifier. You are either performing a gesture, or you are not. On the other hand, you have continuous detection, which relies on RFRProgress and takes advantage of regression analysis. It tells you which stage you are in during the execution of the gesture. A progression value of 0 means you have yet to begin, and 1 means you are at the end stage. The tagging process for each differs, so we will explore them separately.

■ **Note** Many gestures can be modeled using both discrete and continuous detection. It is up to you to determine the best fit for your application. Typically, single-pose gestures are best for discrete and moving gestures are best for continuous. More advanced scenarios mix and match both. Mixing and matching allows you to use discrete gestures to start various progressive stages of continuous gestures; think golf swing, punching, or kicking.

Discrete Gestures

To understand the process of creating a discrete gesture, we will create one of our own. Gestures that are typically easier for VGB to learn are simple ones that involve singular (non-repetitive), large movements, such as sitting down or a single punch. Finer movements, such as a Vulcan salute, or repetitive motions, such as clapping, can be a bit harder to learn. Despite this, we will create a clapping gesture detector to show how powerful VGB can be.

We start off by gathering the data and recording three ~1:30-minute clips in Kinect Studio. These clips should include you clapping in certain portions and you doing other things or nothing at all in the remaining parts. For now, the clapping can be performed in a consistent manner; for example, both hands pointed up and positioned in front of the chest. To perfect the gesture in the long run, you would need to clap at different angles with variable hand positioning. This can be done in future clips; for now, we will focus on a standard interpretation of clapping.

■ **Note** Body, depth, and IR data must specifically be recorded in Kinect Studio for use in VGB. Color is not required and should be turned off if possible to save space. Only use body data where the skeleton is aligned to the actual body of the player and where flickering is limited.

Once you have your clips, search for *Visual Gesture Builder* on your computer and launch it. It might have - PREVIEW affixed to the application title, as in Figure 8-1. Start off by creating a new *Visual Gesture Builder Solution* (.vgbsln) for your Kinect application. This solution will consist of all the separate gestures used for one application. For example, the solution for a fighting game might include the punch, kick, and parry gestures.

Once you have created a new solution, it should show up in the Explorer pane on the right side of the window. This pane details the data involved in your gesture-detection solutions in a hierarchal manner. As in Visual Studio, the solution will contain project files within it. Each *Visual Gesture Project* (.vgbproj) represents exactly one gesture (i.e., clapping). To create a new project, right-click on the solution icon in the Explorer pane and select *Create New Project With Wizard*, as in Figure 8-1.

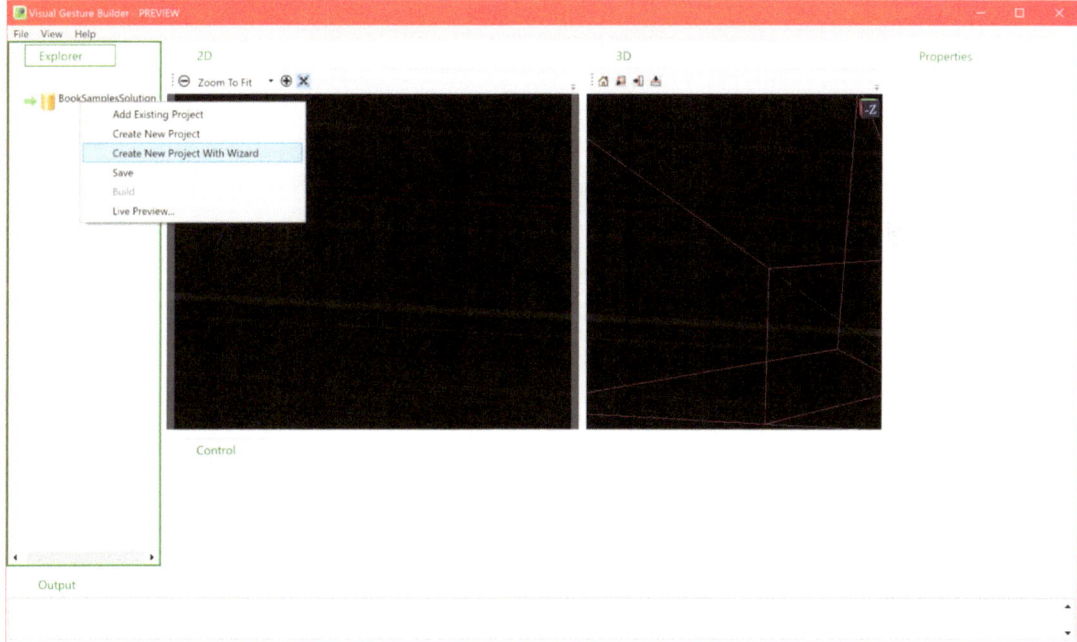

Figure 8-1. *The primary Visual Gesture Builder view. Right-clicking the solution icon presents the option of creating a new project with the wizard. The Explorer window is seen on the left.*

The VGB Gesture Wizard will now take you through the different settings that can be configured for your gesture. The first setting, as seen in Figure 8-2, is none other than the name of the gesture. This should be a base name that does not take into account the handedness of the gesture (i.e,. left or right). Choose an appropriate name, **Clapping** in our case, and then click *Next*.

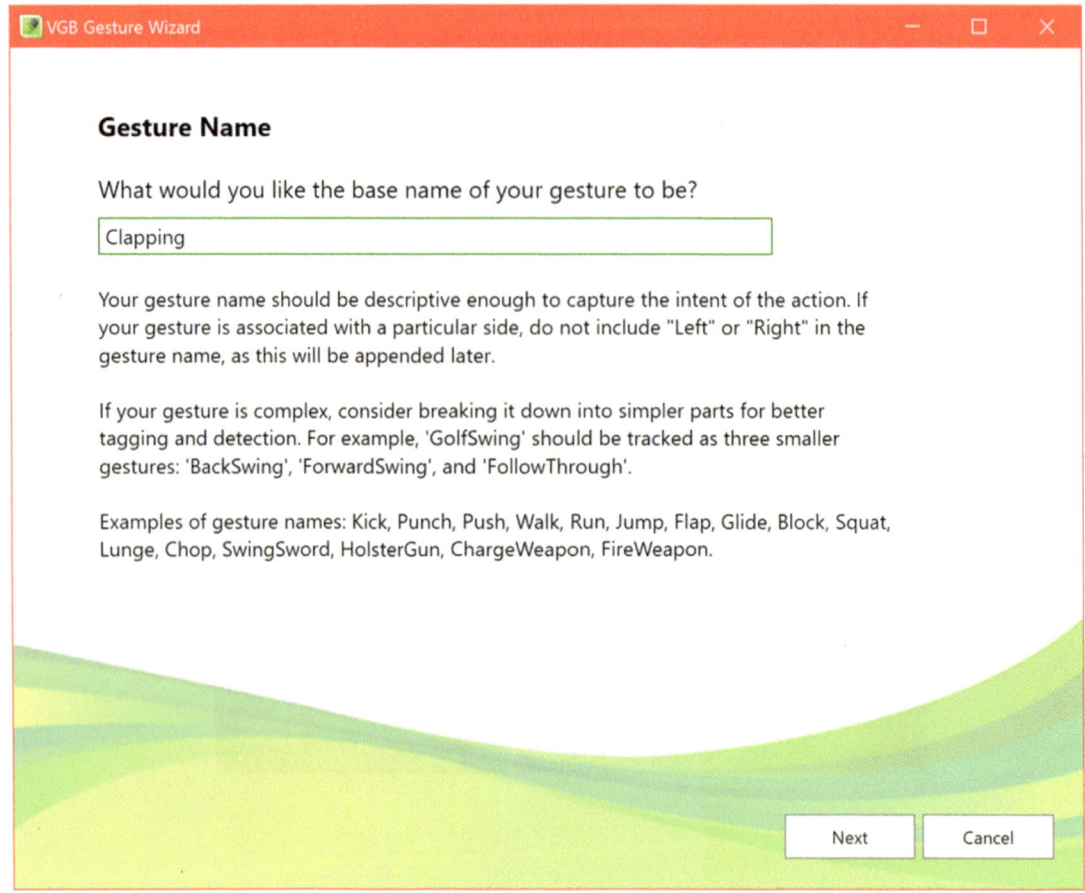

Figure 8-2. *Choosing the name of the gesture in Gesture Wizard*

The next question (Figure 8-3) asks whether we need the joints below the hip. You should choose *No* if you can help it. The more joints the algorithms have to process, the more work it has and the longer it will take. They can also affect the detection accuracy. If all the clips inputted into VGB feature people clapping while standing straight and then in production someone is clapping while squatting, the squatting person might not be identified as clapping because their legs are not standing straight. Since clapping does not involve feet (unless your hands are preoccupied or you are a monkey, in which case we would have to create a new gesture altogether anyway), we can safely choose *No*.

Body Region

Does this gesture rely on joints in the lower body (below the hips)?

○ Yes (default)
● No

Figure 8-3. *Deciding whether to include the lower body region in Gesture Wizard*

Next, we are asked whether we want to support hand states (Figure 8-4). I prefer to keep these options turned off. Hand states can be very jittery if they are detected at all. It is mainly useful if you have two very similar gestures that can use every bit of differentiation available. An example of such would be a Force Push gesture versus a punch gesture. Both involve extending the arm forward. The Force Push requires an open palm whereas punching requires a closed one. In the case of clapping, I opted not to use hand states, though it would be interesting to see if it would work better with hand states.

Figure 8-4. *Deciding whether to rely on hand states in Gesture Wizard*

The following step (Figure 8-5) lets you pick among four possible joint masks for the training process. This basically allows you to eliminate certain inconsequential joints from the training. The top-left joint mask is best suited for clapping, since clapping mainly relies on the arms. The bottom-left joint mask is appropriate for something like a sitting-down gesture. The two on the right are appropriate for handed gestures, such as a right kick or a left slap.

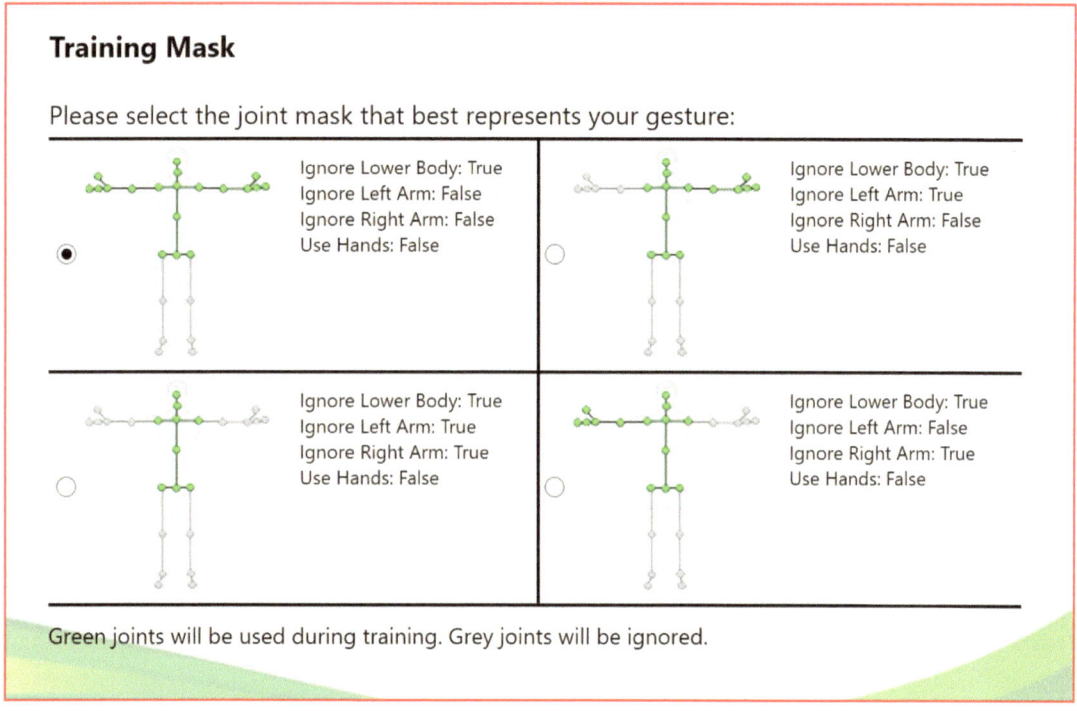

Figure 8-5. *Choosing the joint mask in Gesture Wizard*

We are then asked whether our gesture depends on handedness (Figure 8-6). This is so that VGB can affix the side at the end of the project name. This is not the case for the clapping gesture.

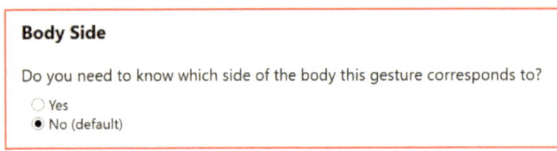

Body Side

Do you need to know which side of the body this gesture corresponds to?

○ Yes
● No (default)

Figure 8-6. *Choosing the body side of the gesture in Gesture Wizard*

If our gesture is symmetrical, we can choose to duplicate our data so that we have an extra set where everything is mirrored (Figure 8-7). This essentially halves the time required to record gestures. A symmetrical gesture refers to one where flipping all the joints along the horizontal and depth axes would result in the gesture's still being valid, (though it might be considered an opposite-handed gesture in the process). Although I chose *Yes* for the clapping gesture, choosing *No* may also be a valid option. This is because the gesture may end up being over-indexed on one side. For example, if you are doing a right-punch gesture and no left-punch gesture, you do not want to replicate the gesture for the left side of your skeleton.

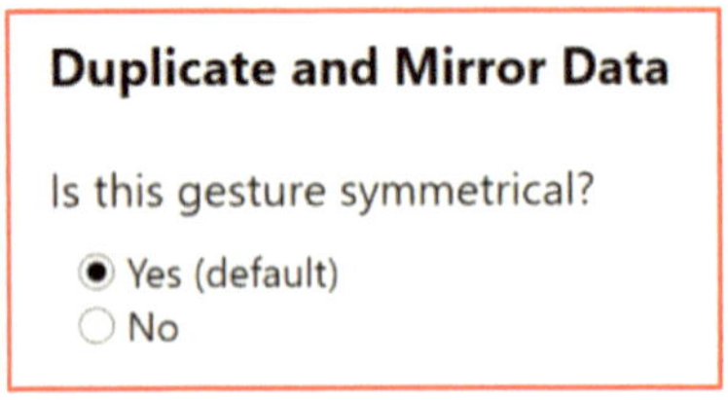

Duplicate and Mirror Data

Is this gesture symmetrical?

● Yes (default)
○ No

Figure 8-7. *Choosing whether to duplicate and mirror data in Gesture Wizard*

Getting fed up with all the configuration options? Bear with me, because we are almost done. The last thing we need to do is to indicate whether our gesture requires progress detection (Figure 8-8). If you have not skipped through the last few pages, you will know that the answer is *No*, as we are not creating a continuous gesture.

Progress

Will you need to detect progress during this gesture (typically used for animation)?

○ Yes
● No (default)

Figure 8-8. *Choosing whether our gesture is discrete or continuous in Gesture Wizard*

If VGB has not crashed, which it is unfortunately a bit prone to do at times, it should have created two files under your solution file in Explorer, as in Figure 8-9. You might have chosen for these to be in a different folder hierarchy in Windows, but they will still appear as a child of the solution in VGB.

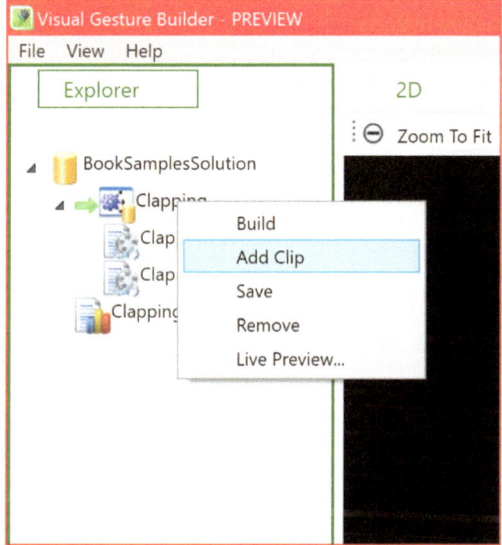

Figure 8-9. *Adding clips to a training project*

The first file created (the one with a blue gear icon) is the *training project*. This is where we add our Kinect Studio clips and perform the labeling process, a.k.a. "tagging," to train our gesture models. The other project, the one with a bar-chart icon and a **.a** file extension, is an *analysis project*. We can use this as the test process to verify the quality of our trained gesture models.

Of our three Kinect Studio clips, we will add two to our training projects. Right-click the training project and click on *Add Clip*, as in Figure 8-9. For convenience, I renamed these files to "**ClappingX,**" where **X** is simply a number I am incrementing per clip. Rename these before you add them to the project, or you will lose your work if you try to do so later.

Once you add a clip, you should see the 2D and 3D visualized feeds from the Kinect, along with a timeline of sorts underneath. This timeline will be labeled **Control**, as in Figure 8-10. This is where all the tagging for the videos takes place. Each frame in the video is represented as a line on the horizontal gray bar.

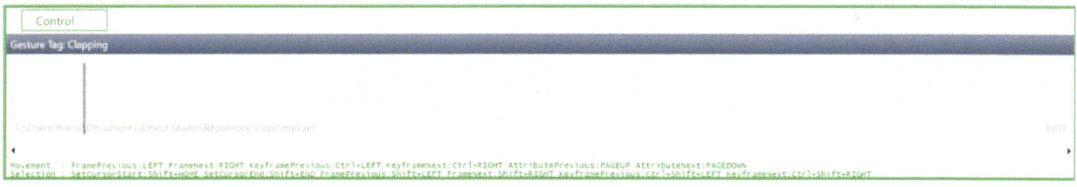

Figure 8-10. *The gesture tagging timeline*

While you can click through the timeline with your mouse, you will probably want to learn the shortcut keys to minimize the effort spent tagging. Depending on your screen size, you might need to resize the Control window to see all of the shortcut keys. They are situated under the gray bar. The most important ones are the **Left** and **Right** arrow keys to move backward and forward through the frames, respectively; the **Shift** key (while moving through frames) to select them; and the **Enter** key to tag a frame as positive. You can also press **Ctrl + Right/Left** arrow key to cycle through the tagged regions.

The process of tagging consists of going through each frame in the clip and deciding whether the gesture is being performed in it or not. This is made easier by the assumption that anything in between the starting frame of a gesture and the ending frame of a gesture is likely a gesture. In our clip, we keep going to the right until we find the first clip of our gesture.

Let us compare the clips in Figures 8-11a and 8-11b to clarify this. In Figure 8-11a, the person is seen to be in the motion of starting a clap, but is not exactly clapping yet. In Figure 8-11b, the person is on the verge of making a clap. These figures are consecutive frames; there are no frames in between.

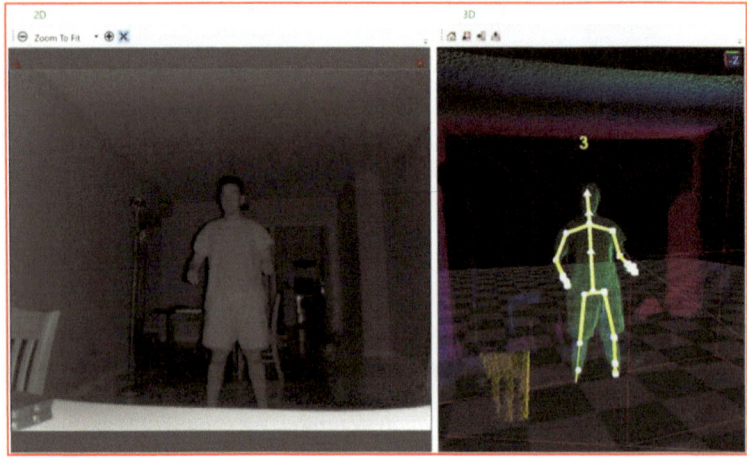

Figure 8-11a. *The person is getting ready to clap, but is not quite there yet*

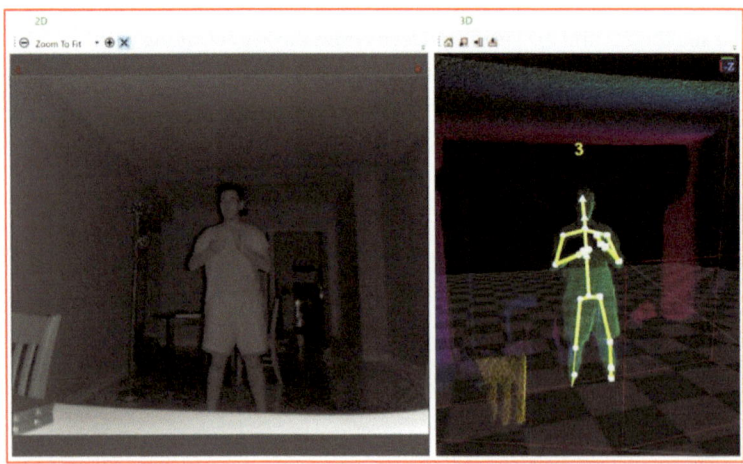

Figure 8-11b. *The person is commencing their clapping gesture*

The subsequent frames are of the person performing a clapping gesture. In this case, Figure 8-11b is the correct frame to tag as true and the start of the gesture. This frame could have conceivably been found in the middle of someone performing a clapping gesture. It looks like they are already clapping in it.

The ending frame is essentially determined in a similar manner. We pick the last clip of the gesture being performed and not the one right after. We have two options on how to handle the gesture's end. We can either end it when an individual clap ends (more work, but more accurate) or just keep it on continuously during a series of claps (much easier, not as accurate). Personally, I just tag a series of claps altogether with no space in between. For certain other gestures, however, you may want to space and delineate the tags more attentively. This is especially the case for clips used in continuous auto-tagging (which we will cover later).

Once all the clips are tagged, the clip timeline should look something like that in Figure 8-12. If you have been following along in VGB until this point, you will know that the blue bars are the regions that have been positively tagged.

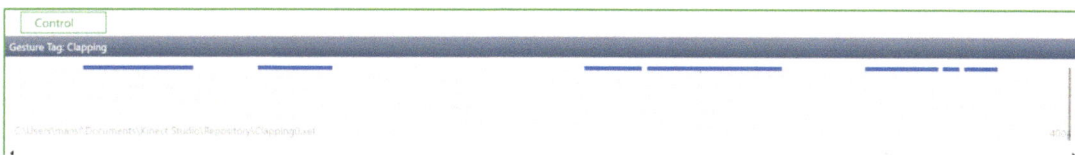

Figure 8-12. *Tags on a clip's timeline*

As mentioned earlier, the untagged parts are considered to be negative, or not consisting of the gesture in question. However, we can also choose to explicitly declare portions of the clip as negative. This is done with the **Backspace** key, used in the same manner as the **Enter** key. If there are explicit negative tags on the clips, untagged regions will no longer be considered negative and will instead not be factored into the learning process at all.

One may wonder why one should explicitly declare a region as negative at all, especially considering all the extra work that this would entail. This feature is useful when you want to purposely provide negative feedback to an undesired behavior in the detection algorithm. Suppose you have a punching gesture that inadvertently also counts pushing with one hand as punching. You can include people pushing in your clips and tag those as negative. With a focus on having those behaviors negated, they should be unlearned more effectively. Theoretically, you could include entire clips where there are no positive tags and only negative tags.

Once you have tagged both clips, right-click on the solution file for the gesture and click *Build*. You will be prompted to save a **.gbd** file, which is the gesture database file that you can use with your Kinect applications. To visually test your new gesture database, go to **File ➤ Live Preview...** and select the .gbd file that you just created. You will see an interface resembling the one shown in Figure 8-13.

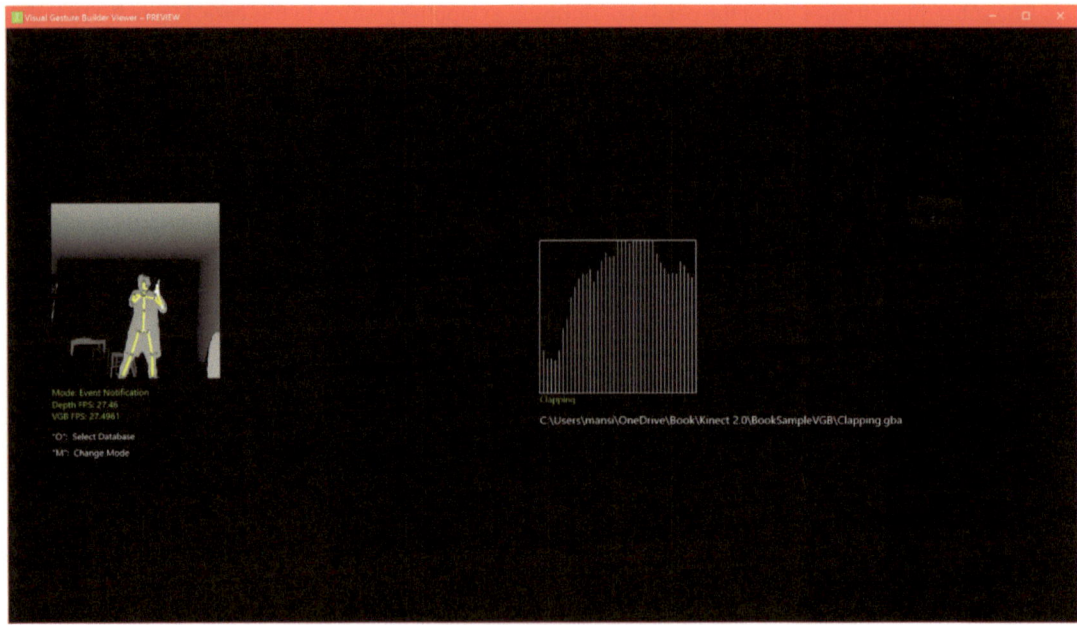

Figure 8-13. *Live preview of the clapping gesture database in VgbView*

■ **Note** It is possible to build either the project or the solution. Building the solution includes of all the gestures in the database, whereas building a gesture project will include just a specific one.

To test your gesture database in a live preview, ensure that your Kinect is plugged in and operational, then perform your gesture. The graph on the right with the white lines indicates the confidence levels for the gesture you are performing. If the white bars are small or nonexistent, no gesture is being detected. If the bars are close to touching the graph's ceiling, the gesture is being detected strongly.

It may be unfeasible to test every variation of every gesture you have trained. There may just be too many, or it may be too hard for a human to reliably determine the accuracy of the gesture detection in realtime. To render this process more convenient, Visual Gesture Builder comes with an analysis tool.

To make use of it, add one or more clips to the analysis project (again, that is the one that ends in `.a`), then right-click the project and select *Analyze*. After the analysis is complete, you will see the average RMS, number of false positives, and number of false negatives for the database used for the analysis project (as seen in Figure 8-14).

Project Analysis Results:

ID	Build	# Frames	Worst Error	Average RMS	False Positives	False Negatives		
1	Clapping.gba-20170311220635	1850	1	0.195904	1	0	Export	Clear

Figure 8-14. *Analysis of errors in the detection capabilities of the selected gesture database*

Tip Average RMS refers to the *average root mean squared error*. It is used to measure the difference between values of the gesture-detection database and the actual observed values in the analysis. An average RMS error of 0 would indicate a perfect score.

If you click on a specific clip used in the analysis, you can see how closely it adheres to the gesture database's definitions. The green line shown in Figure 8-15 represents where the gesture was assumed to be occurring by the gesture-recognition database on the analysis clips. It is natural for it to lag slightly after each (blue) tag portion. If there are areas where it believes there is a gesture, even though you have indicated otherwise, you should rework the training dataset until it shows otherwise. Try including additional clips, and play with the tagging a bit. Creating gesture databases is an iterative process that will involve a lot of fine-tuning. Depending on your use, a database with mild inaccuracies may be fine, but perfecting a database is a rewarding process nonetheless.

Figure 8-15. *Seeing how well the gesture database fits the analysis clips*

Continuous Gestures

Creating a continuous gesture is not too different from creating a discrete gesture. Typically, you will need to create a discrete gesture equivalent for a continuous gesture, but the reverse will not be necessary. This is because continuous detection always gives you a progression value, even if the action is not being performed. Having discrete detection will help determine if the action is being performed at all and whether we can turn off the continuous-gesture detector to save system resources and ignore false progression values. Another use of this technique is that we can stich together multiple simpler gestures in our code to create a more complex gesture.

To create a continuous gesture, open the VGB Gesture Wizard again and create a gesture for clapping (keep the title as **Clapping**), but opt for the progress option this time (see Figure 8-8). At the final dialog, you will notice that the wizard asks to create both a continuous and discrete gesture for clapping. If you already have the discrete version, you can uncheck the checkbox next to the discrete clapping gesture.

Add a clip to the new *ClappingProgress* project. As with the discrete gesture, you will have to tag examples of this gesture. However, the manner in which it is done is a bit different. Instead of a boolean value saying whether a frame in the clip contains the gesture or not, we have a float value describing the progress of the gesture. If you look at Figure 8-16, you will see a line graph in the Control section. It represents the progress of the gesture over time. There are set values on the graph, which are depicted with little gray boxes, but the rest of the lines are interpolated by VGB. If you look at the top-right corner of Figure 8-16, you will notice the ClappingProgress float value (set to 0), which tells us what the value of the progress at the current instant is.

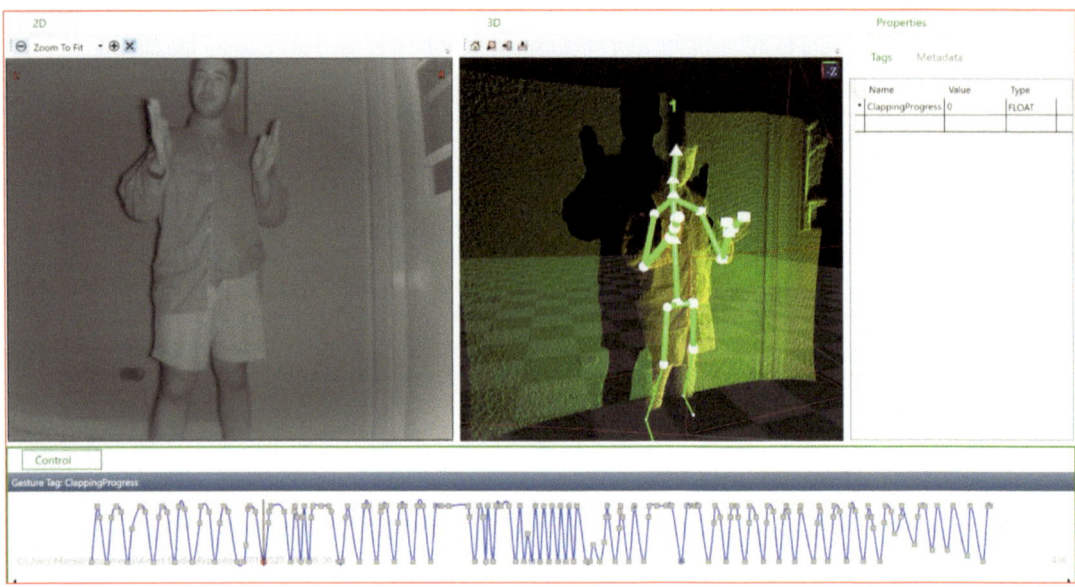

Figure 8-16. *VGB's view for the tagging of continuous gestures*

To tag the progression of the gesture, pick a point on the timeline, then set the `ClappingProgress` value in the *Properties* section to a float value between 0 and 1. The value should represent the progression from 0 percent (0) to 100 percent (1). The more user-defined tags there are on the timeline, the more the gesture detection will correspond to what you have in mind for the gesture. Humans are inherently imprecise, however, so your user-defined tags might not end up representing the true progression of the gesture.

A quick way to add progression tags on the timeline is through keyboard number keys. Each number on the keyboard corresponds to a 10^{-1} progression value. So, for example, pressing the **3** key on your keyboard will result in a frame being tagged with a progression value of 0.3.

After having tagged your clip, you can build your project or solution and visit Live Preview again to test it.

In Figure 8-17, we see the progress detection for the `ClappingProgress` gesture. Unlike for the discrete gesture, the white lines do not represent confidence levels. Instead, they indicate the progression of the gesture. If they reach the roof, the progression is near complete. If they straddle the bottom of the square, the gesture has barely begun. If they take up about half the graph, that might be an indication that no gesture is being detected. It is not a sure sign though (the gesture might just be halfway complete), so it is better to have discrete gesture detection on for clapping beforehand.

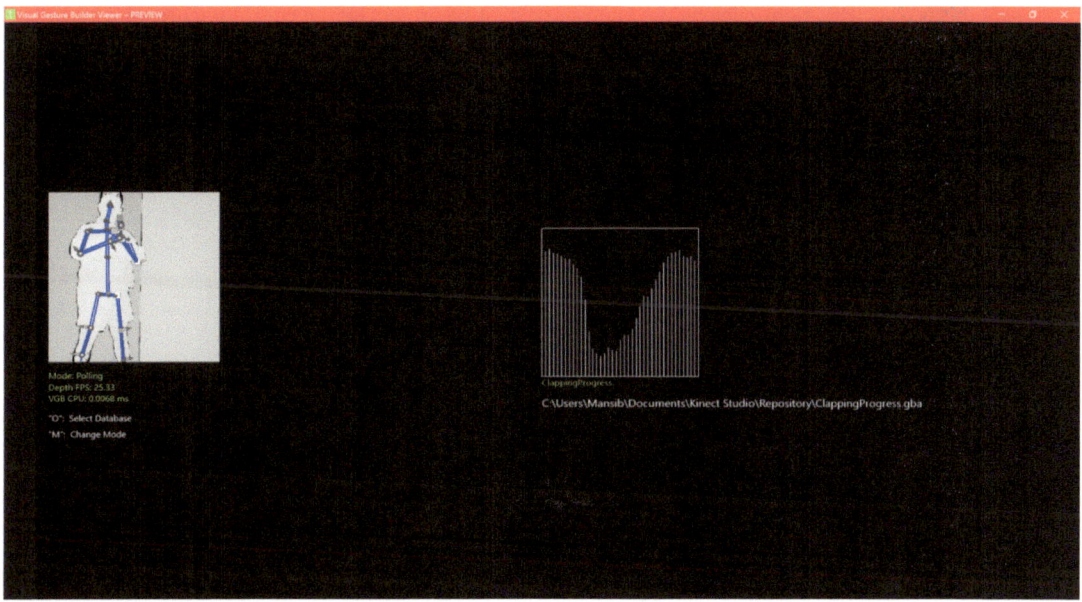

Figure 8-17. ClappingProgress gesture being tested in VgbView

Auto-Tagging

It can be tedious to have to tag hundreds of clips, so fortunately Microsoft went ahead and included an auto-tagger in VGB. For a continuous gesture, what this implies is that we can map discrete tags to progression values. For example, if we have a kick gesture that has been tagged in a clip, the beginning of the kick gesture will be set to have a 0 progression value and the end of the kick gesture will be set to have a 1 progression value.

Moreover, you can stitch multiple gestures together and create a compounded continuous gesture using this technique. For example, let's say you have a single continuous gesture called "PunchKick" in a fighting game, but it is actually composed of a "Punch" discrete gesture followed by a "Kick" discrete gesture. You can set the progression of the PunchKick gesture from 0 to 0.5, mapped to the discrete Punch gesture, and 0.5 to 1 for the discrete Kick gesture.

To perform auto-tagging for continuous gestures, ensure the clips you want to auto-tag have discrete tag labels in their properties. You can achieve this by adding the clip to all the relevant discrete gesture projects. For example, if you have clips for PunchKick, add the same clips to both Punch and Kick as well.

Once this is complete, right-click on the analysis or training project for the continuous gesture and click on *Generate Tags*. The *Link Gestures* dialog will open up as in Figure 8-18.

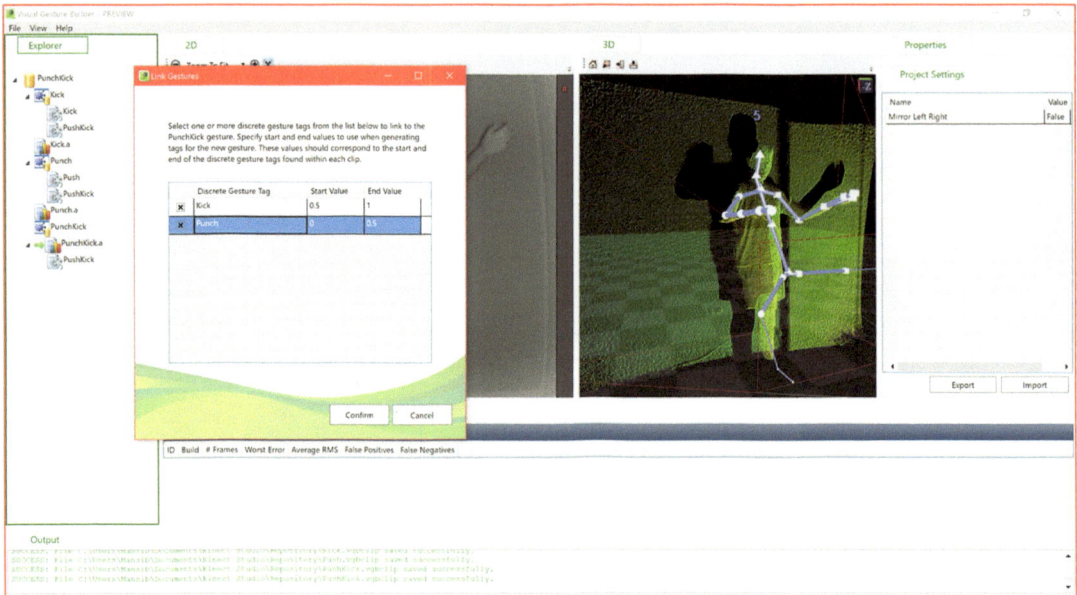

Figure 8-18. *Generating tags and creating a compound gesture (note that certain clips erroneously say Push, but should really say Punch)*

In the dialog, you will have to attribute progression values to the start and end of discrete gestures. Once you are done, click *Confirm*, and the tags will be auto-generated. You will get a pink line on the timelines for your clips, as shown in Figure 8-19, with some explicit progression values set (where the discrete values start and end). This should be fairly accurate, but depending on your gesture, you may need to add or edit midpoint values. In this case, on the neighbor of the frame shown in Figure 8-19, I added a 0 value to indicate that the gesture has not begun. Originally, it was a value around 0.3. To complete the tagging of this clip, I would repeat this on other parts of the timeline and then click on *Accept Tags*, which is right on top of the timeline.

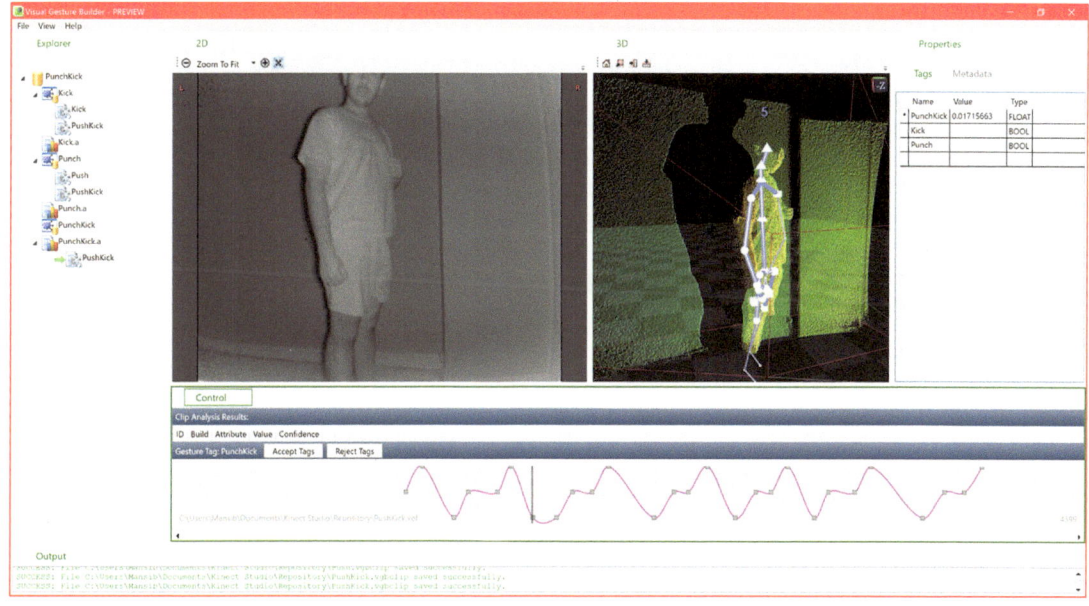

Figure 8-19. *Tags generated for a continuous gesture in VGB*

Including Gesture Databases in a Kinect Project

One you have your .gbd file, we can include it in a Kinect project and perform gesture recognition. You can add it to your Visual Studio project as you would any external file. Simply click on your project in Solution Explorer, press **Shift** + **Alt** + **A** and then select the file. There is a crucial step you have to perform after having imported your file into the project. Find the .gbd file in Solution Explorer, right-click, and then click on *Properties*. You will have to set the *Build Action* value to *Content*, as seen in Figure 8-20.

Figure 8-20. *Setting the Build Action in Visual Studio for the gesture database*

241

The general procedure for detecting gestures follows a paradigm similar to those of all the other Kinect Data Sources. There is a FrameArrived event handler that continuously fires and informs us whether gestures have been detected or not. All we have to do is ensure that a valid body tracking is available for the detector to work with.

To get started, we have to include the Microsoft.Kinect.VisualGestureBuilder namespace and reference. Conveniently, there is a NuGet package for that: Microsoft.Kinect.VisualGestureBuilder.x64 (https://www.nuget.org/packages/Microsoft.Kinect.VisualGestureBuilder.x64). After that is included, we can incorporate the core VGB variables, as shown in Listing 8-1.

Listing 8-1. Gesture-Detection Variable Initializations

```
private VisualGestureBuilderDatabase vgbDb;
private VisualGestureBuilderFrameSource vgbFrameSource;
private VisualGestureBuilderFrameReader vgbFrameReader;
private Gesture gesture;
private Gesture gestureProgress;

private KinectSensor kinect;
private BodyFrameReader bodyFrameReader;
```

The first object, VisualGestureBuilderDatabase vgbDb, will contain the code's association to the .gbd file we generated in VGB. VisualGestureBuilderFrameSource represents VGB's Data Source. VisualGestureBuilderFrameReader will provide us with frames containing detected gestures. Gesture is an object that represents one of the gestures included in vgbDb. We have one for our discrete gesture and another for its continuous counterpart. KinectSensor invariably makes an appearance.

One thing to keep in mind is that like FaceFrameSource, each VisualGestureBuilderFrameSource only tracks one body. We thus have to keep track of the proper body-tracking ID in projects using VGB, so we also include BodyFrameReader. To track multiple bodies, start multiple VisualGestureBuilderFrameSources and VisualGestureBuilderFrameReaders.

Listing 8-2. Initialization of VGB Variables

```
kinect = KinectSensor.GetDefault();

bodyFrameReader = kinect.BodyFrameSource.OpenReader();
bodyFrameReader.FrameArrived += bodyFrameArrived;

vgbDb = new VisualGestureBuilderDatabase(@"..\Gestures\Gesture.gbd");
vgbFrameSource = new VisualGestureBuilderFrameSource(KinectSensor.GetDefault(), 0);

foreach (var g in vgbDb.AvailableGestures)
{
    if (g.Name.Equals("gesture"))
    {
        gesture = g;
    }
    if (g.Name.Equals("gestureProgress"))
    {
        gestureProgress = g;
    }
    vgbFrameSource.AddGesture(g);
}
```

```
vgbFrameReader = vgbFrameSource.OpenReader();
vgbFrameReader.FrameArrived += vgbFrameArrived;
kinect.Open();
```

In Listing 8-2, we initialize our Kinect variables, load our database, extract the gestures from it, and open our readers. vgbDb = new VisualGestureBuilderDatabase(@"..\Gestures\gesture.gbd"); loads our .gbd file from the filesystem and initializes VisualGestureBuilderDatabase with its reference. The input represents the path of the database file.

■ **Tip**   The @ character declares a string (starting and ending in double quotation marks) as a verbatim string literal. A verbatim string literal ignores escape characters, which facilitates the writing of certain types of strings. In our case, we can write a path without having to use double backslashes (\\).

Using vgbDb.AvailableGestures, we can get a list of available gestures in our gesture data store. We loop through it with a foreach loop and add gesture definitions to our Gesture variables based on their names. We also have to add those gestures to VisualGestureBuilderFrameSource so that our Data Source knows what to look for.

Listing 8-3. Body Frame Event Handler for VGB Project

```
private void bodyFrameArrived(object sender, BodyFrameArrivedEventArgs e)
{
    if (!vgbFrameSource.IsTrackingIdValid)
    {
        using (BodyFrame bodyFrame = e.FrameReference.AcquireFrame())
        {
            if (bodyFrame != null)
            {
                Body[] bodies = new Body[6];
                bodyFrame.GetAndRefreshBodyData(bodies);
                Body closestBody = null;

                foreach (Body b in bodies)
                {
                    if (b.IsTracked)
                    {
                        if (closestBody == null)
                        {
                            closestBody = b;
                        }
                        else
                        {
                            Joint newHeadJoint = b.Joints[JointType.Head];
                            Joint oldHeadJoint = closestBody.Joints[JointType.Head];
                            if (newHeadJoint.TrackingState == TrackingState.Tracked &&
                            newHeadJoint.Position.Z < oldHeadJoint.Position.Z)
                            {
                                closestBody = b;
                            }
```

```
                }
              }
            }
            if (closestBody != null)
            {
                vgbFrameSource.TrackingId = closestBody.TrackingId;
            }
          }
        }
      }
}
```

Most of the code in Listing 8-3 should be self-evident by now. We are simply trying to find the body closest to the Kinect, or, more specifically, the head closest to the Kinect. The interesting bit is that we assign the tracking ID to vgbFrameSource.TrackingID. This focuses the gesture detector on that specific body.

Listing 8-4. VGB Gesture Frame Event Handler

```
private void vgbFrameArrived(object sender, VisualGestureBuilderFrameArrivedEventArgs e)
      {
          using (var vgbFrame = e.FrameReference.AcquireFrame())
          {
            if (vgbFrame != null && vgbFrame.DiscreteGestureResults != null)
            {
            var result = vgbFrame.DiscreteGestureResults[gesture];
              if (result.Detected)
              {
                  float progressResult = vgbFrame.ContinuousGestureResults[gestureProgress];
                  //Do something with progressResult
              }
              else
              {
                  //set progress in front end to 0
              }
            }
          }
      }
```

In Listing 8-4, we have the event handler for our gesture frames. Our frame provides us with a list of discrete and continuous gestures with the DiscreteGestureResults and ContinuousGestureResults lists, respectively. If our gesture result has been detected, we can decide to check for a progress value and execute any relevant code after this point.

■ **Note** Although we are only checking for a progress value after having first checked for a discrete result, the detection for the continuous gesture is still ongoing. The benefit to checking for a discrete value first is that our progress value will be meaningful since the gesture is actually present.

If we want to turn off continuous gesture detection altogether and re-enable it when needed, we can use the VisualGestureBuilderFrameSource.SetIsEnabled(Gesture gesture, bool isEnabled) method.

Unity

VGB is utilized much the same way in Unity as in WPF. The primary peculiarity is how the gesture database is imported into the project. We have to include it in a *StreamingAssets* folder in Unity to make use of it. This enables access to the database through the filesystem as opposed to building it into Unity. To create a *StreamingAssets* folder, simply create a folder under the *Assets* folder named *StreamingAssets*. You can then copy your .gbd file into there (Listing 8-5).

Listing 8-5. Importing a Gesture Database in Unity

```
VisualGestureBuilderFrameSource vgbFrameSource;
VisualGestureBuilderFrameReader vgbFrameReader;

KinectSensor kinect = null;

void Start()
{
    kinect = KinectSensor.GetDefault();
    kinect.Open();

    vgbFrameSource = VisualGestureBuilderFrameSource.Create(kinect, 0);
    vgbFrameReader = vgbFrameSource.OpenReader();

    var databasePath = Path.Combine(Application.streamingAssetsPath, "gesture.gbd");
    using (VisualGestureBuilderDatabase vgbDb = VisualGestureBuilderDatabase.
    Create(databasePath))
    {
        foreach (Gesture gesture in vgbDb.AvailableGestures)
        {
            //add gestures to frame source as in prior example
        }
    }
    ...
```

The code for Unity is pretty similar to that for its WPF counterpart. Instead of using the WPF lifecycle, we have to use the Unity one. This means using Start() instead of a MainWindow constructor. The other important difference is with the constructors of VisualGestureBuilderFrameSource and VisualGestureBuilderDatabase. There is no public constructor for them in Unity, so you have to use their .Create() methods. VisualGestureBuilderFrameSource.Create() takes a KinectSensor argument, and the initial tracking ID (0 if there is none) and VisualGestureBuilderDatabase.Create() takes the path to the gesture database.

Kinect Fusion

Kinect Fusion is one the coolest Kinect applications, but also one of the least utilized. This is partially a result of the fact that though its v1 incarnation was technically impressive, it was still somewhat uninspiring when it came to performance. A series of optimizations was made in v2, and, although not perfect, it is arguably useful for production right now.

For those who did not have the privilege of being wowed in person for the public demo of "KinectFusion" at SIGGRAPH 2011 or have not followed up since, Kinect Fusion is software (accessible through the Kinect APIs) that allows you to capture 3D objects and their environment and reconstruct them digitally. Yes, this means you can use it to scan a real world object and import it into a video game or 3D print it.

Kinect Fusion integrates depth data over time to recreate a 3D volumetric representation. The orientation and location of the camera is tracked so that the software can relate their positions from one frame to another. This creates a *voxel* reconstruction, which can later be exported as a triangle mesh.

■ **Note** Voxels are the equivalent of pixels in 3D. It can be convenient to think of them as "volumetric pixels." Each voxel in Kinect Fusion contains a value indicating the distance to the nearest surface. This value can be positive or negative, depending on which side of the surface the voxel is found.

■ **Cool Fact** *Voxel* is often thought to be a portmanteau of *volume* and *pixel*, but it is in fact a contraction of *volume* and *element* (i.e., an element of volume.) Likewise, the *el* in *pixel* also derives from the word *element*.

The limit to how much space you can record with Kinect Fusion will depend on your GPU and/or CPU. Most GPUs can only typically allocate 1 GB of contiguous memory, which in theory should suffice for 640^3 voxels. At the lowest resolution, you can capture a volume of 8m³ (the higher the resolution, the more this volume shrinks). With parallelization, you may be able to develop more-complex tracking applications that capture larger spans of volume, but you can also just record objects in parts and then fuse them together.

Kinect Fusion Explorer

Kinect Fusion is available as a part of the Kinect API (which you can explore at https://msdn.microsoft.com/en-us/library/microsoft.kinect.fusion.aspx), but Microsoft has produced some applications that you can use to try it out without any coding. Kinect Fusion Explorer is one of them.

A note on specs before we dive in further. As you might imagine, Kinect Fusion works best with a beefy graphics card. While I have gotten it to work on devices with no dedicated video card, a card with at least 2 GB would be ideal for basic usage, and 4 GB+ is recommended. DirectX 11 (with C++ AMP) is also paramount.

To get started, visit the **Kinect SDK Browser** and find the *Kinect Fusion Explorer-WPF* project, then click *Run*.

No configuration is required to get started with Kinect Fusion. As soon as you start the application, it should start recording the surrounding scene and reconstructing a voxel representation for you to view. Figure 8-21 contains some parameter alterations, but the 3D reconstruction should look similar to yours overall.

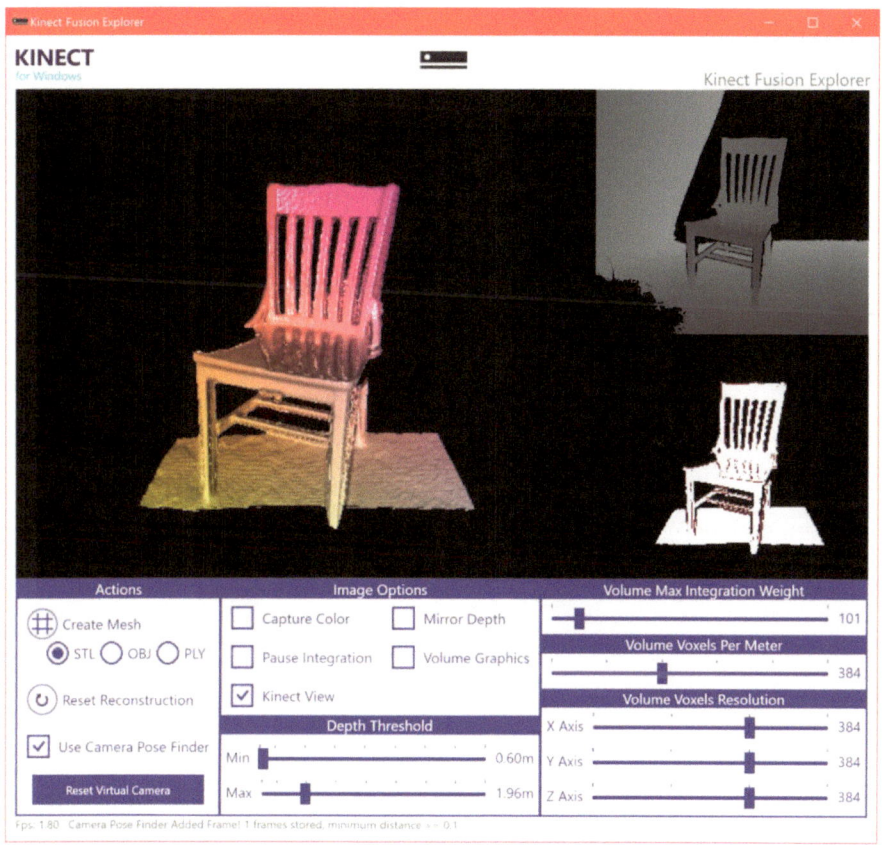

Figure 8-21. *Kinect Fusion Explorer capturing a 3D scan of my dining room*

The larger visualization on the left is that 3D reconstruction. The one on the top-right is the Kinect's depth feed. Finally, the bottom-right one is the delta from the reference image. The red lines on it represent the hard edges that the Kinect detects and uses for its camera-pose tracking.

Comparing the Fusion-generated object with the real world counterpart in Figure 8-22, we can see that Kinect Fusion replicates it with reasonable fidelity. There is some graininess, but that has more to do with the resolution we set for the voxels than it does with the Kinect's capabilities.

Figure 8-22. *Chair as scene from a traditional camera*

It is interesting to note the difficulty that the Kinect has when dealing with reflective surfaces. The chair's glossy flat surface subtly acts as a mirror, trumping the Kinect's IR-based depth sensors. The incidental sunlight illuminates the chair's slats' reflections onto the sitting area, which are then portrayed as projections on the Fusion-generated image. The Kinect also has difficulty with the wooden floor tiles. While they are perfectly flat in reality, they appear dented when reconstructed. This is because the Kinect does perceive the minute, millimeter-wide slits in between each tile. The voxel resolution is too low, however, and the algorithm prefers to perceive curves instead of tiny slits, so Fusion has to show a depression on the surface to make a sensible reconstruction. It generally also has trouble with really flat surfaces.

In Figure 8-23, the *Capture Color* setting is ticked and the color feed is consequently projected onto the Fusion reconstruction. The graphics make the reflective surfaces more palatable.

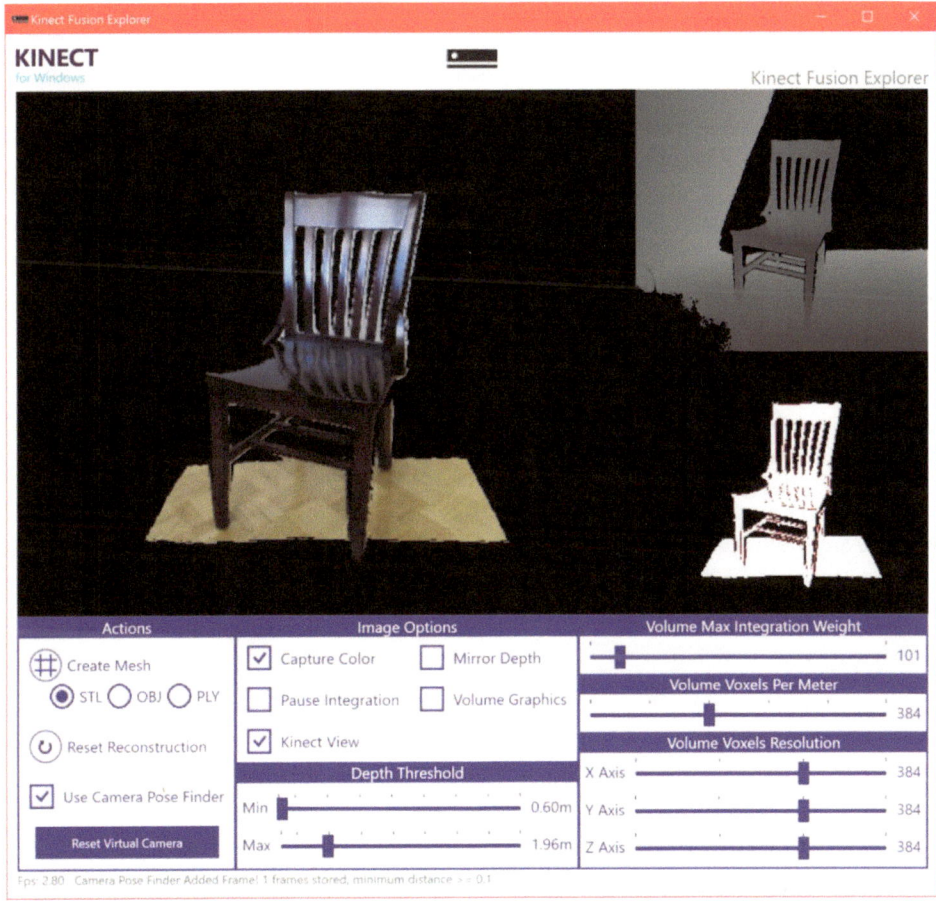

Figure 8-23. *Kinect Fusion Explorer capturing a 3D scan of my dining room chair, in color*

There are some other interesting options to tinker with in Fusion Explorer. One that I had altered beforehand is the *Depth Threshold*. This basically dictates the depth range of the scan. In Figure 8-23, anything in between 60cm and 1.96cm is detected, and everything else is occluded. It was set this way to block anything but the chair from being detected.

Volume Max Integration Weight does not change anything in a static scene, but is useful when there is motion or jitter in the scene. The larger its value, the less susceptible the Fusion reconstruction is to change in the scene. For example, if I kick the chair, the 3D reconstruction will likely only shift a bit. This tends to increase the accuracy of the reconstruction over time. Of course, if you want Fusion to react to change in depth (at the expense of noise), a lower value is appropriate. A value of 100 or 200 is reasonable, but try playing around on your own to see what you prefer.

Volume Voxels Per Meter and *Volume Voxels Resolution* will decide how much space your reconstruction will take up. At a density of 384 voxels/m and a resolution of 384 voxels × 384 voxels × 384 voxels, our reconstruction will take up 1m³ and will contain 56,623,104 voxels. If our resolution were 768 × 512 × 256 and our voxels per meter were 128, our volume would be 6m × 4m × 2m. The more voxels and the higher the resolution, the lesser the performance.

If you ever screw up your scan, the *Reset Construction* button will delete it and restart from scratch. *Pause Integration* will stop using depth data in order to add more content to the scene, but will allow you to still track the scene in 3D space with the camera. *Mirror Depth* will flip the camera image (and reset the reconstruction in the process). *Volume Graphics* will indicate the bounds of the volume graphically. *Reset Virtual Camera* will have the camera attempt to reorient itself in 3D space. This is useful when the tracking has gone awry. *Use Camera Pose Finder* will allow the Kinect to attempt to recover its camera pose more reliably when it has been lost.

The final and possibly the most interesting feature is the ability to export Kinect Fusion reconstructions. Simply click on the *Create Mesh* button. You can choose between STL, OBJ, and PLY file formats, depending on the application.

3D Scan

3D Scan is a very simple Windows Store application that has been built by Microsoft using the Kinect Fusion API. You can move the Kinect around an object, and a 3D scan will be recorded. To get started, simply set the desired depth, width, and height with the slider and enable a timer or rotate the camera viewer if desired, then click *Scan*. On the bottom-right corner of the 3D Scan app (as seen in Figure 8-24), you can see the compounded depth images and see how that reinforces a 3D reconstruction in realtime. If a timer has not been enabled, click the *Stop* button when you are done.

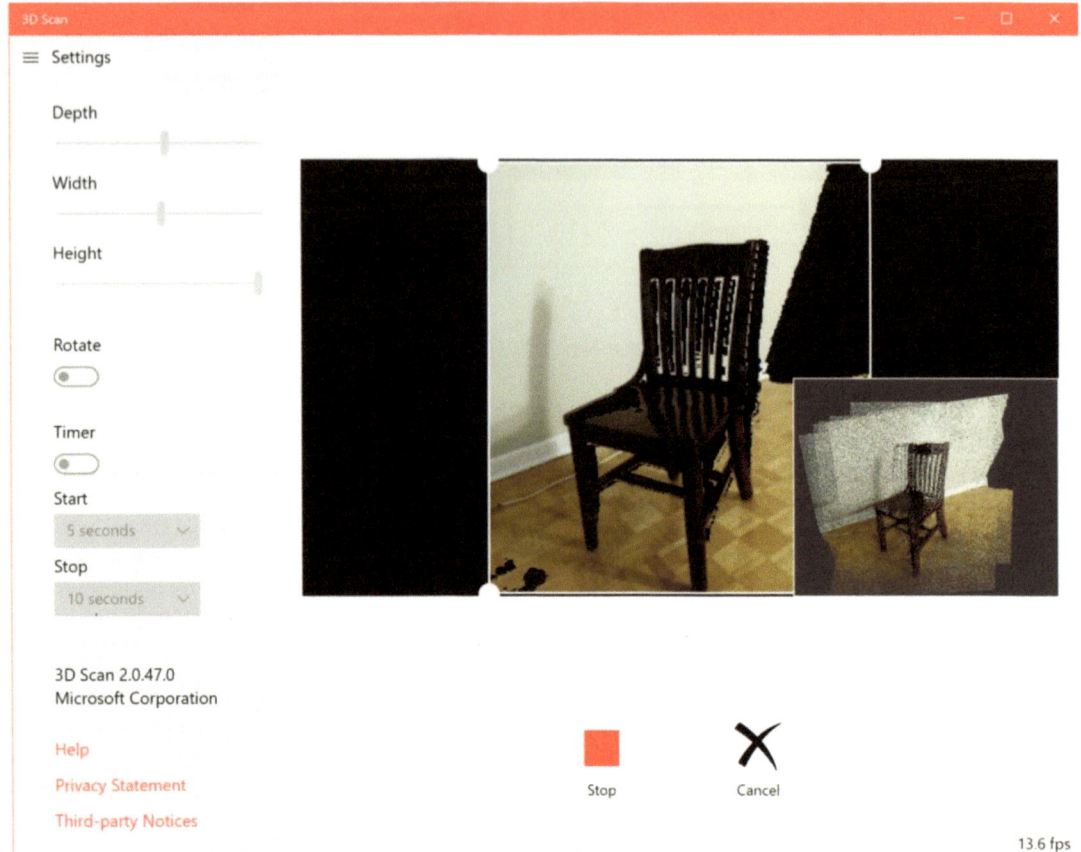

Figure 8-24. *3D Scan attempting to scan my dining room chair*

After the recording has been stopped, 3D Scan will build and texture a mesh using the depth point cloud captured by the Fusion API. The mesh can then be saved to the disk and imported into **3D Builder**. The scan shown in 3D Builder in Figure 8-25 was performed really quickly and on underperforming hardware and thus did not give accurate results for the whole chair. Even then, most of the chair is still there, and we can rotate around it. 3D Builder gives an invalid definition error because not all the surfaces of the chair have been fully scanned and stitched together. If the scan were to be taken more carefully, we would not encounter this error.

Figure 8-25. *A hastily scanned chair visualized in 3D Builder*

Kinect Ripple

Kinect Ripple is an open source, dual projection-based infotainment system that users can interact with by standing on predefined areas on a projected surface. The system consists of a projector mounted overhead pointing downward; a projector ideally mounted overhead (as seen in Figure 8-26), aimed at a surface such as a projection screen; a Kinect positioned above or below the screen and facing the floor projection; and a computer to power and manage the other devices. Only one person can interact with the system at a time (the person closest to the Kinect), and the only available gestures are up, up with both hands, and left and right swipes (I know what you are thinking, but the world can survive without Tinder for Kinect). While not ground-breaking, it is a nifty way of presenting information and making simplistic games.

Figure 8-26. *Ideal projector setup for Kinect Ripple*

I have to say, I was quite impressed to see Microsoft come out with Kinect Ripple. The project, while very cool, is not exactly device-selling and was hardly promoted outside of India (where it was developed), and, consequently, it seems charitable that they even put in the effort to build it. They had several software engineers working on it full-time and even flew them abroad to demo the system at a handful of hackathons (I had the chance to meet two of them in New York City for a Kinect hackathon). Regardless, my respects go to Microsoft for expending resources on this project; it is surely being used in many niche scenarios.

Physical Setup

The hardware setup of the Kinect Ripple is not too demanding, though the documentation adds some weird unnecessary requirements (like an RFID card reader and 16 GB of RAM. . . For HTML and JavaScript? I knew that the web was bloated in this day and age, but not this badly!). The crucial thing to have is a computer that can support two projectors running web apps and any overhead from the Kinect. You will need a decent graphics card and a handful of ports to accomplish this. Additionally, your projector should have a sufficient number of lumens to project strongly onto the desired surfaces. The recommended quantity is 5,000 lumens, though I am sure you can get away with weaker projectors.

As shown in Figure 8-27, the floor projection should measure 8 feet by 6 feet. Other sizes are acceptable, as long as the ratio remains 4:3. The front-facing screen can be any size, though for aesthetics I recommend that its width matches that of the floor projection.

Figure 8-27. *Ideal Kinect Ripple floor size and screen setup*

The ideal Kinect position is 5 feet off the ground, directly beneath the front-facing screen, and 2.6 feet from the floor projection. In practice, I have seen the Kinect placed above the screen and angled down, so you can try other positions, though your mileage may vary.

The resolution for the monitor screen pointed downward should be 1600 x 1200, and the front-facing screen can be anything else, though it is recommended that it be 1600 x 1200 as well. If your computer has another display/is a laptop, you should turn off the main display while Ripple is active. PowerPoint display settings may need to be configured if you choose to play a PowerPoint presentation using Ripple.

Configuring Content

░ **Note**   Support and documentation for the Kinect Ripple project is weak, so I recommend that you command a decent knowledge of WPF and HTML5/JavaScript before setting up Kinect Ripple.

The first thing to do is download the Kinect Ripple project from GitHub. It is available at `https://github.com/Microsoft/kinect-ripple`. Unpack the ZIP and then open the `Ripple` solution in the *Ripple-V2* folder. You will be asked to convert your unsupported solution, which you should say yes to. You may also require the WiX toolset to support the **RippleSetup** project. It can be found at `https://marketplace.visualstudio.com/items?itemName=RobMensching.WixToolsetVisualStudio2015Extension`.

Attempt to build the solution in Release mode. Your solution will probably throw a few errors. These exist because the `Microsoft.Kinect Extended.Wpf.Toolkit` references have not been included in the project. You can grab `Extended.Wpf.Toolkit` on NuGet on `https://www.nuget.org/packages/Extended.Wpf.Toolkit`. After the project has successfully compiled, run the MSI found in the *RippleSetup* folder on the machine that you wish to have Ripple on.

■ **Note** *Microsoft Speech Platform – Runtime (Version 11)* will be required to run the MSI. You can find it in the Kinect SDK 2.0 Browser.

After Ripple has been installed, you have to run the **Ripple Calibration** application in `Ripple-V2\`
`RippleCalibration\bin\Release`. Alternatively, you can find it in `Program Files (x86)\Ripple\`
`RippleCalibration`. It will walk you through the process of calibrating the projectors and Kinect for the Ripple, as skewed projectors can negatively impact your user experience.

Ripple Editor

Ripple Editor is the tool used to configure the content to be displayed on Ripple. It collates all the content into a folder named *Assets*, which is found in the parent folder of the *RippleEditor* folder. The content will be organized in a generated XML file that Ripple will use to determine how the content should be presented.

The Ripple Editor can be found at `Program Files (x86)\Ripple\RippleEditor` or in `Ripple-V2\`
`RippleEditor\bin\Release`. When you start Ripple Editor, you will be greeted with a UI that delineates the floor and screen content. To start a new Ripple experience, click on **File ➤ New**. You will be asked to select a template, as in Figure 8-28. The template indicates how many grid squares will be available on the floor for the user to walk on.

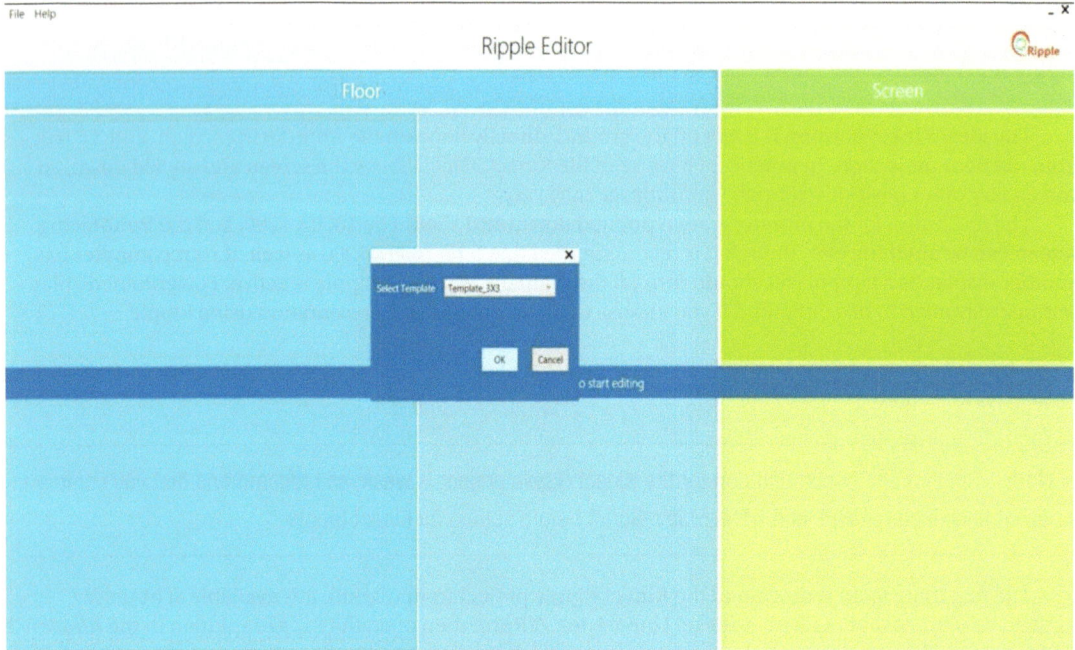

Figure 8-28. *Template selection dialog in Ripple Editor*

You should be presented with a Ripple floor representation on the left side of the window. If you select a title, the other half of the window will fill with properties for the tile and screen, as in Figure 8-29.

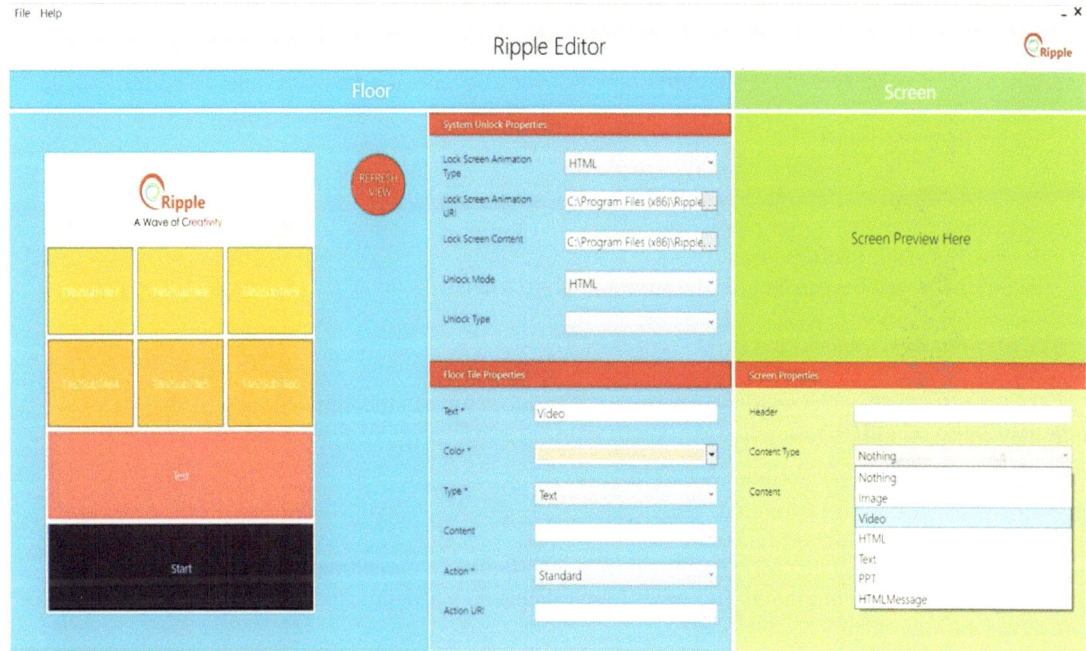

Figure 8-29. *Editing tiles and content in Ripple Editor*

You can set the content and color of the tile in the *Floor Tile Properties* section. If properties are changed, the red *REFRESH VIEW* button under the heading for the **Floor** panel should be clicked. If the Standard *Action* is set, then the front-facing screen will change its content to what is defined under *Screen Properties* when a user is standing on the floor tile. The screen content can be of numerous types, such as video, HTML, or PowerPoint. If the content is PowerPoint, the user can go through it by swiping left and right. If the HTML Action is set, the screen and floor will display the associated HTML files. This can essentially be used to create interactive web applications that play on both screens. The Ripple SDK in the next section will show us how to take advantage of this.

Ripple SDK

Ripple's screens are essentially browser controls rendered in a pair of WPF applications so you can develop HTML5/CSS/JS content for it. Every time you walk on the floor projection, the mousemove DOM event is fired. The mouse's position is where you are currently positioned in relation to the floor projection. The SDK has a handful of events and a helper to facilitate development for the platform.

Getting Started

To get started with Ripple development, create two HTML files representing the floor and front-facing screens. For each, you have to add the JavaScript SDK files found in `kinect-ripple-master\RippleSDK_With_Example\RippleSDK` (sample HTML pages are also found in there). To do so, you can include

```
<script src="rippleFloor.js"></script>
```

for the floor page and

```
<script src="rippleScreen.js"></script>
```

for the front-facing screen page.
Additionally, you have to include the appropriate `meta` tag:

```
<meta http-equiv="X-UA-Compatible" content="IE=10">
```

This is to let the browser window know to use HTML5 features in the Ripple app. Content can alternatively be set to `IE=11` as well.

Ripple Helper

Ripple Helper is an object that contains the brunt of the Ripple SDK functionality.
First, it contains a debug mode. This prints Ripple debug messages onto the browser console. To activate it, simply call the following:

```
rippleHelper.setDebugMode(true);
```

There is also an emulator mode. This lets you run Ripple from your own browser instead of having to set up all the hardware, as seen in Figure 8-30. You can even emulate gestures using button clicks. To turn it on, include the following line:

```
rippleHelper.setEmulatorMode(true);
```

You also have to include the emulator HTML file that you open on your browser. This will be used to display the other screens. There is a sample one in `kinect-ripple-master\RippleSDK_With_Example\RippleSDK` (Figure 8-30).

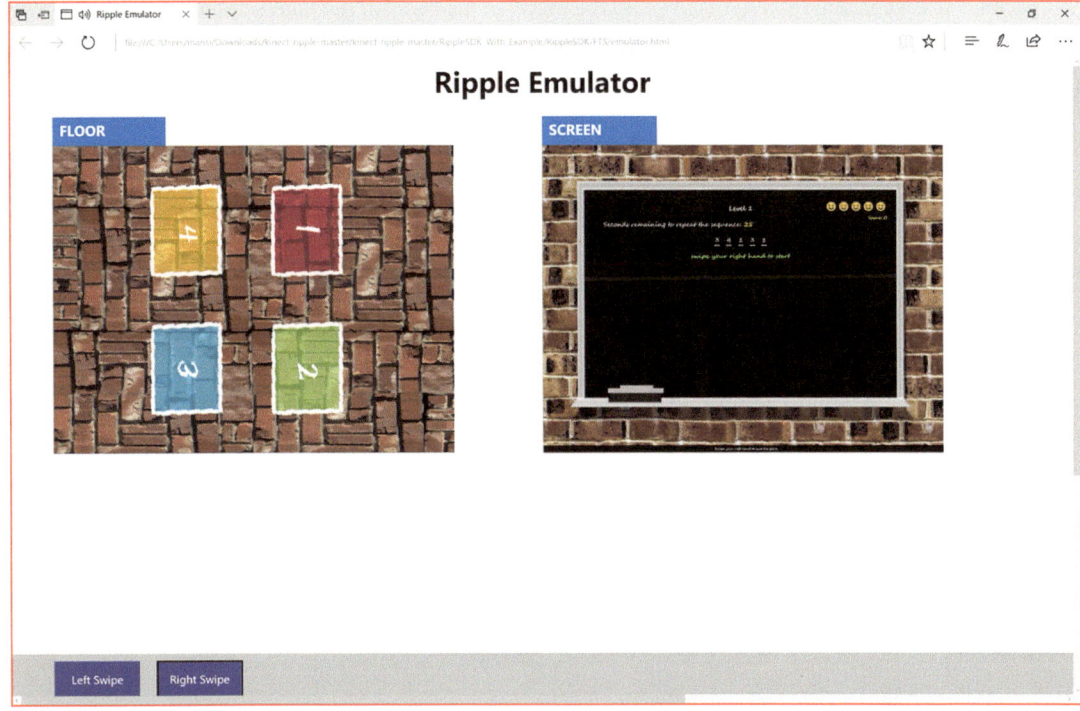

Figure 8-30. *Playing a game on Ripple Editor*

Gesture Detection

Ripple exposes gesture detection through document-level events. There are only a few gestures currently detectable, but since the code is open source and built on top of the Kinect SDK, you can probably manage to include your own. Gestures are always handled in the code associated with the floor.

To detect a gesture, simply register an event listener on the gesture event:

```
document.addEventListener("LeftSwipe", function(e) {
    //code to execute when discrete gesture is detected
}
```

Communicating Between Screens

Ripple Helper has a set of methods to allow communication between screens. The helper methods send a command to the other screen, which is reacted to with an event listener. An array of string parameters can also be included.

The relevant methods are as follows:

```
rippleHelper.sendCommandToFrontScreen("Command", ["Parameter1"]);
rippleHelper.sendCommandToFloor("OrderPizza", ["Cheese", "Tomatoes", "Arugula"]);
```

Let's say you wanted to react to the `OrderPizza` event; in the code for the floor screen, you would include the following:

```
document.addEventListener("OrderPizza", function(e) {
    console.log(e.commandParameters); //do something with command parameters
}
```

The `commandParameters` array contains all the string array parameters sent from the other screen.

Text-to-Speech

Ripple Helper exposes a text-to-speech engine for you to make vocal prompts with. Usage is trivial. You simply have to enter your text as a string in the relevant method:

```
rippleHelper.textToSpeech("Eat my shorts");
```

Timer

There are often times when you will want a user to confirm a selection by standing on a tile for a certain number of seconds. Ripple Helper can create a timer object for you to measure seconds in simple manner.

The timer object can be instantiated with the following:

```
var t = rippleHelper.timer();
```

To start and stop the timer, you can use the following:

```
t.startTimer();
t.stopTimer();
```

Finally, to get the current time of the timer, use the following:

```
t.getTime();
```

Conclusion

In this chapter, we covered some of the very cool tools every Kinect developer should have in their arsenal. While they may not have the best support or documentation, it is clear that they are very powerful and have the capacity to greatly extend our applications. We saw how Visual Gesture Builder facilitates gesture detection to the point that a novice developer can incorporate such functionality in their Kinect app very easily. We learned about Kinect Fusion, a one-of-a-kind technology that can help us digitalize the real world in an unprecedented three dimensions. Finally, we covered Kinect Ripple, a novel system that allows us to give our presentations a degree of interactivity that has not been seen to date. All in all, though these technologies could stand to be used more in production, we can see how they could serve as the basis and inspiration for more advanced and custom applications that could not be achieved with the traditional means of computing.

Conclusion, for Real

And now it has come time to bid adieu. To think, we got to the end! I just finished writing a whole book and you just finished reading a whole book. In the process, we both learned a *lot* about the Kinect; hundreds of pages' worth! Yet, there is still so much more to uncover. It was absolutely unfeasible to include everything there is to know about the Kinect. In fact, I could easily write another couple of hundred pages! But then, the book would never get published.

So, instead, I hope I have been able to instill within you the foundations of Kinect development and an understanding of the philosophies and technologies behind the Kinect. The Kinect is definitely one of the most enjoyable software and hardware platforms to develop for, as there is just so much you can do with it and always something new to learn about it. With these parting words, I hope you will always have a positive and ambitious outlook on Kinect development, one that will help you bring to life your wildest dreams and effect a positive change in your world, just as the Kinect did for me. Keep Kinecting!

APPENDIX A

Windows 10 & Universal Windows Platform

By Dwight Goins

Foreword by Mansib Rahman

It may not be a surprise to many that the current state of affairs for the Kinect is in flux. While the core APIs and drivers are relatively stable, Microsoft has somewhat dropped the ball on further development. That is not necessarily a bad thing. They have taken their learnings and applied them to the development of newer technologies, such as HoloLens and other mixed-reality headsets. The Kinect for Windows v2 still offers a strong entry point for hobbyists, developers on a budget, and research/commercial developers looking for a proven depth-sensing technology.

So, when Microsoft (somewhat quietly) announced Kinect support for Windows 10, you can imagine I was pretty ecstatic. I had no expectations, so anything at all was welcomed by me. I really wanted to ensure that this book included what was essentially the start of the last chapter of Kinect for Windows v2, even at the risk of delaying publication by several months. Unfortunately, I simply did not have enough experience with the new programming paradigms for Windows 10 UWP and Kinect. I happened to know someone who did. Dwight, the author of this appendix, is a Kinect MVP and had the fortune of learning about Kinect development for Windows 10 before it was announced by Microsoft publicly. Thus, I did not hesitate to enlist his expertise and have him draft the following pages. Aside from some grammar and syntactic edits, I let him have purview of this entire appendix.

A lot of the theory and backstory in this appendix may be confusing for those not experienced with the Microsoft development stack. The story from Microsoft is not clear in itself, and Dwight was compelled to work with what he had at his disposal. If you find it difficult to grasp, do not hesitate to jump right to the code.

Additionally, I should warn you that the content in this appendix reflects the current state of affairs, which is subject to change by Microsoft at any time, depending on their overarching Windows 10 and mixed-reality strategies. This is the uncool part about being a developer (though admittedly, it does help justify our paychecks). As a result, some of the content here may not reflect what is current at the time of publishing.

© Mansib Rahman 2017
M. Rahman, *Beginning Microsoft Kinect for Windows SDK 2.0*, DOI 10.1007/978-1-4842-2316-1_9

Introduction

There are two different approaches to building applications for Windows 10. The first approach is to utilize a software development kit targeting Windows 10 traditional Win32 desktop-based applications running on x86 and x64-bit platforms, which have been around for the past 30 years. The second approach is to utilize an SDK targeting Windows 10 Universal Windows Platform (UWP). The second approach has only been around more recently.

To target the Windows 10 Win32 desktop, everything you've learned up to this point is how you would go about it. You would utilize the Kinect for Windows v2 SDK in both .NET and C++ libraries; there is nothing special you need to do. Just follow the steps mentioned in the previous chapters, and your application will work. This brings up a very important point: applications that you built using the Kinect for Windows v2 on Windows 8 will work unchanged on Windows 10 desktops.

On the other hand, to target the Windows 10 UWP and its store, there is a new paradigm you must learn. The architecture has changed, the design philosophy has changed, and the programming model is different. You can pretty much say it's a new SDK and framework for building Windows 10 UWP applications altogether, which follow the Windows 10 UWP development mantra.

This section will take you through this new paradigm, the history of the Kinect v2 for UWP, architecture, and programming model, and show step by step the building blocks for how to build a Windows 10 UWP store application with Kinect.

Windows 10 UWP Paradigm

One Billion Devices

One of the main goals of Windows 10 UWP is to make sure it runs on as many Windows devices as possible. The goal is to reach one billion-plus devices over the next year. This includes device form factors such as IoT, hobby or prototyping programmable boards such as **Raspberry Pi**, **Particle**, and **Arduino**, mobile phones, tablets, desktops, laptops, and wearables. Even the more specific form factors such as **HoloLens** (head-mounted displays), **Surface Hub** (wall-mounted, large, shared-screen experience), and **Xbox** gaming and entertainment consoles are a part of this grouping. This means that, as a developer, when you create your UWP application it has the potential to run on over one billion devices, which means the potential to reach over one billion customers.

Reaching this many customers and running on this many types of devices requires a change in thinking and a change in development strategies. This is especially the case when it comes to display and form factors, as well as how an application runs and how an application closes down. The latter is typically known as lifecycle management. While the form factor is addressed through a fast and fluid design philosophy, there are many books and articles on how this is accomplished within the Windows 10 SDK. The lifecycle management, on the other hand, needs some more attention, as it relates to Kinect for Windows v2 development.

Lifecycle Management

The lifecycle management of a UWP application compared to legacy Win32 desktop applications is a lot like your father's root beer, where anything may go into the drink and anything can happen if he stays on it too long. Traditional Win32 applications are typically controlled by the user, and anything can go in it. The application starts when the user wants, and it can stay running as long as the user wants; anything is likely to happen. The user terminates and closes down the application when the user wants by pressing the x-close button.

UWP applications, however, do not necessarily behave this way. The lifecycle of a UWP application is modeled as a state machine. A state machine, in this example, is a UWP application where states are maintained and can transition to different states based on events and data flowing between the states. A user can initiate the state machine, but there are many things outside the user's control and within the UWP engine that can change the state of the application. A UWP application contains three main states: **Running**, **Suspended**, and **Not Running**. An application can go from Running to Suspended to Not Running. An application can also go from Suspended back to Running again without going to Not Running. Figure A1-1 shows UWP states an application can be in. As a Kinect UWP developer, these states must be accounted for in your application. The following is a quick diagram showing the possible states and events you have access to for determining which state you came from or are about to go into.

Figure A1-1. *Diagram of UWP states*

System Environment Changes

Traditionally, legacy-based applications have free rein as to what resources are available for use. When running legacy applications with administrative permissions, the application can access and modify registry settings, connected devices, and any file on the system, including system files, and can basically communicate with any other running applications. These legacy applications took the concepts of standard C—trust the programmer and don't prevent the programmer from doing what needs to be done—and ran with it. Legacy applications became kings of the computer. Well, in the hands of untrustworthy developers, these king-like legacy applications led to viruses, and eventually led to a concept called *secure computing*.

Fast forward to 30 years later, when trustworthy developers and secure applications are at the top of everyone's list. As a developer now, we must be security conscious. Along with this responsibility comes the idea of secure applications running on one billion devices, and the philosophy is now the anti-C theorem: **do not** "trust the programmer"; **you must** prevent the programmer from doing whatever he or she wants. This, for the most part, is what Windows 10 and the UWP is all about—providing a safe environment and giving control back to the user and safe engine and taking it out of the hands of the untrustworthy programmer. In UWP applications, a user must decide if they trust the programmer. By default, nearly all capabilities within the application are restricted from running. The user has the ability to allow various capabilities to run within an app. As a developer, if you don't request the permissions for the capabilities you need for your application from the user, it will run with the bare minimum of features, basically just displaying some hard-coded text and embedded pictures. Keep in mind that the user also has the ability to say no, which means as a developer your application must account for this.

A UWP application provides a manifest file that contains declared capabilities that a developer requests from a user in order to run on a target system. These capabilities can be declared in the package manifest file or programmatically. The capabilities are grouped into general capabilities, device capabilities, and special capabilities.

General capabilities apply to the most common UWP application scenarios and contain features such as access to the *Music, Pictures, Videos,* and *Removable Storage* folders and libraries. If your application requires network connectivity, there's the Internet and public networks capabilities, as well as home and work networks. There is also support for appointments, contacts, code generation, AllJoyn, phone calls, VOIP calling, 3D objects, chat messaging, and a few more.

If your application requires a scenario for accessing peripheral and internal devices, you must include device capabilities. The Kinect, for example, is a peripheral device that has many sensors and capabilities.

If I may, I'd like to go on a tangent for a second. In the new Windows 10 and beyond world, we must think about the Kinect for Windows v2 as a set of capabilities instead of as just a device. Just for a second, you could re-imagine the development strategy of UWP and Kinect. Microsoft has always been more focused on the software side rather than the hardware side. For this one- billion-device plan to work, they have to build software and frameworks for device-agnostic capabilities that have depth, perception, color, sound, and location. Well, that's exactly what the UWP SDK provides–a generic pattern where we as developers can now build device-agnostic applications. We can build a suite of applications that take advantage of capabilities rather than tie ourselves to specific devices. These capabilities can be embedded in apparel (wearables), glasses (HMD), TVs, phones, laptops, desktops, household appliances, cars, and other creative mediums. Now, hopefully, you can imagine how emerging technologies can lead to emerging experiences that explode like tiny DNA explosions, changing your molecular structure into a new being.

Sorry for the aside; let's get back to the point. A Kinect UWP application needs to explicitly request one or more of these capabilities. For example, if you want to access any generic microphone connected either internally or externally (such as the Kinect for Windows v2 microphone array), you must add the Microphone capability to the package manifest. There are device manifests for Location-aware scenarios (GPS sensors), Proximity (Bluetooth/Wi-Fi or Internet), Webcam (i.e., Kinect's web camera), USB, Human Interface Device (HID), Bluetooth GATT, Bluetooth RFCOMM, and Point of Sale (POS) features. Even though these capabilities seem specific and special enough, there's actually one last unique category left: the special capabilities.

The special capabilities are meant to be for specific scenarios that include Enterprise authentication, Share User certificates, and a special use case for the documents library when accessing files offline or in OneDrive, or when your UWP can't use a file-picker user interface control for selecting files in the folder of your choosing.

Now that you have a general understanding of capabilities, it's time to talk about how packaging and installation works in a UWP application.

Packaging and Store Submission

Packaging is where a developer takes all the files necessary to run the application-included configuration entries and combines them into one file or folder for the purpose of transferring the package to the user. Back in the day, this was done with a packaging utility or installer software, such as **InstallShield** or **WISE**. In more modern applications such as .NET, we have the **Wix Toolset** and set up projects included inside of Visual Studio .NET. These utilities package the application and make it easy to transfer the application to a user system. One of the issues with packaging has always been deciding where you should store the package for high visibility, and also how you know what version of the software is inside the package. Along with answering these questions, it would be great to also have a rating and/or ranking system and comments for other users. Taking a lesson from other vendors, Microsoft realized a repository of trusted packages and versioning was needed, and while they were at it, they added a way to allow developers to monetize the application, along with a way for users to show their trust in the developer, or their distrust. This repository is called the Windows App Store. For UWP applications, you can easily create an App Store–ready package, and Microsoft provides a PowerShell script to make it easy to install the package on a user system for development and testing purposes if you don't care to submit your application to the store. This is known as *sideloading* your application. However, if you do decide to submit your application, it will go through a certification process. This certification process will run automatic tests to check a system shutdown scenario, residual entries upon installation and uninstallation steps, OS Version check, compatibility issues, crashes, user-access control, writing to correct folders, multi-user sessions, and other tests. Once the tests have been passed, a few poor souls at Redmond then have the task of manually reviewing the application and testing it to make sure there's no offensive content and that it is in line with Microsoft store policies. If the manual test is approved, the application is published to the Microsoft App Store.

Now, with the Windows 10 paradigm out of the way, let's take a trip down memory lane for a second to understand how the Kinect's new UWP APIs made it into being.

History of the Kinect and Windows 10

I've been fortunate enough to be a Microsoft Most Valuable Professional (MVP) for Kinect and Emerging Experiences for the past few years. Being an MVP gives me and my fellow MVP colleagues the privilege of sitting down at the table with key members of the Analog team at Microsoft, many of whom either built the Kinect drivers or Kinect SDK and samples. We were even fortunate to participate in multiple future Kinect meetings of the minds and whiteboard talks. While most of those conversations are laced with non-disclosure agreements, I can say that the then-present Analog team members eventually realized the importance of building a Windows-based sensor-agnostic library and SDK, the reason being that the future is teeming with new devices and new capabilities. These new devices are being created in new form factors. They are getting smaller, faster, and better equipped to handle a wide range of scenarios.

If you take a look a *Moore's law* during 1990–2010, you could say that we were in a 2D mindset based on the type of software and applications that were in existence. Everything was focused on the X, Y plane. We as an industry were focused on the color pixels on a flat screen. Figure A1-2 shows the number of transistors, clock speed, and power available to run these types of applications. The experiences were all about using either a mouse or a keyboard to interact with a 2D application.

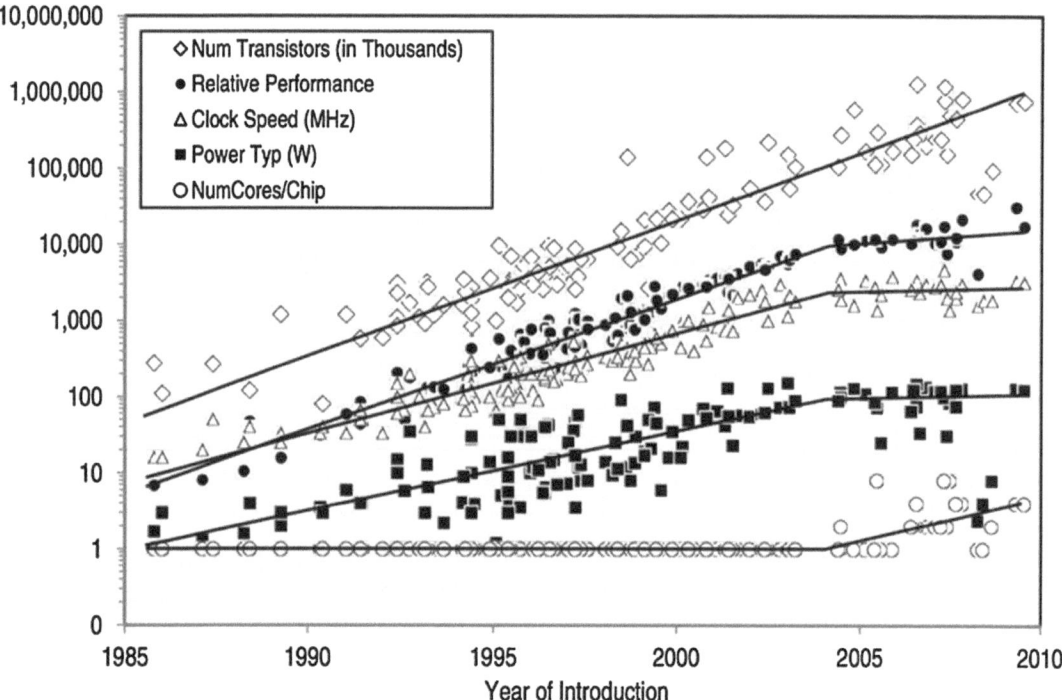

Figure A1-2. *Moore's Law, count of transistors from 1985 to 2010. Source: "2 What Is Computer Performance?." National Research Council. 2011. The Future of Computing Performance: Game Over or Next Level?. Washington, DC: The National Academies Press. doi: 10.17226/12980.*

I ask, how many transistors does it take to display a pretty image on a flat screen? Apparently according to the chart above: 1 million.

When the Kinect v1 device came out in the later part of 2010, this introduced the 3D (3 Dimensions) plane. Figure A1-3 shows the more current Moore's Law count of transistors from 2010 to 2016. What you should notice is that from 2010 to 2012 we saw an increase in about 3 billion transistors.

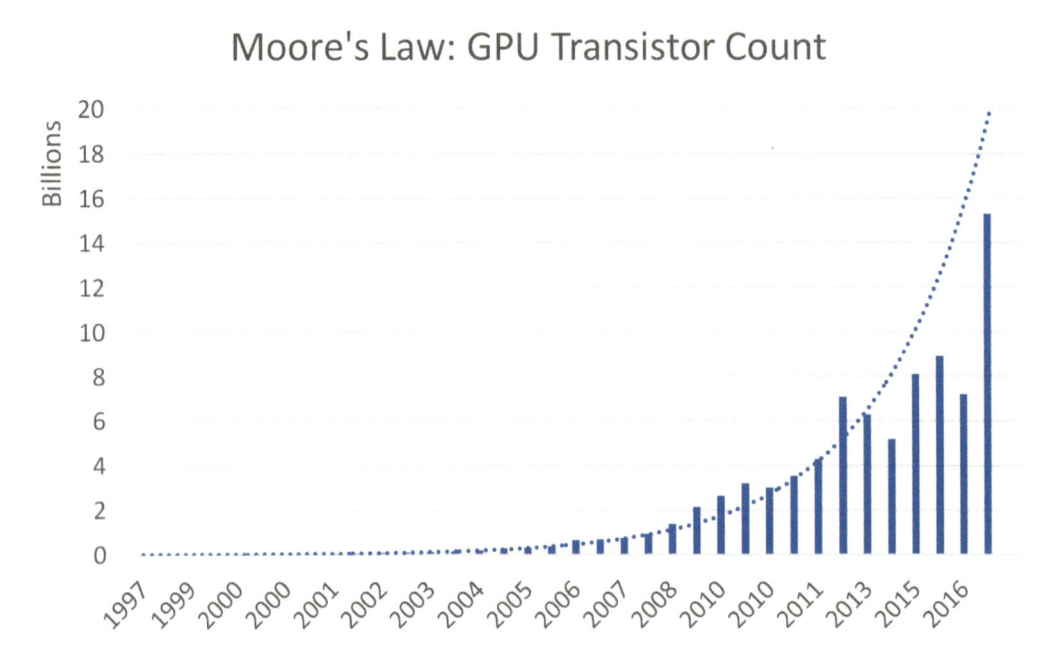

Figure A1-3. *GPU transistor count from 1997 to 2016*

Now we have devices and software to bring about new experiences that deal with X, Y, and now Z, also known as depth, or how far something is from us. We have depth sensors such as the Kinect device, **Intel RealSense**, **Orbbec Astra**, and **Leap Motion**. We also have 3D printers, 3D scanners, 3D monitors, and 3D glasses all on the market today.

■ **Note**　A good comparison of the available depth sensors today can be found at `https://stimulant.com/depth-sensor-shootout-2/`.

All this and we've only begun to take advantage of 3D applications and thought patterns, and all the while more and more transistors are being brought to life. I ask the same question today: how many transistors does it take to create a 3D experience? Again, if we consult the modern chart from Figure A1-3, the answer is apparently three billion.

However, let's not stop there. With new transistors come new capabilities. If three billion more transistors go toward 3D processing, looking at Figure A1-3 going past 2016 into 2017 we see a total of at least fifteen billion transistors. That's seven billion more than we had in 2015. What are we as developers going to use the remaining seven billion transistors for?

I have a suggestion. We are now moving into, dare I say it, out-of-the-box experiences. Out-of-the-box as in literally out of the concept of a three-dimensional box. We are now talking out of the 3D spectrum. We are talking *1D* and *4D* experiences.

The *1D* concept is conversation as a platform. The idea behind this concept is that of an automated service taking requests and returning automated responses, a.k.a. a *Bot*. It's considered one-dimensional because there is no user interface. There are no X and Y planes, only input and output. Let's call this the W plane. Now, for the geeky people, we could argue that's still 2D, as output is a function of the input, but we're specifically talking about the user interface and how that relates to the user experience.

On the other side of the box, let me be one of the first to introduce you to the *4D* concept. This is the idea that we can model a 3D object and print it out. However, with this object, we create changes over time, such that it's no longer what we printed out, but rather it itself changes into the best-shaped object for its purpose. We do this with the instructions we give it. Envision an architect creating the blueprint for a toy car and printing out that car with a 3D printer. This 3D car also has an embedded interpreter, sensor, printer, and compiler, such that if that car senses water, it can recompile its makeup and print out floatable buoys, thus making it behave like a boat. Researchers at MIT are already trying out biochemical components for CPUs that are embedded with sensors and instruction sets so that these components can be used for biological experiments. Think: mechanical implants that grow into place.

With the 1D to 4D spectrum, it's easy to see we that have new devices and many more transistors, and they are being coupled with integrated computers (a.k.a. Orbbec Persee). They are also getting integrated into wearables such as glasses (a.k.a. the Microsoft HoloLens). They are now also being embedded into laptops, cellphones, and drones (Intel RealSense). Thus, the questions for Microsoft started piling in.

Just how, exactly, is any future version of the Kinect device going to fare in this new wave of crazy gadgets? Microsoft led the way, now should it continue? We asked the same and even asked about the focus. There were many ideas that floated around, and one thing sounded loud and clear. Microsoft is a great software company, and I'd say a decent hardware company. Many, including myself, felt Microsoft's focus shouldn't be on the device itself, but rather on the device's capabilities. Fast forward into 2017, and the focus is indeed on building the best platform for developers so they are able to take advantage of all these potential Windows-based devices. The key is *Windows-based devices*. As long as all the aforementioned new-wave devices are Windows based–and this simply means they can work on Windows, or even better *run* Windows 10–then expect a Microsoft development platform that can target it: UWP.

Now, the next obvious question became how was the Analog team going to do their part? The details weren't quite ironed out, and the question of which patterns they were going to use was still up in the air, but one thing was certain: it all had to align with the current Windows OS and development platform during that time.

Enter the first attempt: Windows 8 Run Time (WinRT). At the very core, a new engine was brought to light, rewritten from the ground up. I'll spare all the details about its coolness and wackiness and only shed light on how it affected the Kinect for Windows. Microsoft released a WinRT Runtime library (WRL) that wrapped the Kinect's core functionality. It was a somewhat futile attempt **(Mansib's edit: Hey!)** to keep the Kinect device working with its current design and architecture, yet follow the programming model of WinRT. This component was released as the `WindowsPreview.Kinect` runtime library.

As a developer, to utilize this library you first had to create a project that targeted the Windows 8 platform SDK and WinRT engine. Most of the time, you couldn't just convert or retarget an existing project, simply because the WinRT engine followed a different programming paradigm as far as access, APIs, and libraries were concerned. Additionally, it only had a subset of .NET's assemblies. This meant you had to create a new project from scratch and bring over concepts, code, and implementation judiciously.

Once you targeted Windows 8 WinRT, the next step was to add a reference to the Kinect Windows 8 WinRT Extensions; in other words, the Windows Preview Kinect (WRL). You could download this WRL preview from Microsoft, and once downloaded and installed these extensions were physically found on your computer at the default path: `Program Files (x86)\Microsoft SDKs\Windows\v8.0\ExtensionSDKs\WindowsPreview.Kinect\2.0\References\CommonConfiguration\neutral\WindowsPreview.Kinect.winmd`. There were also WRL extensions for Kinect Face, Kinect Tools, and Visual Gesture Database. The upside behind this WRL was that the classes, methods, and events remained exactly the same as was what was exposed in .NET. In other words, you could take your C# code from a WPF application and copy and paste it into a WinRT Windows 8 Store application. The only thing you had to focus on was the differences in asynchronous design and behavior for Windows 8 RT as opposed to .NET.

One thing to note is the name of the WRL, `WindowsPreview.Kinect`. The word *Preview*, in my opinion, was a slight admission of their not being ready. This WRL wasn't quite ready for prime time, nor was it a true rewritten-from-scratch component. It was just a wrapper with the purpose of supporting Windows 8 Store applications.

The Windows 8 Kinect Preview WRL and its popularity lasted just about as long as Windows 8/8.1 did. During this time, there were concerns that the Kinect as a brand or product was in its sunset years, but that its capabilities would emerge in a new form factor. Enter Windows 10 and UWP.

Windows 10 UWP Kinect Architecture

The Kinect for Windows UWP support is a complete rewrite of components that adhere to the Windows 10 mantra. It is based on *Microsoft Media Foundation* (MF), a `Windows.Devices.Perception` sensor provider, a new hardware driver, and the notion that Windows 10 UWP applications provide an underlying set of contracts known as the *Application Binary Interface* (ABI) that dictates how a device and software can interact with Windows 10 and each other. Any device that adheres to this set of contracts can be considered a Windows-based device and provide implemented features exposed by the contracts.

Some of the features provided by these contracts are security and authentication through facial recognition, standard camera capabilities, and speech recognition through the microphone. These features make the Kinect device's capabilities available to all UWP applications on Windows 10. The Kinect for Windows UWP support effectively conforms to the various ABIs for a web camera, microphone, IR, and even a custom contract for body tracking. In other words, the Kinect is no longer perceived as a specific device with its own SDK by Windows, but rather as a series of sensors that afford data through the various ABI contracts. In Figure A1-4, you can see the Kinect camera being exposed as a web camera in Skype.

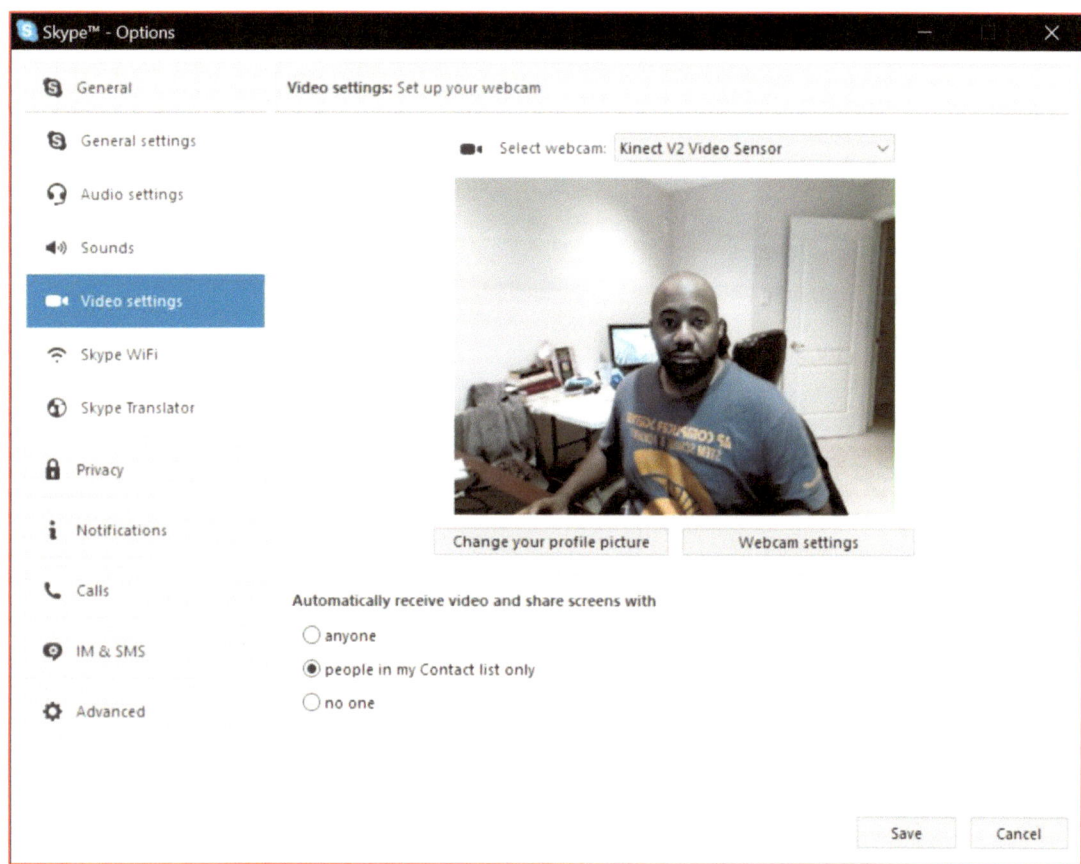

Figure A1-4. *Kinect Color Camera exposed as a web camera for Skype*

How Do the New Components Work Together?

The Kinect UWP support implements the ABI contracts through three major components: the *Windows 10 Kinect Driver version 2.2*, the *Kinect UWP Device Perception Provider*, and a new *Microsoft Media Foundation COM library*. Figure A1-5 shows the details and date of the most updated Kinect driver installed on a Windows Anniversary 10 updated system.

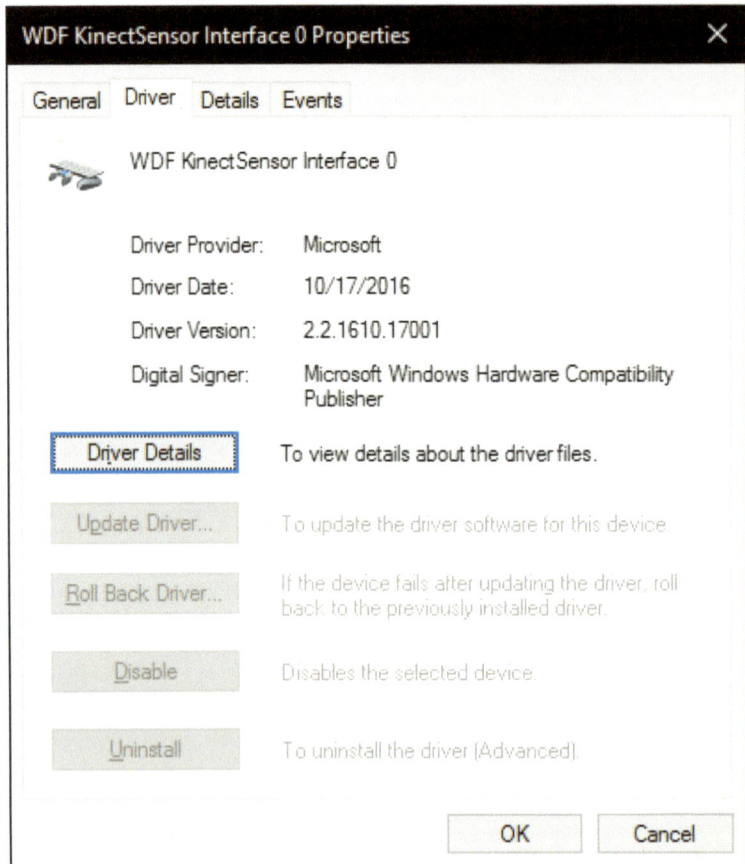

Figure A1-5. *Kinect updated UWP driver*

The purpose of the driver is to communicate with the specific device and send the appropriate device hardware commands to turn on the cameras, read and process the gigabytes of data that is generated from the device and its transistors, and eventually send the interpreted results to the software layer through a service called the `SensorDataService` or the Microsoft Media Foundation (MF) library. The `SensorDataService` is a Windows service that loads up device perception providers that are registered for use with Windows Hello. It allows for interacting with various devices that want to provide perception and depth through infrared cameras to adhere to the security standards for Windows Hello. The device perception provider is the software layer that provides access to the various UWP generic classes for accessing perception frames for the IR camera. The MF library is a set of components and functions that contain a custom *Media Foundation Source* that adheres to the *Media Foundation Pipeline*. It is the core component that provides the color, depth, and IR feeds to standard Windows 10 UWP applications.

Microsoft Media Foundation (MF)

Microsoft Media Foundation is a technology that has been around since Windows Vista, and its core purpose is to be a low-level set of APIs for media development. It is the replacement for **Microsoft DirectShow**, **Microsoft Direct X Media Objects (DMO)**, and **Microsoft Video for Windows** frameworks. MF provides a developer with the ability to create videos and audio, enhance them, and create audio and video editing–type applications and other advanced media-based filters. It is the core functionality behind Windows 8/8.1 Media Capture XAML elements and is used to provide Internet Explorer and Microsoft Edge with their video rendering capabilities. Another way of stating it is as follows: MF is the hard-core media pundit. All things go through MF, 'cause it's one bad MF. In previous releases of MF, there was no ability to treat videos as frames or pixel buffers except through the MF Pipeline, also known as the *topology*. The MF topology is a graph, or a series of tasks handled by individual COM-based components that take some input stream, process the data, and return 0 or more output streams.

There are several different types of components, such as *sources*, *sinks*, and *transforms*. Sources provide the source of the data-input stream. Sinks provide a destination (think file on a drive, or endpoint to a network stream). Transforms are middleware that aggregate, divide, remix, modify, encode, decode, compress, decompress, convert, and anything else you can think of that happens in the middle of processing media files.

Microsoft created a Media Foundation Source that adheres to the Media Foundation Pipeline specifically for the Kinect v2. This source captures all the video and audio streams that the Kinect device offers. There is a video source input for Color, Depth, and IR. This can be seen in the Figure A1-6 using the MF topology editor (`topoedit.exe`). I won't go into detail about the MF topology, as that is beyond the scope of this book. However, I will say that the MF topology provides a low-level architecture to expose the Kinect device as a set of media sources for video and audio to the Windows OS. This mechanism is what allows the Kinect device to be used with Skype, the Windows 10 Camera, and other applications. It is also what allows the Kinect to be exposed to the Windows 10 UWP *MediaCapture API* and XAML controls.

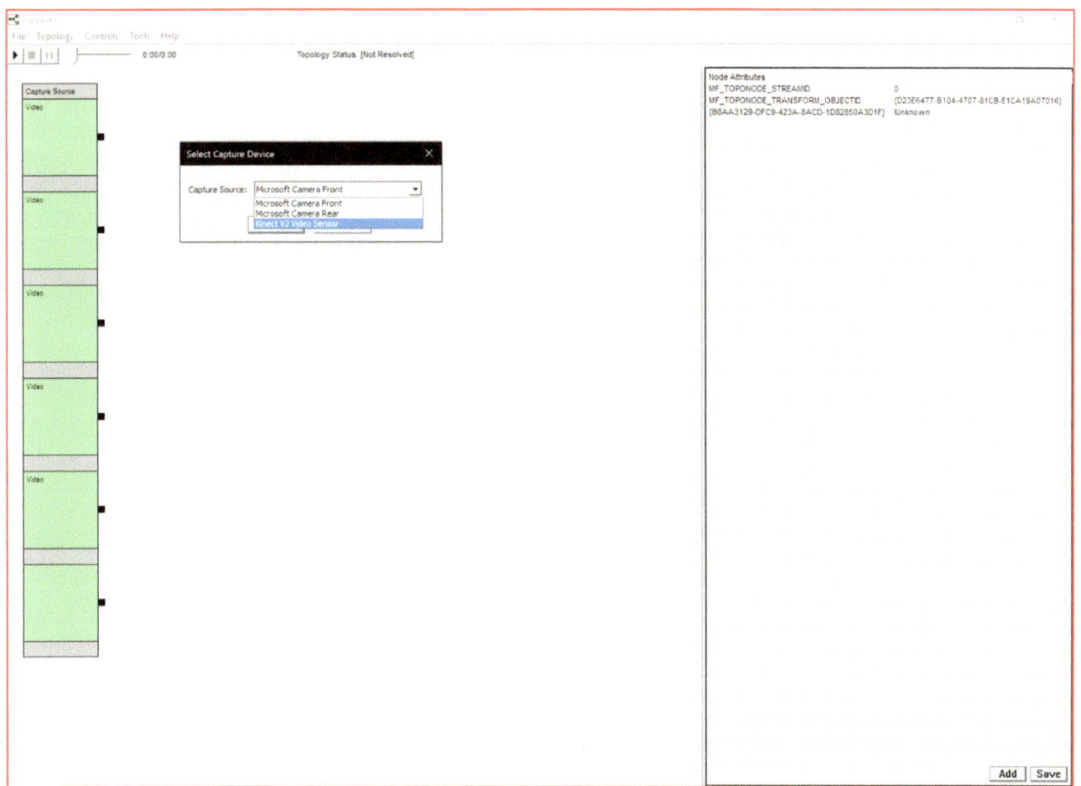

Figure A1-6. *Kinect MF Custom Source in TopoEdit*

The Kinect Media Foundation library can be found at `C:\Windows\System32\Kinect\KinectMFMediaSource.dll`

■ **Note** `KinectMFMediaSource.dll` is the core component that was missing in previous releases of Kinect for Windows that makes it an actual media device for Windows 10.

In previous versions of MF, it was very difficult to get to the underlying audio and video frames and buffers. With the Windows 10 Anniversary update, Microsoft provided a set of new APIs in the `Windows.Media.Capture` and `Windows.Media.Capture.Frames` namespaces that makes it very easy to access video frames and buffers for doing computer vision.

Device Perception Provider

Windows 10 also provides a set of APIs in the `Windows.Devices.Perception` and `Windows.Devices.Perception.Providers` namespaces that allows hardware vendors to expose depth and infrared streams of data from their devices into a generic class for perception: the perception frame. This is made possible by the *Device Perception Provider*. If a device wants to provide access to its data streams to Windows 10 as perception frames, it must conform to the DevicePerception Provider ABIs. In previous versions of this provider, color, depth, and infrared streams were made available; however, this provider was purposed only for use with Windows Hello. Windows Hello only utilizes the infrared camera for its functionality, and thus the Analog team removed the color and depth streams from this provider.

By default, you can find the device providers at `Computer\HKEY_LOCAL_MACHINE\SOFTWARE\Microsoft\Analog\Providers`.

In Figure A1-7, you can see some of the default providers available on a Windows 10 Anniversary Update.

Figure A1-7. *Registry hive showing default perception providers*

■ **Note** I have a Surface Book (first generation) with the Kinect and Intel RealSense cameras.

Specifically, the Kinect perception provider can be found at `%SystemDrive%\windows\system32\Kinect.PerceptionFrameProvider.dll`.

■ **Note** In Figure A1-7, I also have a device perception provider for Surface. This is because in addition to the Kinect device, the Surface Book also contains an embedded infrared camera that is used for Windows Hello face authentication.

As an Independent Hardware Vendor (IHV), Original Equipment Manufacturer (OEM), or hobbyist maker, you can implement your own PerceptionFrameProvider. To do so, you have to create a driver for your device that can generate data to send to the SensorDataService through the IPerceptionFrameProvider and IPerceptionFrameProviderManager interfaces. Implementing these interfaces and the exact steps required is beyond the scope of this book.

Updated Kinect Driver

The Kinect for Windows v2 hardware driver was updated to support sending data to the sensor provider so that the provider could create perception frames to send to the SensorDataService. The Kinect Device Perception Provider is what allows the Kinect to also support Windows Hello. This driver was also updated to support Media Foundation for sending color, depth, IR, and audio streams into the MF Pipeline. With all these new updates, one would think that the old functionality would take a hit and not work anymore, but that is not the case. Microsoft did not modify the old behavior of the driver. This means it still works the same way for previous applications. It is backward-compatible with the pre–Windows 10 UWP SDK and framework. This means that applications built with the .NET framework and C++ libraries will still continue to function as normal. Also, as an added bonus, the new UWP framework does not hinder the old framework and actually can work side by side simultaneously. Figure A1-8 shows an application running the DirectX Depth Sample found in the previous SDK and the UWP Camera Frames sample simultaneously.

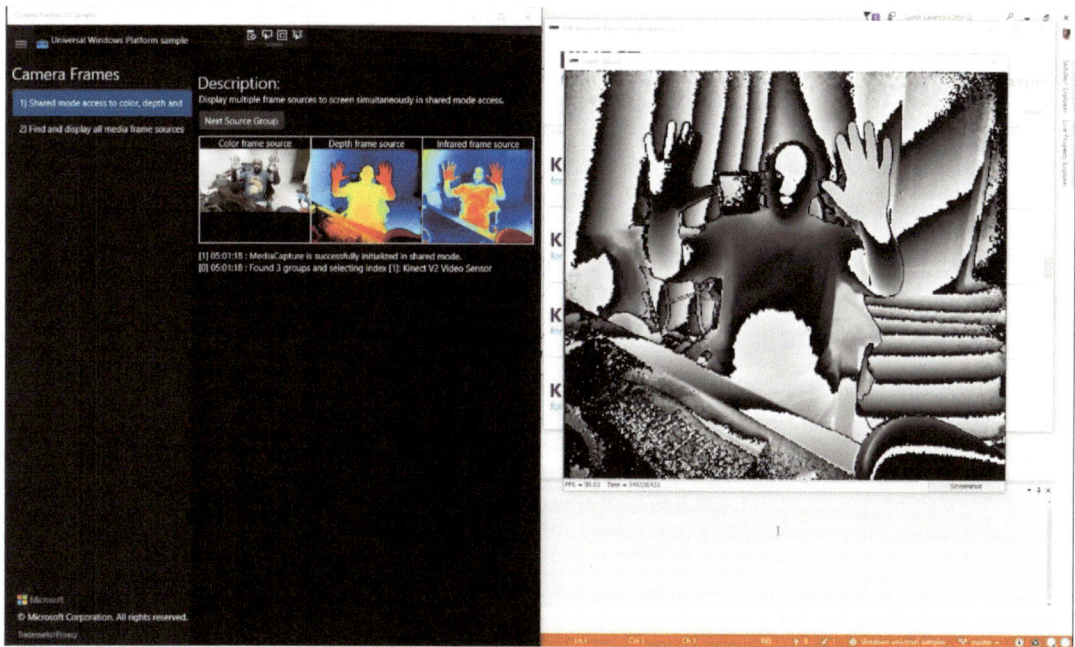

Figure A1-8. *.NET and UWP Kinect applications running simultaneously*

If you didn't catch this the first time I mentioned it, this means you can run your old .NET or C++ Kinect-based applications at the same time as you run newer UWP store Kinect-based applications. Along with running the applications side by side, as you start to develop with UWP Kinect-based applications, you'll notice a similar programming pattern when it comes to asynchronous behavior and listening for events and frames. Let's discuss that next.

Similarities in the Architecture

I know that it was stated earlier that developing a Kinect application for UWP is a completely new paradigm, and this leaves the impression that there will be a steep learning curve. However, to our advantage, the learning curve isn't very steep at all. This is because if you've ever done any Kinect programming in the past, even by reading through this book up to this point, you'll be comforted to know that there are some very similar development patterns with UWP. For example, Windows 10 UWP applications are based on an asynchronous design pattern, so if you've worked with the `await` keyword in C#, nothing has changed. Next, there is the concept of frames. If you remember, in previous chapters we had `ColorFrames`, `DepthFrames`, and `InfraredFrames`. In UWP, we have Media Capture frames and perception frames. Similarly, if you remember, each type of frame was able to generate various events, such as the `FrameArrived` event. In UWP, we have the same pattern. Perception and Media Capture frames also generate an arrived event, and as developers we can subscribe to these events to listen for data. Another similarity we have is the concept of disposing of a frame as fast as we possibly can once we're done with it. In UWP Kinect-based applications, we follow the same principle. As you can see, the learning curve isn't as steep as it may appear when it comes to media frames. With the architecture out of the way, let's start creating our UWP Kinect-based Store app.

Creating a Windows 10 UWP Kinect-Based Store App

To start developing with Windows 10 UWP and Kinect, you are going to need to get the latest Kinect for Windows v2 drivers. The first thing you want to do is make sure Kinect for Windows v2 is running with version **2.0.xx+** and Windows 10. If not, download and install the Kinect for Windows v2 SDK as described in the first chapter of this book. Once your system is up and running with the Kinect, we can move forward.

How to Get the Latest Drivers

The next and easiest, but probably most time-consuming, step is to upgrade your Windows 10 to the Anniversary edition: version **10.0.14394.xxx** or above. If you're not sure what version of Windows 10 you're running, click the Start button or press the Windows key and type *version*. In Figure A1-9, you'll see the **System Information** application reveal what version of Windows 10 you are running. If you're on version **10.0.14394** or above, you're good to go. If not, go ahead and head over to your system settings, navigate to **Windows Update,** and update to the Anniversary edition.

Figure A1-9. *Windows 10 System Information and version details*

If you don't have the Anniversary edition, you can still get the latest driver, which is the preview "test" version but is the latest none the less. To get this version, you will also have to be a Windows 10 Insider and opt in for test drivers. Search how to opt in for test drivers on Windows 10. It's a registry setting. Turn on Windows Hello, and you should be ready to go. I won't go into the specifics simply because I recommend you go the safer route of updating to the Anniversary edition.

Once you are on the Anniversary edition, plug your device in and open up **Device Manager** by clicking on the Start button and typing *device manager*. Locate the Kinect device, expand it, right-click on *WDF KinectSensor Interface 0*, and select *Update driver*. Let Windows choose the best and most updated driver. Figure A1-10 shows the menu item to select.

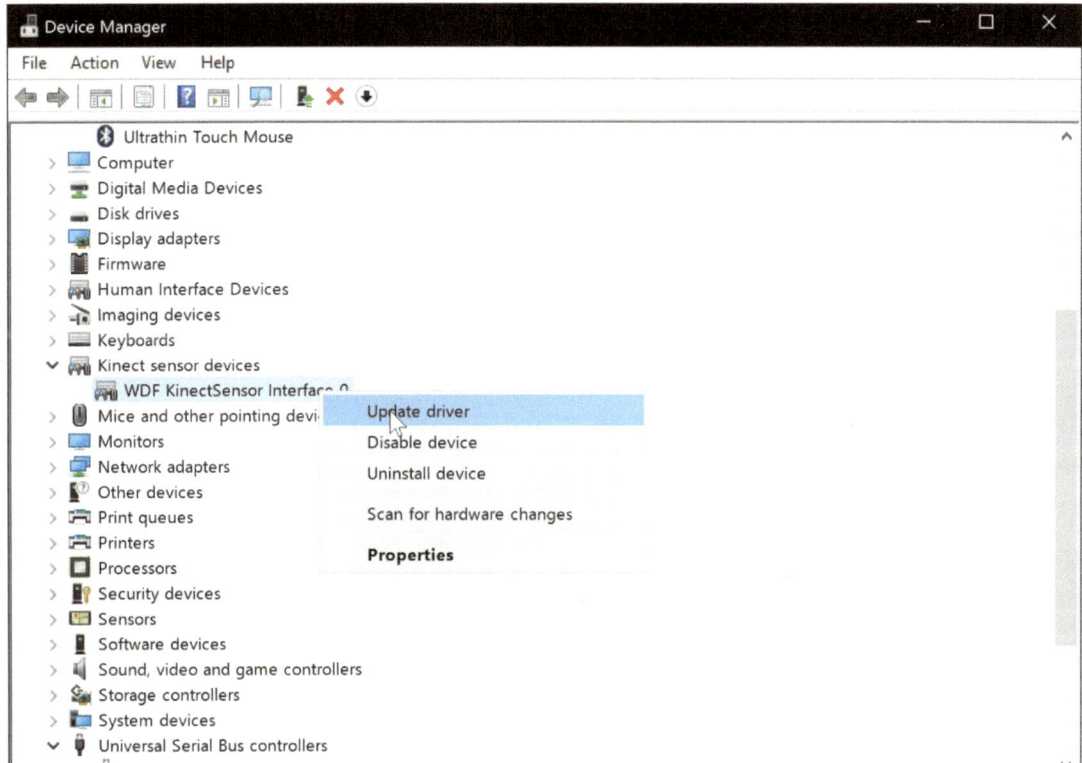

Figure A1-10. *Selecting Update driver in Device Manager*

If everything goes well without an error, then your driver should be updated to the latest version. Figure A1-11 shows the properties of the latest Kinect UWP driver from Device Manager.

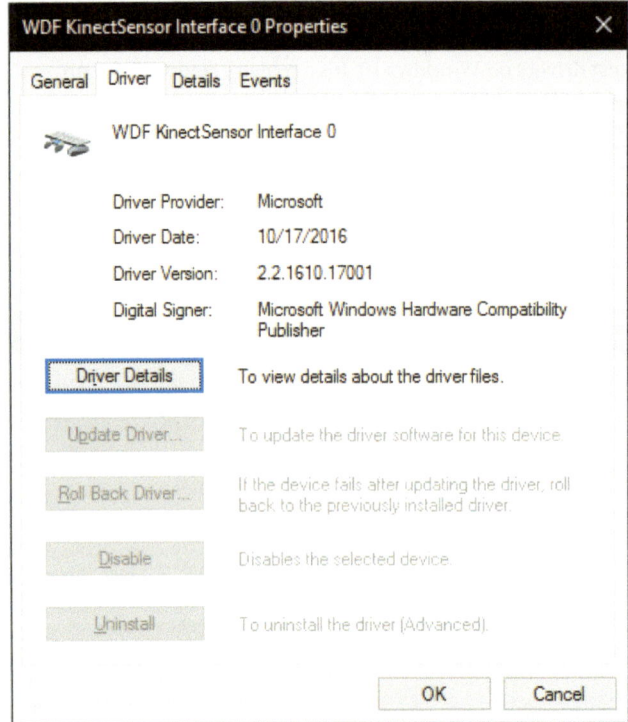

Figure A1-11. *Kinect's latest driver details*

To verify your Kinect driver is working with UWP, run the Windows 10 Camera application and cycle through your cameras, verifying that your Kinect turns on and that the Kinect v2 is listed as the camera. Figure A1-12 shows the Kinect device as a camera available for selection in the UWP Camera application. No camera feed might show.

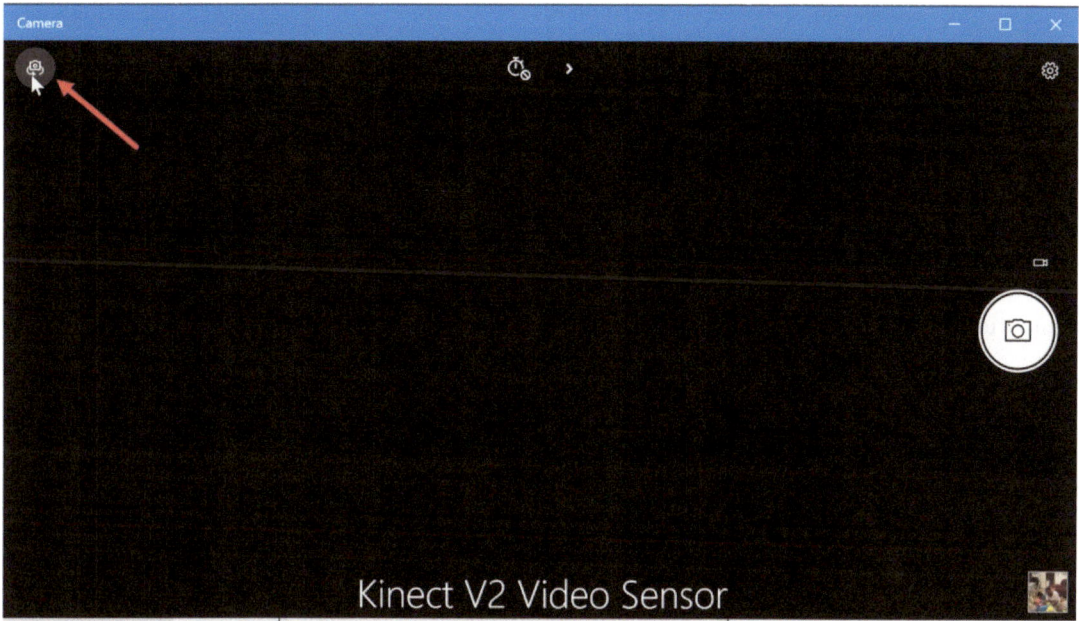

Figure A1-12. *Kinect Device selected in the Camera application*

Note After updating your driver, if your Kinect camera does not turn on make sure the `KinectMonitor.exe` window service is running from within your services in the system settings control panel.

To run the standard Camera application, click the Start button and type *camera*.

Building a UWP Store application

We will start up by building a very simple UWP Kinect-based application that can show the color web camera stream and allow you to take a picture of the color stream. The application will make use of the *Media Frame Source Group, Media Frame Reference, Media Frame, Software Bitmap, XAML Image Element,* and *XAML Writeable Bitmap.*

Let's start from scratch building a Visual Studio 2015 C# UWP project. Name the project `WorkingWithMediaCaptureFrames`.

Tip You can find the solution on my GitHub page at `https://github.com/dngoins/KinectUWPApps.git`.

Add the namespaces seen in Listing A1-1 to the `MainPage.xaml.cs` file.

Listing A1-1. Media Capture Namespaces

```
using Windows.Media.Capture;
using Windows.Media.Capture.Frames;
```

These are the core namespaces required for the Media Capture and Media Capture Frames classes.

Add the remaining namespaces to include asynchronous behavior, software bitmaps, and file storage functionality, as shown in Listing A1-2.

Listing A1-2. Namespaces for Async, Imaging, and File Storage

```
using System.Threading.Tasks;
using Windows.Graphics.Imaging;
using Windows.UI.Xaml.Media.Imaging;
using Windows.Storage;
using Windows.Storage.Pickers;
using Windows.Storage.Streams;
```

Add the private class member variable seen in Listing A1-3.

Listing A1-3. MediaCapture Variable Declaration

```
// MF Capture device to start up
        MediaCapture _mediaCapture;
```

This variable represents the Media Foundation Media Capture device. Basically, this represents the Kinect Device as a Media Foundation UWP object. Now, add the remaining private member class variables that will be used to help us display and save the color image, as shown in Listing A1-4.

Listing A1-4. Imaging Variable Declarations

```
private enum FileFormat
{
    Jpeg,
    Png,
    Bmp,
    Tiff,
    Gif
}

// Constants that let us decide on which frame source kind we want
const int COLOR_SOURCE = 0;
const int IR_SOURCE = 1;
const int DEPTH_SOURCE = 2;
const int LONGIR_SOURCE = 3;
const int BODY_SOURCE = 4;
const int BODY_INDEX_SOURCE = 5;

// UWP BitmapSource
private SoftwareBitmapSource source = new SoftwareBitmapSource();

// XAML Bitmap for updating bitmaps in XAML
private WriteableBitmap extBitmap;
```

The preceding variables are used to help us determine which file format we'd like to use to save the color image, which Media Capture frame source types we are going to display, the UWP bitmap source, and the XAML bitmap source, respectively. The remaining steps deal more with how to access various data streams and work with the individual streams (i.e. color, depth, and infrared).

How to Access Data Streams

To get access to the various data streams from a Media Capture device, we must request permission to access the device. We can do this declaratively using the `package.appxmanifest` file. Double-click on the package manifest and click on the Capabilities tab. Select *Webcam* and *Pictures Library*. Figure A1-13 shows the `package.appmanifest` file opened inside Visual Studio 2015.

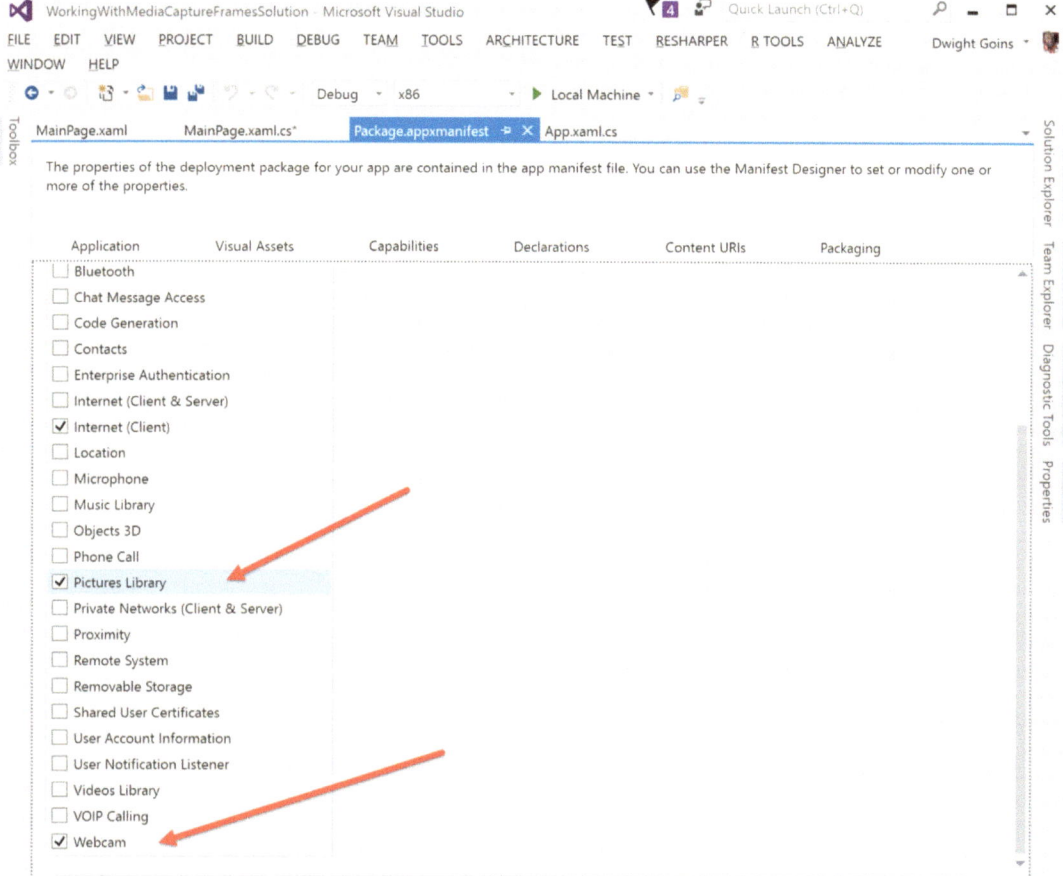

Figure A1-13. *Declaring capabilities with package manifest*

Selecting these options will have the application asking or requesting permission to read and write to the Pictures Library, as well as asking for permission to show and use the device camera (the Kinect, in our case).

Once we have requested the proper capabilities, we focus our attention on how to access them. UWP treats a Media Capture device as a set of features. This set or collection is known as a `MediaFrameSourceGroup`. For each MF device, if it can send at least one type of data stream, then it is represented as one of this group. This is where things start to differ from traditional Kinect development. Normally, with the Kinect device you'd call `GetDefault()`, which will provide you with the default device that represents the Kinect.

In the UWP paradigm, you query for a group, the MediaFrameSourceGroup. The concept is that one or more devices will have a group of sensors. Plus, as a developer you may want to access these sensors synchronously. Basically, each device (a.k.a. MediaFrameSourceGroup) will return with data that contains one or more sources of information regarding each data stream that may be exposed by that device (a.k.a. MediaFrameSourceGroup). This group is more akin to a single device with a collection of capabilities. The MediaFrameSourceGroup is an abstraction over that aforementioned idea. A MediaFrameSourceGroup provides a collection of sources. These are the Data Sources. They are represented in the UWP platform as a class called SourceInfos. The SourceInfos is a collection that contains a collection of data streams. A data stream is of a particular type; for example, it could be a color stream, depth stream, infrared stream, or custom stream. We determine which type of stream it is by the property of source info called the SourceKind. SourceInfos also contain information about the source's DisplayName, Location, and CameraIntrinsics. It's here that you can filter out which device group you want to listen to for frame events. Let's add the method shown in Listing A1-5 to the class.

Listing A1-5. Color Frame Initialization

```
async void InitializeColorFrames()
{
    // Devices come grouped together as a set of capabilities
    // Find all the media frame source devices - as a collection of groups
    var allGroups = await MediaFrameSourceGroup.FindAllAsync();

    // Let's filter using LINQ, based on a device group with a DisplayName of "Kinect"
    var eligibleColorGroups = allGroups
        .Where(g => g.SourceInfos.FirstOrDefault(info => info.SourceGroup.DisplayName.
        Contains("Kinect") ) != null).ToList();

    // Check to see if we found a device
    if (eligibleColorGroups.Count == 0)
    {
        //No kinect camera found
        return;
    }

    //Only one Kinect camera supported, so always take the first in the list
    var kinectColorGroup = eligibleColorGroups[0];
    try
    {
        // Initialize MediaCapture with selected group.
        // This can raise an exception if the source no longer exists,
        // or if the source could not be initialized.
        await InitializeMediaCaptureAsync(kinectColorGroup);
    }
    catch (Exception exception)
    {
        //Release any resources if something goes wrong
        await CleanupMediaCaptureAsync();
        return;
    }
```

```
// Let's get a Device Capability; in this case, let's get the color source stream
MediaFrameSourceInfo colorInfo = kinectColorGroup.SourceInfos[COLOR_SOURCE];

if (colorInfo != null)
{
    // Access the initialized frame source by looking up the the ID of the source found above.
    // Verify that the ID is present, because it may have left the group
    // while we were busy deciding which group to use.
    MediaFrameSource frameSource = null;
    if (_mediaCapture.FrameSources.TryGetValue(colorInfo.Id, out frameSource))
    {
        // Create a frameReader based on the color source stream
        MediaFrameReader frameReader = await _mediaCapture.CreateFrameReaderAsync(frameSource);

        // Listen for Frame Arrived events
        frameReader.FrameArrived += FrameReader_FrameArrived;

        // Start the color source frame processing
        MediaFrameReaderStartStatus status = await frameReader.StartAsync();

        // Status checking for logging purposes if needed
        if (status != MediaFrameReaderStartStatus.Success)
        {

        }
    }
    else
    {
        // Couldn't get the color frame source from the MF Device

    }
}
else
{
    // There's no color source

}
}
```

The preceding method uses the static MediaFrameSourceGroup.FindAllAsync() method to get back a collection of device "groups" that have a synchronous collection of data streams. A LINQ statement follows, which filters the collection of groups based on the SourceInfo's DisplayName. If the groups' DisplayName contains the word *Kinect*, then the group represents the Kinect's collection of synchronized capabilities.

Note In the previous programming model, we had two modes of retrieving frames: each data feed individually without synchronization and the multisource frame, which synchronizes all the frames together. In UWP, we don't have the non-synchronous option.

SourceInfo Class

There are various other values in the `SourceInfo` metadata that allow you to further filter the data streams you're interested in. The `SourceInfo` is kind of a mixture between the `FrameDescription` class and the actual xSource (`ColorFrameSource`, `DepthFrameSource`, `InfraredFrameSouce`) we had in previous frameworks. For example, you can look at metadata properties for a source stream using the `SourceInfo.Properties` map. You can access the device that the individual source stream using the `SourceInfo.DeviceInformation` property. There's also the `SourceInfo.SourceKind` mentioned earlier. This is an enumeration that specifies the different types of MF sources, such as `Color`, `InfraRed`, `Depth`, and `Custom` (used for body and bodyIndex or hand tracking). There's also the `MediaStreamType`, which allows you to determine if the source stream is for photo, video, or audio, and, last but not least, there is the `CoordinateSystem` property. This is a `Windows.Perception.SpatialCoordinateSystem` class that allows you to get to a matrix representing where the device thinks it's positioned and transform it to other coordinate systems for advanced projection mapping and 3D positioning and rotation mechanisms. In summary, the `SourceInfo` class will be utilized heavily in your applications, so get comfortable with all its properties and methods. It will make UWP Kinect-based development easy.

Initializing the Media Capture Object

After we've retrieved the `MediaFrameSourceGroup` we're interested in from MF, we need to initialize the Media Capture device. This is accomplished with the call to `InitializeMediaCaptureAsync`, as shown in Listing A1-6.

Listing A1-6. Media Capture Initialization

```
private async Task InitializeMediaCaptureAsync(MediaFrameSourceGroup sourceGroup)
{
    if (_mediaCapture != null)
    {
        return;
    }

    // Initialize mediacapture with the source group.
    _mediaCapture = new MediaCapture();
    var settings = new MediaCaptureInitializationSettings
    {
        SourceGroup = sourceGroup,

        // This media capture can share streaming with other apps.
        SharingMode = MediaCaptureSharingMode.SharedReadOnly,

        // Only stream video and don't initialize audio capture devices.
        StreamingCaptureMode = StreamingCaptureMode.Video,

        // Set to CPU to ensure frames always contain CPU SoftwareBitmap images
        // instead of preferring GPU D3DSurface images.
        MemoryPreference = MediaCaptureMemoryPreference.Cpu
    };

    await _mediaCapture.InitializeAsync(settings);
}
```

This method initializes the _mediaCapture private variable using a MediaCaptureInitializationSettings class. We use the IntializationSettings class to specify the device we want MF to use. In this case, it's the Kinect device, represented as the SourceGroup of capabilities we queried and filtered on with the MediaFrameSourceGroup.FindAllAsync() earlier. The next property we must set on the settings class is whether we want MF to initialize the device in ExclusiveControl or ShareReadOnly mode. ExclusiveControl gives our application sole control over the device. Using this mode allows us to change settings on the device, such as the white balance, frame height and width, and other advanced properties we didn't have with the previous versions of the Kinect SDK. There's one catch, however. If another application is currently using the Kinect, either UWP or a previous framework, then when we request exclusive control it will throw an error. You will have to wait until the other application has finished usage. This option is great for doing more creative things with your device, such as changing the white balance of the sensors and changing resolution and modes. However, with great power comes great responsibility. Using exclusive mode blocks other applications and could potentially cause other side effects if you're not careful.

For the most part, SharedReadOnly mode will be the mode we choose. This mode allows multiple applications to use the Kinect device in a sharing, read-only fashion. This means you can't modify metadata or camera intrinsics.

After setting the SharingMode property, we must set the StreamingCaptureMode. This is an enumeration that accepts three possible values: Video, Audio, AudioAndVideo. These three values are self-explanatory; if the device SourceGroup supports audio and video, you have a choice of which capture you want to work with. You can choose audio only, video only, or both.

The last property of the MediaCaptureInitializationSettings class we set is the MemoryPreference value. This value tells the Media Foundation how and where it should construct the pixel buffer array. The options are Cpu and Auto. Cpu instructs MF to create SoftwareBitmap classes on the CPU. Auto instructs MF to use Hardware Acceleration where available and check for a DirectX-compatible GPU if present. If a GPU is available, then create an ID3DSurface object in which to store the pixel buffer array. When rendering, D3DSurface provides the best performance; however, the use of DirectX and Direct3D is outside the scope of this book. There are many more settings you can apply to the MediaCaptureInitializationSettings class, but these are some of the more common you'll use for Kinect-based UWP applications.

Once all the settings are populated with their respective values, the MediaCapture instance invokes InitializeAsync(settings), passing in the settings object. Add the helper method shown in Listing A1-7 for cleanup.

Listing A1-7. Media Capture Cleanup

```
/// <summary>
/// Unregisters FrameArrived event handlers, stops and disposes frame readers
/// and disposes the MediaCapture object.
/// </summary>
private async Task CleanupMediaCaptureAsync()
{
    if (_mediaCapture != null)
    {
        using (var mediaCapture = _mediaCapture)
        {
            _mediaCapture = null;

        }
    }
}
```

The `CleanupMediaCaptureAsync()` is used to free up resources of the MF `mediaCapture` instance. This instance represents the actual device, and it needs to be freed from memory if anything goes wrong, or you're finished using it. We do this simply by setting the private member variable to null. Now, with all the MF setup and clean up out the way, let's focus our attention on getting the color frames and processing individual color buffers.

How to Work with Color

Up to this point, we've discussed the basic setup for accessing the Kinect device through Media Foundation. We have the `mediaCapture` object, and we have the Kinect source data streams through two different objects: `mediaSourceGroup` (a synchronized grouping of all the source streams, similar to the `MultiSourceFrameReader` in the previous SDK) and `SourceInfo` object, which represents the individual source stream and description. Let's revisit a section of code in the `InitializeColorFrames()` method. There's a section of code we haven't discussed yet, and this section of code is how we work with the core source data streams. See Listing A1-8.

Listing A1-8. Handling the Color Data Source

```
//...
// Let's get a Device Capability, in this case let's get the color source stream
MediaFrameSourceInfo colorInfo = kinectColorGroup.SourceInfos[COLOR_SOURCE];

if (colorInfo != null)
{
    // Access the initialized frame source by looking up the ID of the source found above.
    // Verify that the ID is present, because it may have left the group
    // while we were busy deciding which group to use.
    MediaFrameSource frameSource = null;
    if (_mediaCapture.FrameSources.TryGetValue(colorInfo.Id, out frameSource))
    {
        // Create a frameReader based on the color source stream
        MediaFrameReader frameReader = await _mediaCapture.CreateFrameReaderAsync(frameSource);

        // Listen for Frame Arrived events
        frameReader.FrameArrived += FrameReader_FrameArrived;

        // Start the color source frame processing
        MediaFrameReaderStartStatus status = await frameReader.StartAsync();

        // ...
```

The preceding section of code queried all the source groups and filtered down the specific device source group we're interested in. If you remember from earlier, it was the source group that had a display name of "Kinect" contained inside its string's full name. Once we've selected the source group, we can get to the capability list, better known as the `SourceInfo` collection. In the preceding code I use an indexed array, as follows:

```
//...
// Let's get a Device Capability, in this case let's get the color source stream
        MediaFrameSourceInfo colorInfo = kinectColorGroup.SourceInfos[COLOR_SOURCE];
```

Where COLOR_SOURCE is a constant representing the positional index that is retrieved from the MF Kinect's color stream representation. However, if you're working with other source groups for other devices, this may not be the best way to go, because each device has the option of changing how it presents the collection of *source streams* and in what order. Again, remember the SourceInfo collection allows you to determine which source stream you want using SourceKind, MediaStreamType, Device Information, and many other properties. I chose the simplest way for this demo.

The next lines of code should be somewhat familiar:

```
MediaFrameSource frameSource = null;
if (_mediaCapture.FrameSources.TryGetValue(colorInfo.Id, out frameSource))
{
    // Create a frameReader based on the color source stream
    MediaFrameReader frameReader = await _mediaCapture.CreateFrameReaderAsync(frameSource);

    // Listen for Frame Arrived events
    frameReader.FrameArrived += FrameReader_FrameArrived;

    // Start the color source frame processing
    MediaFrameReaderStartStatus status = await frameReader.StartAsync();
```

These lines use the mediaCapture.FrameSources.TryGetValue() method to get a generic MediaFrameSource. This MediaFrameSource represents a generic version of the ColorFrameSource, DepthFrameSource, and InfraredFrameSource found in previous versions. Once we have the frame source, we use the mediaCapture object to create a MediaFrameReader. This frame reader is used to listen for FrameArrived events. The MediaFrameReader also supports the polling design. There is a TryAcquireLatestFrame() method that you can use to get the latest frame in a polling pattern if you need to use that design. The MediaFrameReader also contains the methods to StartAsync() and StopAsync() the frames from being processed on the device.

Let's now add the FrameReader_FrameArrived method to our class, as shown in Listing A1-9.

Listing A1-9. FrameArrived Event Handler

```
private void FrameReader_FrameArrived(MediaFrameReader sender, MediaFrameArrivedEventArgs args)
{
    // Try to get the FrameReference
    using (var frameRef = sender.TryAcquireLatestFrame())
    {
        if(frameRef != null)
        {
            if (frameRef.SourceKind == MediaFrameSourceKind.Color)
            {
                // Process the frame drawing the color pixels onto the screen
                ProcessColorFrame(frameRef);
            }
        }
    }
}
```

This method provides us with a signature that takes in a MediaFrameReader, as well as MediaFrameArrivedEventArgs parameters. Normally in .NET programming the EventArgs parameter provides context information about the event being generated; however, in this framework EventArgs are not used at all, so you can safely ignore that parameter. In this method, we use the MediaFrameReader parameter to try to get the latest frame. This method returns a MediaFrameReference, similar to in the previous SDK. Once we have the media frame reference, we can access the specific frame we're interested in. Some of you may have noticed that it's generic. In the previous framework, we had a specific frame reference–one for Color, one for Body, one for Depth, and one for Infrared. In this new UWP framework, there's only one. This means we have to check for which type of frame reference it is by using the SourceKind enumeration property.

Now, let's add the ProcessColorFrame method, as shown in Listing A1-10.

Listing A1-10. Processing Color Data

```
private void ProcessColorFrame(MediaFrameReference  clrFrame)
{
    try
    {
        // Get the Individual color Frame
        var vidFrame = clrFrame?.VideoMediaFrame;
        {
            if (vidFrame == null) return;

            // create a UWP SoftwareBitmap and copy Color Frame into Bitmap
            SoftwareBitmap sbt = new SoftwareBitmap(vidFrame.SoftwareBitmap.BitmapPixelFormat,
            vidFrame.SoftwareBitmap.PixelWidth, vidFrame.SoftwareBitmap.PixelHeight);
                    vidFrame.SoftwareBitmap.CopyTo(sbt);

            // PixelFormat needs to be in 8bit BGRA for Xaml writeable bitmap
            if (sbt.BitmapPixelFormat != BitmapPixelFormat.Bgra8)
            sbt = SoftwareBitmap.Convert(vidFrame.SoftwareBitmap, BitmapPixelFormat.Bgra8);

            if (source != null)
            {
                // To write out to writeable bitmap which will be used with ImageElement, it needs to run
                // on UI Thread thus we use Dispatcher.RunAsync()...
                var ignore = Dispatcher.RunAsync(Windows.UI.Core.CoreDispatcherPriority.Normal, () =>
                    {
                        // This code runs on UI Thread
                        // Create the writeableBitmap for ImageElement display
                        extBitmap = new WriteableBitmap(sbt.PixelWidth, sbt.PixelHeight);

                        // Copy contents from UWP software Bitmap
                        // There are other ways of doing this instead of the double copy,
                            1st copy earlier
                        // this is a second copy.
                        sbt.CopyToBuffer(extBitmap.PixelBuffer);
                        extBitmap.Invalidate();
```

```
                // Set the imageElement source
                var ig = source.SetBitmapAsync(sbt);
                imgView.Source = source;

            });
        }
      }
    }
    catch (Exception ex)
    {

    }
}
```

The process color frame method is used to retrieve either the VideoMediaFrame or a BufferMediaFrame. The BufferMediaFrame is used to retrieve body skeletal data through an IBuffer interface. The VideoMediaFrame is used for all other data. We use the VideoMediaFrame to determine if we have data in the Direct3DSurface or SoftwareBitmap, get access to the depth- and infrared-specific properties, get access to camera intrinsics, and determine video format. The video media frame is the key to processing and displaying frames.

Once we have the video media frame, we use the frame's SoftwareBitmap to copy the contents of the bitmap to another SoftwareBitmap, which we can use inside a writeable bitmap for saving out to a file and also for re-copying to a SoftwareBitmapSource class for use with the XAML Image element.

Note   There are more efficient ways to share the bitmap contents. These can be found in the UWP samples located at https://github.com/Microsoft/Windows-universal-samples/tree/master/Samples/CameraFrames. However, these samples deal with locking strategies and multiple dispatcher calls that require a lot more explanation than what is possible for the scope of this book.

As an aside, because we are updating an XAML user interface control (the Image element), we are confined to only updating this control on the user interface thread. The way to do this is to use the Dispatcher.RunAsync() method call. This is the reason for all the SoftwareBitmap copy code living inside the RunAsync call.

Finally, let's add the final helper methods and user interface so we have something to show. Add the helper methods shown in Listing A1-11 to your class.

Listing A1-11. Helper Methods for Screenshotting

```
private async Task<StorageFile> WriteableBitmapToStorageFile(WriteableBitmap WB, FileFormat
fileFormat, string fileName = "")
{
    try
    {
        string FileName = string.Empty;
        if (string.IsNullOrEmpty(fileName))
        {
            FileSavePicker savePicker = new FileSavePicker();
            savePicker.SuggestedStartLocation = PickerLocationId.PicturesLibrary;
```

```
    // Drop-down of file types the user can save the file as
    savePicker.FileTypeChoices.Add("jpeg", new List<string>() { ".jpg", ".jpeg" });

    // Default file name if the user does not type one in or select a file to replace
    savePicker.SuggestedFileName = "WorkingWithMediaCapture.jpg";
    fileName = (await savePicker.PickSaveFileAsync()).Name;
}

FileName = fileName;

Guid BitmapEncoderGuid = BitmapEncoder.JpegEncoderId;
switch (fileFormat)
{
    case FileFormat.Jpeg:
    //  FileName = string.Format("{0}.jpeg", fileName);
    BitmapEncoderGuid = BitmapEncoder.JpegEncoderId;
    break;

    case FileFormat.Png:
    //  FileName = string.Format("{0}.png", fileName);
    BitmapEncoderGuid = BitmapEncoder.PngEncoderId;
    break;

    case FileFormat.Bmp:
    //  FileName = string.Format("{0}.bmp", fileName);
    BitmapEncoderGuid = BitmapEncoder.BmpEncoderId;
    break;

    case FileFormat.Tiff:
    //  FileName = string.Format("{0}.tiff", fileName);
    BitmapEncoderGuid = BitmapEncoder.TiffEncoderId;
    break;

    case FileFormat.Gif:
    //  FileName = string.Format("{0}.gif", fileName);
    BitmapEncoderGuid = BitmapEncoder.GifEncoderId;
    break;
}

var file = await Windows.Storage.KnownFolders.PicturesLibrary.
CreateFileAsync(FileName, CreationCollisionOption.GenerateUniqueName);
using (IRandomAccessStream stream = await file.OpenAsync(FileAccessMode.ReadWrite))
{
    BitmapEncoder encoder = await BitmapEncoder.CreateAsync(BitmapEncoderGuid, stream);
    Stream pixelStream = WB.PixelBuffer.AsStream();
    byte[] pixels = new byte[pixelStream.Length];
    await pixelStream.ReadAsync(pixels, 0, pixels.Length);

    encoder.SetPixelData(BitmapPixelFormat.Bgra8, BitmapAlphaMode.Ignore,
        (uint)WB.PixelWidth,
        (uint)WB.PixelHeight,
```

```
                96.0,
                96.0,
                pixels);
            await encoder.FlushAsync();
        }
        return file;
    }
    catch (Exception ex)
    {
        return null;
        // throw;
    }
}

private void btnExternalCapture_Click(object sender, RoutedEventArgs e)
{
    if (extBitmap != null)
    {
        var ignore = Dispatcher.RunAsync(Windows.UI.Core.CoreDispatcherPriority.Normal, async () =>
        {
            var file =
                await WriteableBitmapToStorageFile(extBitmap, FileFormat.Jpeg,
                "WorkingWithMediaCaptureFrames.jpg");
            fileLocation.Text = file.Path;
        });
    }
}
```

The preceding two methods take the writeable bitmap and save the contents out to a file in the *Pictures* library. The second method is an event handler that listens for the button client event for the External Capture button in the XAML shown in Listing A1-12.

Listing A1-12. MainPage.xaml Markup

```
<Grid Background="{ThemeResource ApplicationPageBackgroundThemeBrush}">
    <StackPanel  Margin="5" VerticalAlignment="Center" HorizontalAlignment="Center">
        <TextBlock >External Cameras View:</TextBlock>
        <Image x:Name="imgView" Stretch="UniformToFill" MaxWidth="600" ></Image>
        <Button x:Name="btnExternalCapture" HorizontalAlignment="Right"
         Click="btnExternalCapture_Click">Save External Camera Shot</Button>
        <TextBlock>FilePath:</TextBlock>
        <TextBox x:Name="fileLocation" TextWrapping="Wrap" />
    </StackPanel>
</Grid>
```

Compile your application and run it. The final result is shown in Figure A1-14.

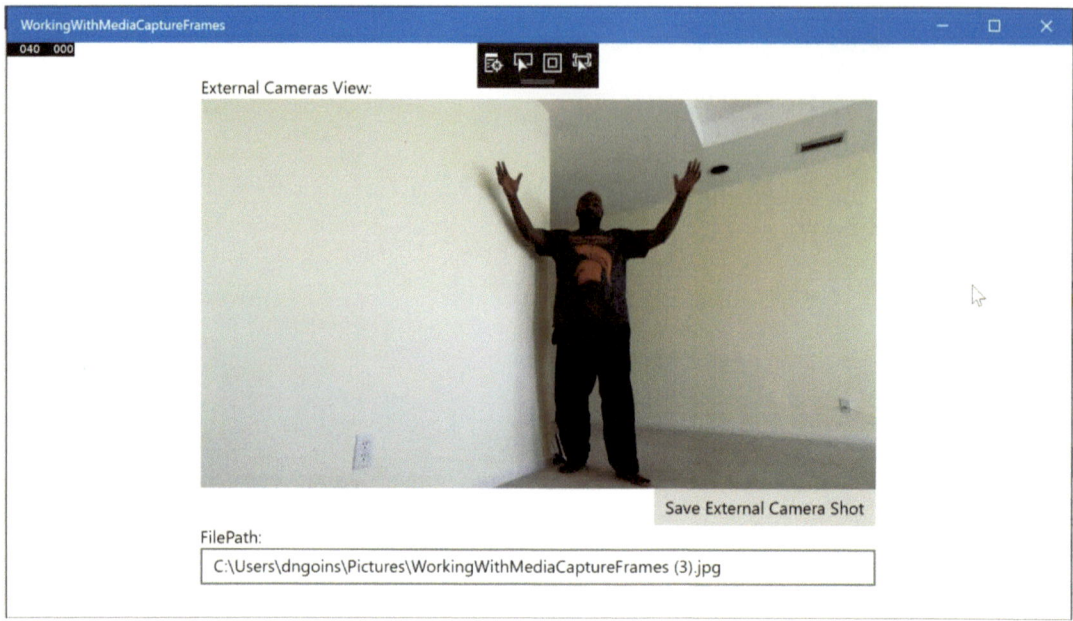

Figure A1-14. *Using Color stream with Kinect in UWP application*

How to Work with Depth

Working with Depth is exactly the same as working with Color frames. Let's revisit the code to retrieve Color Source Info, which gives us access to the color frames:

```
//...
// Let's get a Device Capability, in this case let's get the color source stream
MediaFrameSourceInfo colorInfo = kinectColorGroup.SourceInfos[COLOR_SOURCE];
```

To use Depth, we simply change out which `SourceInfo` item we want. We can simply change this to `DEPTH_SOURCE`, where `DEPTH_SOURCE` is a constant representing the positional index that is retrieved from MF Kinect's Depth stream representation.

```
//...
// Let's get a Device Capability, in this case let's get the depth source stream
MediaFrameSourceInfo depthInfo = kinectColorGroup.SourceInfos[DEPTH_SOURCE];
```

Again, remember that if you're working with other `SourceGroups` (i.e. other devices), this may not be the best way to do this, because each device has the option to change how it presents the collection of source streams and in what order. The `SourceInfo` collection allows you to determine which source stream you want by using the `SourceKind`, `MediaStreamType`, Device Information, and many other properties. I chose the simplest way for this demo.

How to Work with IR

Working with Infrared is exactly the same as working with Color frames. Let's revisit the code to retrieve Color Source Info, which gives us access to the color frames:

```
//...
// Let's get a Device Capability, in this case let's get the color source stream
MediaFrameSourceInfo colorInfo = kinectColorGroup.SourceInfos[COLOR_SOURCE];
```

To use either the default IR or the LongIR, we simply change out which SourceInfo item we want. We can simply change this to IR_SOURCE or LONGIR_SOURCE, where IR_SOURCE or LONGIR_SOURCE is a constant representing the positional index that is retrieved from the MF Kinect's Infrared stream representation.

```
//...
// Let's get a Device Capability, in this case let's get the IR source stream
MediaFrameSourceInfo irInfo = kinectColorGroup.SourceInfos[IR_SOURCE];
MediaFrameSourceInfo longIRInfo = kinectColorGroup.SourceInfos[LONGIR_SOURCE];
```

How to Work with Custom Data Streams (Preview)

The Media Capture frames API supports devices' being able to yield custom data streams. This means devices like the Kinect can send body, body index, and other data through the custom data streams pattern. This also means other devices, like Leap Motion, can send hand-tracking data. There is no limit to the types of data that can be sent through the custom data streams.

As of the time of this writing, the Kinect device is the only device that makes use of this custom data stream, and its current implementation is in preview, which means it's likely to change by the time this feature is finalized.

Working with Kinect Body Streams

The current implementation of Body frames is through the Custom SourceKind data stream. This data stream defines a very unique buffer layout and set of properties that outline the various body joint data, similar to the previous Kinect SDK. This implementation contains the same number of joints, 25, in pretty much in the same physical areas of the body. The joints also expose the joint position and orientation. Body-tracking statuses of **tracked**, **not tracked**, and **inferred** are also exposed; thus, nothing has changed when comparing to the previous framework's functionality.

What has changed, however, is how you access the body joint information. Body joints are accessed through custom SourceInfo as a form of enum from the WindowsPreview.Media.Capture.Frames.BodyPart enumeration. The position and orientation come from the WindowsPreview.Media.Capture.Frames. TrackedPose class. The tracking status comes from the WindowsPreview.Media.Capture.Frames.TrackingStatus enumeration.

■ **Note**   These enums and classes are currently a part of the **PoseTrackingPreview** project, which can be found on GitHub at https://github.com/Microsoft/Windows-universal-samples/tree/master/Samples/ CameraStreamCorrelation.

The custom data stream has a specific structure. This structure is basically a byte array of various components that stream the BodyPart, TrackedPost, TrackingStatus, and another object called the TrackingEntity. See Listing A1-13.

Listing A1-13. Structure of Pose-Tracking Entity

```
// The layout of a pose frame buffer is:
//    1. PoseTrackingFrameHeader: Common structure for all providers.
//        1. CustomDataTypeGuid: Guid corresponding to specific provider (16 bytes).
//        2. EntitiesCount: How many entities are in this frame (4 bytes).
//        3. EntityOffsets: Offsets of entity data from buffer start (4 * EntitiesCount bytes).
//
//    2. Provider-specific, per-frame data.
//
//    3. PoseTrackingEntityData #0: First entity data. Common structure for all providers.
//        1. DataSizeInBytes: Size of entire entity (PoseTrackingEntityData + custom data) in
//           bytes (4 bytes).
//        2. EntityId: Allows correlation between frames (16 bytes).
//        3. PoseSet: Guids for Body (such as Kinect), Handtracking, etc.: defines the meaning
//           of each Pose (16 bytes).
//        4. PosesCount: Count of poses in this entity data (4 bytes).
//        5. IsTracked: Whether or not this entity is being tracked (1 byte).
//        6. Poses: Array of common structure TrackedPose (sizeof(TrackedPose) * PosesCount bytes).
//        7. Customer-specific data for this entity (DataSizeInBytes - sizeof(PoseTrackedEntityData) -
//           sizeof(TrackedPose) * (PosesCount -1) bytes)
//
//    4. Provider-specific data for entity #0 in this frame.
//
//    5. PoseTrackingEntityData #1: Second entity data. Common structure for all providers.
//
//    6. Provider-specific data for entity #1 in this frame.

struct PoseTrackingFrameHeader
{
    // This Guid allows the consumer of the buffer to verify that
    // it has been written by the expected provider
    GUID CustomDataTypeGuid;
    uint32_t EntitiesCount;
    uint32_t EntityOffsets[1]; // actual length is EntitiesCount
};

struct PoseTrackingEntityData
{
    uint32_t DataSizeInBytes;
    GUID EntityId;
    GUID PoseSet;
    uint32_t PosesCount;
    bool IsTracked;
    TrackedPose Poses[1]; // actual length is PosesCount
};
```

The easiest way to access the Body information is to add a reference to the **PoseTrackingPreview** project and utilize its logic. This project contains all the code to unpack the stream and convert the data into programmable objects in code for easy access.

You can review the sample listed in the aforementioned link to see exactly how it's done. One note of caution: the sample is rather complex and involved and is likely to change to a more encapsulated and basic design when this feature is released.

How to Use Audio

There is nothing special about using Kinect audio within a UWP application. As a matter of fact, we lose some capabilities here. In previous frameworks, we had the ability to do beamforming and to determine where sound was coming from using the Microphone array. We could even figure out which sound was coming from which tracked user. I'm afraid to say these features are not implemented in UWP audio for Kinect. I'm also afraid to say I haven't heard any rumors about whether there will even be an update to include these features in the near future. However, if you know Microsoft, if we as a community provide enough vocal feedback and devise a real usage scenario for them, they will consider it. After all, that is how we got the exclusive mode option in the Kinect, as well as many other features.

Using audio in a UWP Kinect-based application requires working with the Audio graph class. The Kinect microphone array is treated like any other standard microphone. This means that the standard audio samples that come with the UWP samples will work for Kinect. I won't go into detail on this since no special settings or functionality are required to get it working.

■ **Tip**   You can find one Audio sample here: `https://github.com/Microsoft/Windows-universal-samples/tree/master/Samples/AudioCreation` and many more here: `https://github.com/Microsoft/Windows-universal-samples`.

What About Kinect Fusion, Face, HD Face, Visual Gesture Builder, and Kinect Studio Tools?

I'm sad to say that in this current release there are no features or capabilities to work with Kinect Face, HD Face, Fusion, Visual Gesture Builder, or Kinect Studio tools for unit testing and automating replay mechanisms. These topics suffer the same fate as the more specific Kinect Audio capabilities. I do offer some advice and alternatives though.

Kinect Fusion is all about volumetric rendering and can be accomplished by extrapolating on the basic Data Sources. We have all the core components in the Media Capture APIs to establish a generic framework for doing volumetric rendering.

Kinect Face can be accomplished using the `Windows.Media` classes. There are classes inside of UWP that provide the ability to locate a face and various face points using any camera, not just Kinect. More advanced face features can be accomplished using OpenCV and other computer-vision tools. HD Face can specifically be accomplished using volumetric rendering on top of `Windows.Media` classes and OpenCV.

Visual Gesture Builder and Kinect Studio tools, on the other hand, are separate applications that will require more involved effort. Their current versions may be able to work with UWP applications as is, but it will require some thought.

Summary

Microsoft changed the game plan when it comes to developing Kinect-based applications for the UWP Store. They have a new paradigm, framework, and philosophy for dealing with device-agnostic apps. This change does not put Kinect in the tomb, but rather forces the developer to treat the device as a group of capabilities that can be used across Windows 10 devices. Learning the new paradigm doesn't mean you have to start from scratch, especially since a lot of the patterns and design dealing with frames, events, and asynchronous behavior are practically the same. Where Microsoft will take the Kinect from here on out is left to be seen.

Index

© Mansib Rahman 2017
M. Rahman, *Beginning Microsoft Kinect for Windows SDK 2.0*, DOI 10.1007/978-1-4842-2316-1

Get the eBook for only $5!

Why limit yourself?

With most of our titles available in both PDF and ePUB format, you can access your content wherever and however you wish—on your PC, phone, tablet, or reader.

Since you've purchased this print book, we are happy to offer you the eBook for just $5.

To learn more, go to http://www.apress.com/companion or contact support@apress.com.

Apress®